DICTIONARY OF BIOLOGY

DERIVED FROM THE CONCISE SCIENCE DICTIONARY

DICTIONARY OF
BIOLOGY

DERIVED FROM THE CONCISE SCIENCE DICTIONARY

Sphere Reference
LONDON & SYDNEY

First published by Sphere Books Ltd., 1985.
30–32 Gray's Inn Road, London WC1X 8JL

This dictionary has been derived from the
Concise Science Dictionary, first published
in hardback by Oxford University Press, 1984.

TRADE
MARK

Printed and bound in Great Britain by
Cox & Wyman Ltd, Reading

Preface

This dictionary is derived from the *Concise Science Dictionary*, published by Oxford University Press in 1984. It consists of all the entries relating to biology and biochemistry in this dictionary, together with those entries relating to geology that are required for an understanding of palaeontology and a few entries relating to physics and chemistry that are required for an understanding of the physical and chemical aspects of biology. This dictionary also includes a selection of the words used in medicine and palaeoanthropology.

The more chemical aspects of biochemistry and the chemistry itself will be found in the Sphere *Dictionary of Chemistry*; this and the Sphere *Dictionary of Physics* are companion volumes to this dictionary.

SI units are used throughout this book and its companion volumes.

EAM, 1985

Editor

Elizabeth Martin M.A.

Advisors

B. S. Beckett B.Sc., B.Phil., MA (Ed.)
R. A. Hands B.Sc. Michael Lewis MA

Contributors

Tim Beardsley BA
Lionel Bender B.Sc.
W. M. Clarke B.Sc.
Derek Cooper Ph.D., FRIC
E. K. Duintith B.Sc.
D. E. Edwards B.Sc., M.Sc.
Malcolm Hart B.Sc., M.I.Biol.
Robert S. Hine B.Sc., M.Sc.
Ann Lockwood B.Sc.
J. Valerie Neal B.Sc., Ph.D.
R. A. Prince MA
Jackie Smith BA
Brian Stratton B.Sc., M.Sc.
Elizabeth Tootill B.Sc., M.Sc.
David Eric Ward B.Sc., M.Sc., Ph.D

A

abdomen The posterior region of the body trunk of animals. In vertebrates it contains the stomach and intestines and the organs of excretion and reproduction. It is particularly well defined in mammals, being separated from the *thorax by the *diaphragm. In many arthropods, such as insects and spiders, it may be segmented.

abiogenesis The origin of living from non-living matter, as by *biopoiesis. *See also* spontaneous generation.

ABO system One of the most important human *blood group systems. The system is based on the presence or absence of *antigens A and B on the surface of red blood cells and *antibodies against these in blood serum. A person whose blood contains either or both these antibodies cannot receive a transfusion of blood containing the corresponding antigens as this would cause the red cells to clump (*see* agglutination). The table illustrates the basis of the system: people of blood group O are described as 'universal donors' as they can give blood to those of any of the other groups. *See also* immune response.

abscisic acid A naturally occurring plant *growth substance that promotes leaf ageing, leaf fall (*see* abscission), and *apical dominance and induces *dormancy in seeds and buds.

abscission The separation of a leaf, fruit, or other part from the body of a plant. The process is controlled by growth substances, notably *abscisic acid; it involves the formation of an *abscission zone*, at the base of the part, within which a layer of cells (*abscission layer*) breaks down.

absorption The movement of fluid or a dissolved substance across a cell membrane. In animals, for example, soluble food material is absorbed into the circulatory system through cells lining the alimentary canal. In plants, water and mineral salts are absorbed from the soil by the *roots. *See* osmosis.

absorption spectrum *See* spectrum.

abyssal zone The lower depths of the ocean (below approximately 2000 metres), where there is effectively no light penetration. Abyssal organisms are adapted for living under high pressures in cold dark conditions.

accommodation 1. (in animal physiology) The process by which the focal length of the *lens of the eye is changed so that clear

Group	Antigens on red cell surface	Antibodies in serum	Blood group of people donor can receive blood from	Blood group of people donor can give blood to
A	A	anti-B	A,O	A, AB
B	B	anti-A	B,O	B, AB
AB	A and B	none	A,B, AB, O	AB
O	neither A nor B	anti-A and anti-B	O	A, B, AB, O

The ABO blood group system

images of objects at a range of distances are displayed on the retina. In man and some other mammals accommodation is achieved by reflex adjustments in the shape of the lens brought about by relaxation and contraction of muscles within the *ciliary body. **2.** (in botany) The ability of a plant to adapt itself to gradually changing environmental conditions. **3.** (in animal behaviour) The psychological adjustment made by an animal in response to continuously changing environmental conditions.

acellular Describing tissues or organisms that are not made up of separate cells but often have more than one nucleus. Examples of acellular structures are muscle fibres. *Compare* unicellular.

acetic acid (ethanoic acid) A carboxylic acid, CH_3COOH, that, when combined with *coenzyme A (to form *acetyl coenzyme A*), plays a crucial role in energy metabolism.

acetone *See* ketone body.

acetylcholine A substance that is released at some (*cholinergic*) nerve endings (*see* neurotransmitter). Its function is to pass on a nerve *impulse to the next nerve (i.e. at a *synapse) or to initiate muscular contraction. Once acetylcholine has been released it has only a transitory effect because it is rapidly broken down by the enzyme *acetylcholinesterase*.

achene A dry indehiscent fruit formed from a single carpel and containing a single seed. An example is the feathery achene of clematis. Variants of the achene include the *caryopsis, *cypsela, *nut, and *samara. *See also* etaerio.

acid–base balance The regulation of the concentrations of acids and bases in blood and other body fluids so that the *pH remains within a physiologically acceptable range. This is achieved by the presence of natural *buffer systems, such as the haemoglobin, bicarbonate ions, and carbonic acid in mammalian blood. By acting in conjunction, these effectively mop up excess acids and bases and therefore prevent any large shifts in blood pH. The acid–base balance is also influenced by the selective removal of certain ions by the kidneys and the rate of removal of carbon dioxide from the lungs.

acidic stains *See* staining.

acid rain *See* pollution.

acquired characteristics Features that are developed during the lifetime of an individual, e.g. the enlarged arm muscles of a tennis player. It is a basic tenet of current evolutionary thought that such characteristics are not genetically controlled and cannot be passed on to the next generation. *See also* Lamarckism; neo-Lamarckism.

acrosome *See* spermatozoon.

ACTH (adrenocorticotrophic hormone, corticotrophin) A hormone produced by the anterior *pituitary gland in response to stress that controls secretion of certain hormones (the *corticosteroids) by the adrenal glands. It can be administered by injection to treat such disorders as rheumatic diseases and asthma, but it only relieves symptoms and is not a cure.

actinomorphy *See* radial symmetry.

actinomycetes A group of Gram-positive mostly anaerobic nonmotile bacteria. All species are fungus-like, with filamentous cells producing reproductive spores on aerial branches similar to the spores of certain moulds. The group includes bacteria of the genera *Actinomyces*, some species of which cause disease in animals and man; and *Streptomyces*, which are a source of many important antibiotics (including streptomycin).

action potential The change in electrical potential that occurs across a cell membrane during the passage of a nerve *impulse. As an impulse travels in a wavelike manner along the *axon of a nerve, it causes a localized and transient switch in electrical potential across the cell membrane from −60 mV (millivolts) to +45 mV. Nervous stimulation of a muscle fibre has a similar effect.

action spectrum A graphical plot of the efficiency of electromagnetic radiation in producing a photochemical reaction against the wavelength of the radiation used. For example, the action spectrum for photosynthesis using light shows a peak in the region 670–700 nm. This corresponds to a maximum absorption in the absorption *spectrum of chlorophylls in this region.

active site The site on the surface of an *enzyme molecule that binds the substrate molecule. The properties of an active site are determined by the three-dimensional arrangement of the polypeptide chains of the enzyme and their constituent amino acids. These govern the nature of the interaction that takes place and hence the degree of substrate specificity and susceptibility to *inhibition.

active transport The movement of substances through membranes in living cells, often against a concentration gradient: a process requiring metabolic energy. Organic molecules and inorganic ions are transported into and out of both cells and their organelles. It is thought that the substance binds to a carrier protein embedded in the membrane, which carries it through the membrane and releases it on the opposite side. Active transport serves chiefly to maintain the normal balance of ions in cells, especially the concentration gradients of sodium and potassium ions crucial to the activity of nerve and muscle cells. *Compare* facilitated diffusion.

adaptation 1. (in evolution) Any change in the structure or functioning of an organism that makes it better suited to its environment. *Natural selection of inheritable adaptations ultimately leads to the development of new species. Increasing adaptation of a species to a particular environment tends to diminish its ability to adapt to any sudden change in that environment. **2.** (in physiology) The alteration in the degree of sensitivity (either an increase or a decrease) of a sense organ to suit conditions more extreme than normally encountered. An example is the adjustment of the eye to vision in very bright or very dim light.

adaptive radiation (divergent evolution) The evolution from one species of animals or plants of a number of different forms. As the original population increases in size it spreads out from its centre of origin to exploit new habitats and food sources. In time this results in a number of populations each adapted to its particular habitat: eventually these populations will differ from each other sufficiently to become new species. A good example of this process is the evolution of the Australian marsupials into species adapted as carnivores, herbivores, burrowers, fliers, etc. On a smaller scale, the adaptive radiation of the Galapagos finches provided Darwin with crucial evidence for his theory of evolution (*see* Darwin's finches).

adenine A *purine derivative. It is one of the major component bases of *nucleotides and the nucleic acids *DNA and *RNA.

adenosine A nucleoside comprising one adenine molecule linked to a D-ribose sugar molecule. The phosphate-ester derivatives of adenosine, AMP, ADP, and *ATP, are of fundamental biological importance as carriers of chemical energy.

adenosine diphosphate (ADP) *See* ATP.

adenosine monophosphate (AMP) *See* ATP.

adenosine triphosphate *See* ATP.

adenovirus One of a group of DNA-containing viruses found in rodents, fowl, cattle, monkeys, and man. In man they produce acute respiratory-tract infections with symptoms resembling those of the common cold. They are also implicated in the formation of tumours (*see* oncogenic).

ADH (antidiuretic hormone) *See* vasopressin.

adipose tissue A body tissue comprising cells containing *fat and oil. It is found chiefly below the skin (*see* subcutaneous tissue) and around major organs (such as the kidneys and heart), acting as an energy reserve and also providing insulation and protection.

ADP *See* ATP.

adrenal glands A pair of endocrine glands situated immediately above the kidneys (hence they are also known as the *suprarenal glands*). The inner portion of the adrenals, the *medulla*, secretes the hormones *adrenaline and *noradrenaline; the outer *cortex* secretes small amounts of sex hormones (*androgens and *oestrogens) and various *corticosteroids, which have a wide range of effects on the body. *See also* ACTH.

adrenaline (epinephrine) A hormone, produced by the medulla of the *adrenal glands, that increases heart activity, improves the power and prolongs the action of muscles, and increases the rate and depth of breathing to prepare the body for 'fright, flight, or fight'. At the same time it inhibits digestion and excretion. Similar effects are produced by stimulation of the *sympathetic nervous system. Adrenaline can be administered by injection to relieve bronchial asthma and reduce blood loss during surgery by constricting blood vessels.

adrenergic Describing a nerve fibre that either releases *adrenaline or *noradrenaline when stimulated or is itself stimulated by these substances. *Compare* cholinergic.

adrenocorticotrophic hormone *See* ACTH.

adsorption The formation of a layer of solid, liquid, or gas on the surface of a solid or, less frequently, of a liquid. There are two types depending on the nature of the forces involved. In *chemisorption* a single layer of molecules, atoms, or ions is attached to the adsorbent surface by chemical bonds. In *physisorption* adsorbed molecules are held by the weaker physical forces. The property is utilized in adsorption *chromatography.

adventitious Describing organs or other structures that arise in unusual positions. For example, ivy has adventitious roots growing from its stems.

aerobe *See* aerobic respiration.

aerobic respiration A type of *respiration in which foodstuffs (usually carbohydrates) are completely oxidized to carbon dioxide and water, with the release of chemical energy, in a process requiring atmospheric oxygen. The reaction can be summarized by the equation:

$$C_6H_{12}O_6 + 6O_2 \rightarrow 6CO_2 + 6H_2O + \text{energy}$$

Most organisms have aerobic respiration (i.e. they are *aerobes*); exceptions include certain bacteria and yeasts. *Compare* anaerobic respiration.

aerotaxis *See* taxis.

aestivation 1. (in zoology) A state of inactivity occurring in some animals, notably lungfish, during prolonged periods of drought or heat. Feeding, respiration, movement, and other bodily activities are considerably slowed down. *See also* dormancy. *Compare* hibernation. **2.** (in botany) The arrangement of the parts of a flower bud, especially of the sepals and petals.

afferent Carrying (nerve impulses, blood, etc.) from the outer regions of a body or organ towards its centre. The term is usually applied to types of nerve fibres or blood vessels. *Compare* efferent.

aflatoxin Any of four related toxic compounds produced by the mould *Aspergillus flavus*. Aflatoxins bind to DNA and prevent replication and transcription. They can cause acute liver damage and cancers: man may be poisoned by eating stored peanuts and cereals contaminated with the mould.

afterbirth The *placenta, *umbilical cord, and *extraembryonic membranes, which are expelled from the womb after a mammalian fetus is born. In most nonhuman mammals the afterbirth, which contains nutrients and might otherwise attract predators, is eaten by the female.

agamospermy *See* apomixis.

agar An extract of certain species of red seaweeds that is used as a gelling agent in microbiological *culture media, foodstuffs, medicines, and cosmetic creams and jellies. *Nutrient agar* consists of a broth made from beef extract or blood that is gelled with agar and used for the cultivation of bacteria, fungi, and some algae.

agglutination The clumping together by antibodies of microscopic foreign particles (*antigens), such as red blood cells or bacteria, so that they form a visible pellet-like precipitate. Agglutination is a specific reaction, i.e. a particular antigen will only clump in the presence of its specific antibody; it therefore provides a means of identifying unknown bacteria and determining *blood group. When red cells of incompatible blood groups (e.g. group A and group B – *see* ABO system) are mixed together agglutination of the cells occurs (*haemagglutination*).

aggression Behaviour aimed at intimidating or injuring another animal of the same or a competing species. Aggression between individuals of the same species often starts with a series of ritualized displays or contests that can end at any stage if one of the combatants withdraws, leaving the victor with access to a disputed resource (e.g. food, a mate, or *territory) or with increased social dominance (*see* dominant). It is also often seen in *courtship. Aggression or threat displays usually appear to exaggerate the performer's size or strength; for example, many fish erect their fins and mammals and birds may erect hairs or feathers. Special markings may be prominently exhibited, and *intention movements* may be made: dogs bare their teeth, for example. Some animals have evolved special structures for use in aggressive interactions (e.g. antlers in deer) but these are seldom used to cause actual injury; the opponent usually flees first or adopts *appeasement postures. Fights 'to the death' are comparatively rare. *See* display behaviour; ritualization.

Agnatha A class of marine and freshwater vertebrates that lack jaws. They are fishlike animals with cartilaginous skeletons and well-developed sucking mouthparts with horny teeth. The only living agnathans are lampreys and hagfishes (order Cyclostomata), which are parasites or scavengers. Fossil agnathans, covered in an armour of bony plates, are the oldest known fossil vertebrates. They have been dated from the Silurian and Devonian periods, 440–345 million years ago.

air bladder *See* swim bladder.

alanine *See* amino acid.

albinism Hereditary lack of pigmentation in an organism. Albino animals and human beings have no colour in their skin, hair, or eyes (the irises appear pink from underlying blood vessels). The *allele responsible is *recessive to the allele for normal pigmentation.

albumen *See* albumin.

albumin (albumen) One of a group of globular proteins that are soluble in water but form insoluble coagulates when heated. Albumins occur in egg white, blood, milk, and plants. Serum albumins, which constitute about 55% of blood plasma protein, help regulate the osmotic pressure and hence plasma volume. They also bind and transport fatty acids. α-lactalbumin is one of the proteins in milk.

albuminous cell *See* companion cell.

alburnum *See* sapwood.

aldohexose *See* monosaccharide.

aldose *See* monosaccharide.

aldosterone A hormone produced by the adrenal glands (*see* corticosteroid) that controls excretion of sodium by the kidneys and thereby maintains the balance of salt and water in the body fluids. *See also* angiotensin.

algae A large and diverse group of simple plants that contain chlorophyll (and can therefore carry out photosynthesis) and live in aquatic habitats and in moist situations on land. The plant body may be unicellular or multicellular (filamentous, ribbon-like, or platelike). The algae are not now regarded as a taxonomic group, although they are sometimes grouped with other simple organisms in the kingdom *Protista. Usually, however, the component groups are recognized as taxonomic divisions. Separation into these divisions is based primarily on the composition of the cell wall, the nature of the stored food reserves, and the other photosynthetic pigments present. *See* Bacillariophyta; Chlorophyta; Cyanophyta; Euglenophyta; Phaeophyta; Rhodophyta.

alimentary canal (digestive tract; gut) A tubular organ in animals that is divided into a series of zones specialized for the ingestion, *digestion, and *absorption of food and for the elimination of indigestible material. In most animals the canal has two openings, the mouth (for the intake of food) and the *anus (for the elimination of waste). Simple animals, such as coelenterates (e.g. *Hydra* and jellyfish) and flatworms, have only one opening to their alimentary canal, which must serve both functions.

alkaloid One of a group of nitrogenous organic compounds derived from plants and having diverse pharmacological properties. Alkaloids include morphine, cocaine, atropine, quinine, and caffeine, most of which are used in medicine as analgesics (pain relievers) or anaesthetics. Some alkaloids are poisonous, e.g. strychnine and coniine, and *colchicine inhibits cell division.

allantois One of the membranes that develops in embryonic reptiles, birds, and mammals as a growth from the hindgut. It acts as a urinary bladder for the storage of waste excretory products in the egg (in reptiles and birds) and as a means of providing the embryo with oxygen (in reptiles, birds, and mammals) and food (in mammals; *see* placenta). *See also* extraembryonic membranes.

allele (allelomorph) One of the alternative forms of a gene. In most organisms there are two alleles of any one gene (one from

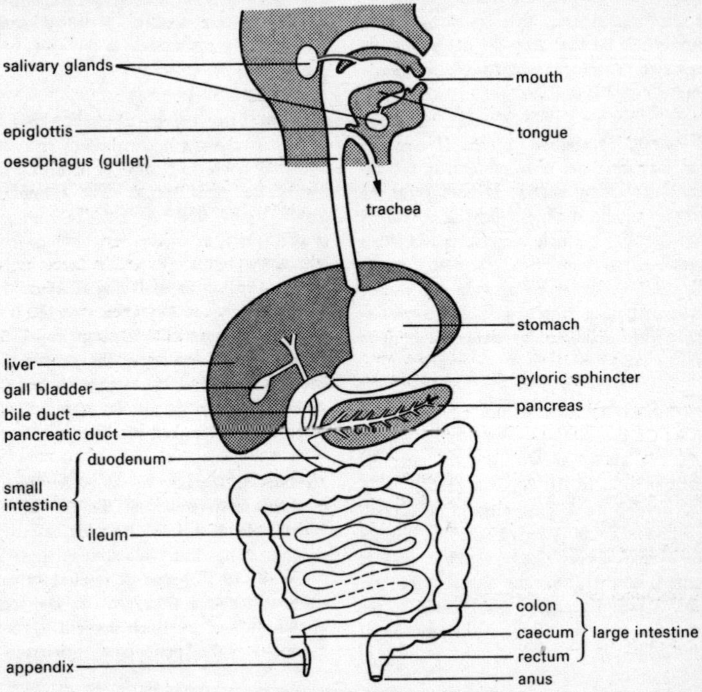

The human alimentary canal

each parent), which occupy the same relative position on *homologous chromosomes. One allele is often *dominant to the other (known as the *recessive), i.e. it determines which aspects of a particular characteristic the organism will display.

allelomorph *See* allele.

allergy A condition in which the body produces an abnormal *immune response to certain *antigens (called *allergens*), which include dust, pollen, certain foods and drugs, or fur. In allergic individuals these substances, which in a normal person would be destroyed by antibodies, stimulate the release of *histamine, leading to inflammation and other characteristic symptoms of the allergy (e.g. asthma or hay fever). This response is a type of *anaphylaxis.

allogamy Cross-fertilization in plants. *See* fertilization.

allometric growth The regular and systematic pattern of growth such that the mass or size of any organ or part of a body can be expressed in relation to the total mass or size of the entire organism according to the allometric equation: $Y = bx^{\alpha}$, where Y = mass of the organ, x = mass of the organism, α = growth coefficient of the organ, and b = a constant.

allopatric Describing or relating to groups of similar organisms that could interbreed but do not because they are geographically separated. *Compare* sympatric. *See* speciation.

all-or-none response A type of response that may be either complete and of full intensity or totally absent, depending on the strength of the stimulus; there is no partial response. For example, a nerve cell is either stimulated to transmit a complete nervous impulse or else it remains in its resting state; a stinging *thread cell of a coelenterate is either completely discharged or it is not.

alpha-naphthol test A biochemical test to detect the presence of carbohydrates in solution, also known as *Molisch's test* (after the Austrian chemist H. Molisch (1856–1937), who devised it). A small amount of alcoholic alpha-naphthol is mixed with the test solution and concentrated sulphuric acid is poured slowly down the side of the test tube. A positive reaction is indicated by the formation of a violet ring at the junction of the two liquids.

alternation of generations The occurrence within the *life cycle of an organism of two or more distinct forms (generations), which differ from each other in appearance, habit, and method of reproduction. The phenomenon occurs in certain lower animals (e.g. coelenterates and parasitic protozoans and flatworms) and in plants. The malaria parasite (*Plasmodium*), for example, has a complex life cycle involving the alternation of sexually and asexually reproducing generations. In plants the generation with sexual reproduction is called the *gametophyte and the asexual generation is the *sporophyte, either of which may dominate the life cycle, and there is also alternation of the haploid and diploid states. Thus in ferns the dominant plant is the diploid sporophyte; it produces spores that germinate into small haploid gametophytes bearing sex organs. In mosses the gametophyte is the dominant plant and the sporophyte is the spore-bearing capsule.

altruism Behaviour by an animal that decreases its chances of survival or reproduction while increasing those of another member of the same species. For example, a lapwing puts itself at risk by luring a predator away from the nest through feigning injury, but by so doing saves its offspring. Altruism in its biological sense does not imply any conscious benevolence on the part of the performer. Altruism can evolve through *kin selection, if the recipients of altruistic acts tend on average to be more closely related to the altruist than the population as a whole. *See also* inclusive fitness.

alveolus 1. The tiny air sac in the *lung of mammals and reptiles at the end of each *bronchiole. It is lined by a delicate moist membrane, has many blood capillaries, and is the site of exchange of respiratory gases (carbon dioxide and oxygen). **2.** The socket in the jawbone in which a tooth is rooted by means of the *periodontal membrane.

amber A yellow or reddish-brown fossil resin. The resin was exuded by certain trees and other plants and often contains preserved insects, flowers, or leaves that were trapped by its sticky surface before the resin hardened. Amber is used for jewellery and ornaments. It also has the property of acquiring an electrical charge when rubbed (the term electricity is derived from *electron*, the Greek name for amber). It occurs throughout the world in rock strata from the Cretaceous to the Pleistocene, but most commonly in Cretaceous and Tertiary rocks.

amino acid Any of a group of water-soluble organic compounds that possess both a carboxyl ($-COOH$) and an amino ($-NH_2$) group attached to the α-carbon atom. Amino acids can be represented by the gen-

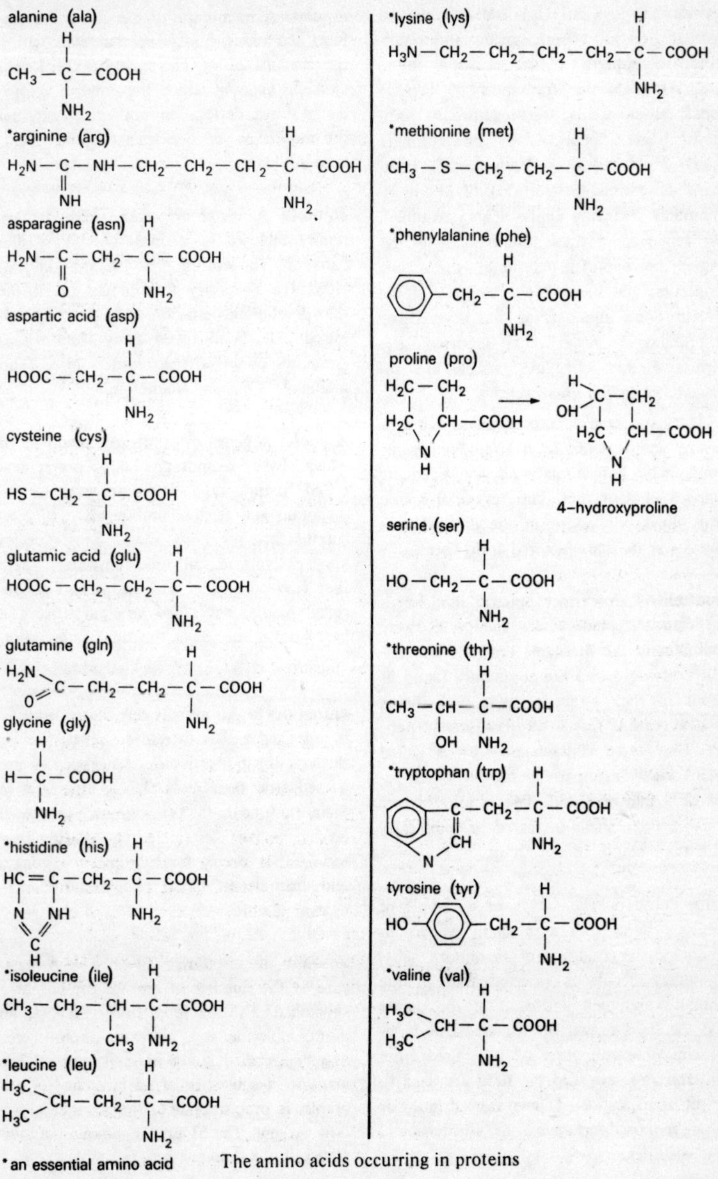

The amino acids occurring in proteins

eral formula $R-CH(NH_2)COOH$. R may be hydrogen or an organic group and determines the properties of any particular amino acid. Through the formation of peptide bonds, amino acids join together to form short chains (*peptides) or much longer chains (*polypeptides). Proteins are composed of various proportions of about 20 commonly occurring amino acids (see table). The sequence of these amino acids in the protein polypeptides determines the shape, properties, and hence biological role of the protein. Some amino acids that never occur in proteins are nevertheless important, e.g. *ornithine and citrulline, which are intermediates in the urea cycle.

Plants and many microorganisms can synthesize amino acids from simple inorganic compounds, but animals rely on adequate supplies in their diet. The *essential amino acids must be present in the diet whereas others can be manufactured from them.

ammonite An extinct aquatic mollusc of the class *Cephalopoda. Ammonites were abundant in the Mesozoic era (225–65 million years ago) and are commonly found as fossils in rock strata of that time, being used as *index fossils for the Jurassic period. They were characterized by a coiled shell divided into many chambers, which acted as a buoyancy aid. The external suture lines on these shells increased in complexity with the advance of the group.

amniocentesis The taking of a sample of amniotic fluid from a pregnant woman to determine the condition of an unborn baby. A hollow needle is inserted through the woman's abdomen and wall of the uterus and the fluid drawn off. Chemical and microscopical examination of cells shed from the embryo's skin into the fluid are used to detect spina bifida, *Down's syndrome, or other serious biochemical or chromosomal abnormalities.

amnion A membrane that encloses the embryo of reptiles, birds, and mammals within the *amniotic cavity*. This cavity is filled with *amniotic fluid*, in which the embryo is protected from desiccation and from external pressure. *See also* extraembryonic membranes.

amniote A vertebrate whose embryos are totally enclosed in a fluid-filled sac – the *amnion. The evolution of the amnion provided the necessary fluid environment for the developing embryo and therefore allowed animals to breed away from water. Amniotes comprise the reptiles, birds, and mammals. *Compare* anamniote.

Amoeba A genus of *Protozoa, members of which have temporary body projections called *pseudopodia. These are used for locomotion and feeding and result in a constantly changing body shape. Most species are free-living in soil, mud, or water, where they feed on smaller protozoans and single-celled plants, but a few are parasitic. The best known species is the much studied *A. proteus*.

amoebocyte An animal cell whose location is not fixed and is therefore able to wander through the body tissues. Amoebocytes are named after their resemblance, especially in their movement, to **Amoeba* and they feed on foreign particles (including invading bacteria). They occur, for example, in sponges and mammalian blood (e.g. some *leucocytes).

amount of substance Symbol n. A measure of the number of entities present in a substance. The specified entity may be an atom, molecule, ion, electron, photon, etc., or any specified group of such entities. The amount of substance of an element, for example, is proportional to the number of atoms present. The SI unit of amount of substance is the *mole.

AMP *See* ATP; cyclic AMP.

Amphibia The class of vertebrates that contains the frogs, toads, newts, and salamanders. Amphibians evolved in the Devonian period (about 370 million years ago) as the first vertebrates to occupy the land, and many of their characteristics are adaptations to terrestrial life. All adult amphibians have a passage linking the roof of the mouth with the nostrils so they may breathe air and keep the mouth closed. The moist scaleless skin is used to supplement the lungs in gas exchange. They have no diaphragm, and therefore the muscles of the mouth and pharynx provide the pumping action for breathing. Fertilization is usually external and the eggs are soft and prone to desiccation, therefore reproduction commonly occurs in water. Amphibian larvae are aquatic, having gills for respiration; they undergo metamorphosis to the adult form.

amphimixis True sexual reproduction, involving the fusion of male and female gametes and the formation of a zygote. *Compare* apomixis.

amylase (diastase) Any of a group of closely related enzymes that degrade starch, glycogen, and other polysaccharides. Plants contain both α- and β-amylases; animals possess only α-amylases, found in pancreatic juice and also (in humans and some other species) in saliva. Amylases cleave the long polysaccharide chains, producing a mixture of glucose and maltose.

amylopectin A *polysaccharide comprising highly branched chains of glucose molecules. It is one of the constituents (the other being amylose) of *starch.

amylose A *polysaccharide consisting of linear chains of between 100 and 1000 linked glucose molecules. Amylose is a constituent of *starch. In water, amylose reacts with iodine to give a characteristic blue colour.

anabolic steroid Any steroid compound that promotes tissue growth, especially of muscles. Naturally occurring anabolic steroids include the male sex hormones (*androgens). Synthetic forms of these are used medically to help weight gain after debilitating diseases; their use by athletes to build up body muscles can cause liver damage and is banned by most athletic authorities.

anabolism The metabolic synthesis of proteins, fats, and other constituents of living organisms from molecules or simple precursors. This process requires energy in the form of ATP. *See* metabolism. *Compare* catabolism.

anaerobe *See* anaerobic respiration.

anaerobic respiration A type of *respiration in which foodstuffs (usually carbohydrates) are partially oxidized, with the release of chemical energy, in a process not involving atmospheric oxygen. Since the substrate is never completely oxidized the energy yield of this type of respiration is lower than that of *aerobic respiration. It occurs in some yeasts and bacteria and in muscle tissue when oxygen is absent (*see* oxygen debt). *Obligate anaerobes* are organisms that cannot use free oxygen for respiration; *facultative anaerobes* are normally aerobic but can respire anaerobically during periods of oxygen shortage. Alcoholic *fermentation is a type of anaerobic respiration in which one of the end products is ethanol.

analogous Describing features of organisms that are superficially similar but have evolved in different ways. The wings of butterflies and birds are analogous organs.

anamniote A vertebrate that lacks an *amnion and whose embryos and larvae must therefore develop in water. Anamniotes comprise the agnathans, fishes, and amphibians. *Compare* amniote.

anaphase The third stage of cell division. In *mitosis the chromatids of each chromosome move .apart to opposite ends of the spindle. In the first anaphase of *meiosis, the paired homologous chromosomes separate and move to opposite ends; in the second anaphase the chromatids move apart, as in mitosis.

anaphylaxis An abnormal *immune response that occurs when an individual previously exposed to a particular *antigen is re-exposed to the same antigen. Anaphylaxis may follow an insect bite or the injection of a drug (such as penicillin). It is caused by the release of *histamine and similar substances and may produce a localized reaction or a more generalized and severe one, with difficulty in breathing, pallor, or drop in blood pressure, unconsciousness, and possibly heart failure and death. *See also* allergy.

anatomy The study of the structure of living organisms, especially of their internal parts by means of dissection and microscopical examination. *Compare* morphology.

androecium The male sex organs (*stamens) of a flower. *Compare* gynaecium.

androgen One of a group of male sex hormones that stimulate development of the testes and of male *secondary sexual characteristics (such as growth of facial and pubic hair in men). *Testosterone* is the most important. Androgens are produced principally by the testes when stimulated with *luteinizing hormone but they are also secreted in smaller amounts by the adrenal glands and the ovaries. Injections of natural or synthetic androgens are used to treat hormonal disorders of the testes and breast cancer and to build up body tissue (*see* anabolic steroid).

anemophily Pollination of a flower in which the pollen is carried by the wind. Examples of anemophilous flowers are those of grasses and conifers. *Compare* entomophily; hydrophily.

Angiospermae The flowering plants: a subdivision of the *Spermatophyta or a class of the *Pteropsida. Angiosperm gametes are produced within *flowers and the ovules (and the seeds into which they develop) are enclosed in a carpel (*compare* Gymnospermae). The angiosperms are the dominant plant forms of the present day. They show the most advanced structural organization in the plant kingdom, enabling them to inhabit a very diverse range of habitats. There are two classes (or subclasses) within this group: the *Monocotyledonae with one seed leaf (cotyledon) in the seed, and the *Dicotyledonae with two seed leaves.

angiotensin Either of two related peptide hormones that raise blood pressure. Angiotensin I is derived from a protein (*angiotensinogen*) secreted by the liver into the bloodstream. As blood passes through the lungs, an enzyme splits angiotensin I, forming angiotensin II. This causes constriction of blood vessels and stimulates the release of the hormones *vasopressin and *aldosterone, which increase blood pressure.

angstrom Symbol Å. A unit of length equal to 10^{-10} metre. It was formerly used to measure wavelengths and intermolecular distances but has now been replaced by the nanometre. 1 Å = 0.1 nanometre. The unit is named after the Swedish pioneer of spectroscopy A. J. Ångstrom (1814–74).

animal Any living organism of the kingdom Animalia, characterized by an inability to manufacture its own food, so that it feeds on other organisms or organic matter (holozoic nutrition; *see* heterotrophism). Animals are therefore typically mobile (to search for food) and have evolved specialized sense organs for detecting changes in the environment; a *nervous system coordinates information received by the sense organs and enables rapid responses to environmental stimuli. Animal *cells lack the cellulose cells walls of *plant cells. For a classification of the animal kingdom, see Appendix.

animal behaviour The activities that constitute an animal's response to its external environment. Certain categories of behaviour are seen in all animals (e.g. feeding, reproduction) but these activities involve different movements in different species and develop in different ways. Some movements are highly characteristic of a species (*see* instinct) whereas others are more variable and depend on the interaction between innate tendencies and *learning during the individual's lifetime. Physiologists study how changes in the body (e.g. hormone levels) affect behaviour, psychologists study the mechanisms of learning, and ethologists study the behaviour of the whole animal: how this develops during the individual's lifetime and how it evolved through natural selection (*see* ethology).

animal starch *See* glycogen.

anisogamy Sexual reproduction involving the fusion of gametes that differ in size and sometimes also in form. *See also* oogamy. *Compare* isogamy.

Annelida A phylum of invertebrates comprising the segmented worms (e.g. the earthworm). Annelids have cylindrical soft bodies showing *metameric segmentation, obvious externally as a series of rings separating the segments. Each segment is internally separated from the next by a membrane and bears stiff bristles (*see* chaeta). Between the gut and other body organs there is a fluid-filled cavity called the *coelom, which acts as a hydrostatic skeleton. Movement is by alternate contraction of circular and longitudinal muscles in the body wall. The phylum contains three classes: *Polychaeta, *Oligochaeta, and *Hirudinea.

annual A plant that completes its life cycle in one year, during which time it germinates, flowers, produces seeds, and dies. Examples are the sunflower and marigold. *Compare* biennial; ephemeral; perennial.

annual ring *See* growth ring.

annulus 1. (in botany) **a.** A ragged ring of tissue that remains on the stalk of a mushroom or toadstool. Also called a *velum*, it is formed from the ruptured membrane that originally covered the lower surface of the cap. **b.** The region of the wall of a fern sporangium that is specialized for spore dispersal. It consists of cells that are thickened except on their outer walls. On drying out, the cells contract and the sporangium ruptures, releasing the spores. The annulus springs back into position when the residual water in the cells vaporizes and any remaining spores are dispersed. **2.** (in zoology) Any of various ring-shaped structures in animals, such as any of the segments of an earthworm or other annelid.

Anoplura *See* Siphunculata.

ANS *See* autonomic nervous system.

antagonism 1. The interaction of two substances (e.g. drugs, hormones, or enzymes) having opposing effects in a system in such a way that the action of one partially or completely inhibits the effects of the other. For example, one group of anti-cancer drugs acts by antagonizing the effects of certain enzymes controlling the activities of the cancer cells. **2.** An interaction between two muscles in which contraction of one (the *antagonist*) prevents that of the other (the *agonist*). The antagonist must relax to enable the agonist to contract and effect movement. **3.** An interaction between two organisms (e.g. moulds or bacteria) in which the growth of one is inhibited by the other. *Compare* synergism.

antenna A long whiplike jointed mobile paired appendage on the head of many arthropods, usually concerned with the senses of smell, touch, etc. In insects, millipedes, and centipedes they are the first pair of head appendages and are specialized and modified in many insects. In crustaceans they are the second pair of head appendages, the first pair (the *antennules*) having the sensory function, while the antennae are modified for swimming and for attachment.

antennule *See* antenna.

anterior 1. Designating the part of an animal that faces to the front, i.e. that leads when the animal is moving. In man and bipedal animals the anterior surface corresponds to the *ventral surface. **2.** Designating the side of a flower or axillary bud that faces away from the flower stalk or main stem, respectively. *Compare* posterior.

anther The upper two-lobed part of a plant *stamen, usually yellow in colour. Each lobe contains two pollen sacs within which are numerous pollen grains, which are released when the anther ruptures.

antheridium The male sex organ of algae, fungi, bryophytes, and pteridophytes. It produces the male gametes (*antherozoids*). It may consist of a single cell or it may have a wall that is made up of one or several layers forming a sterile jacket around the developing gametes. *Compare* archegonium.

antherozoid (spermatozoid) The motile male gamete of algae, fungi, bryophytes, pteridophytes, and certain gymnosperms. Antherozoids usually develop in an *antheridium but in certain gymnosperms, such as *Ginkgo* and *Cycas*, they develop from a cell in the pollen tube.

anthocyanin One of a group of *flavonoid pigments. Anthocyanins occur in the cell vacuoles of various plant organs and are responsible for many of the blue, red, and purple colours in plants (particularly in flowers).

antibiotics Substances obtained from microorganisms, especially moulds, that destroy or inhibit the growth of other microorganisms, particularly disease-producing bacteria and fungi. Common antibiotics include *penicillin, streptomycin, and tetracyclines. They are used to treat various infections but tend to weaken the body's natural defence mechanisms and can cause allergies. Overuse of antibiotics can lead to the development of resistant strains of microorganisms.

antibody A protein (*see* immunoglobulin) produced by certain white blood cells (*lymphocytes) in response to entry into the body of a foreign substance (*antigen) in order to render it harmless. An antibody–antigen reaction is highly specific. Antibody production is one aspect of the *immune response and is stimulated by such antigens as invading bacteria, foreign red blood cells (*see* ABO system), inhaled pollen grains or dust, and foreign tissue *grafts. *See also* immunity; monoclonal antibody.

anticoagulant A substance that prevents the formation of blood clots. *Heparin is a natural anticoagulant, which is extracted to treat such conditions as thrombosis and embolism. Synthetic anticoagulants include *warfarin and dicoumarol.

antidiuretic hormone (ADH) *See* vasopressin.

antigen Any substance that the body regards as foreign and that therefore elicits an *immune response. Antigens may be formed in, or introduced into, the body. They are usually proteins. *Transplantation (histocompatibility) antigens* are associated with the tissues and are involved in tissue or organ *grafts; an example is the *HL-A system. A graft will be rejected if the recipient's body regards such antigens on the donor's tissues as foreign. *See also* antibody.

antihistamine Any drug that inhibits the effects of *histamine in the body and is therefore used to relieve and prevent the symptoms associated with allergic reactions, such as hay fever. Since one of the side-effects produced by antihistamines is sleepiness, some are used to prevent motion sickness and induce sleep.

antiseptic Any substance that kills or inhibits the growth of disease-causing microorganisms but is essentially nontoxic to cells of the body. Common antiseptics include hydrogen peroxide, the detergent cetrimide, and ethanol. They are used to treat minor wounds. *Compare* disinfectant.

anus The terminal opening of the *alimentary canal in most animals, through which indigestible material (*faeces) is expelled.

aorta The major blood vessel in higher vertebrates through which oxygenated blood leaves the *heart from the left ventricle. The aorta branches to form many smaller arteries, which in turn branch many times to supply oxygen and essential nutrients to all living cells in the body.

apatite A complex mineral form of calcium phosphate, $Ca_5(PO_4)_3(OH,F,Cl)$, that is the main constituent of the enamel of teeth.

apical dominance Inhibition of the growth of lateral buds in a plant by the presence of a growing apical bud. It is brought about by the action of auxins (produced by the apical bud) and abscisic acid.

apical meristem A region at the tip of each shoot and root of a plant in which cell divisions are continually occurring to produce new stem and root tissue, respectively. The new tissues produced are known collectively as the *primary tissues* of the plant. *See also* meristem. *Compare* cambium.

apocarpy The condition in which the female reproductive organs (*carpels) of a flower are not joined to each other. It occurs, for example, in the buttercup. *Compare* syncarpy.

apocrine secretion *See* secretion.

apoenzyme An inactive enzyme that must associate with a specific *cofactor molecule or ion in order to function. *Compare* holoenzyme.

apomixis (agamospermy) A reproductive process in plants that superficially resembles normal sexual reproduction but in which there is no fusion of gametes. In apomictic flowering plants there is no fertilization by pollen, and the embryos develop simply by division of a *diploid cell of the ovule. *See also* parthenocarpy; parthenogenesis.

aposematic coloration *See* warning coloration.

appeasement Behaviour that inhibits aggression from another animal of the same species, frequently taking the form of a special posture or *display emphasizing the weakness of the performer. Threatening structures (e.g. antlers) and markings are covered or turned away, and vulnerable parts of the body may be exposed. Appeasement is seen in *courtship, in greeting ceremonies, and often (from the loser) after a fight.

appendix (vermiform appendix) An outgrowth of the *caecum in the alimentary canal. In humans it is a *vestigial organ containing lymphatic tissue and serves no function in normal digestive processes. Appendicitis is caused by inflammation of the appendix.

aqueous humour The fluid that fills the space between the cornea and the lens of the vertebrate eye. In addition to supplying the cornea and lens with nutrients, the aqueous humour helps to maintain the shape of the eye. It is produced and renewed every four hours by the *ciliary body.

Arachnida A class of terrestrial arthropods comprising about 65 000 species and including spiders, scorpions, harvestmen, ticks, and mites. An arachnid's body is divided into an anterior *cephalothorax (*prosoma*) and a posterior abdomen (*opisthosoma*). The prosoma bears a pair of grasping or piercing appendages (the *chelicerae*), a pair of *pedipalps* used for manipulation or as sensory structures, and four pairs of walking legs. The opisthosoma may bear various sensory or silk-spinning appendages (*see* spinneret). Arachnids are generally carnivorous, feeding on the body fluids of their prey or secreting enzymes to digest prey externally. Spiders immobilize their prey with poison injected by the fanglike chelicerae, while scorpions grasp their prey in large clawed pedipalps and may poison it using the posterior stinging organ. Ticks and some mites are parasitic but most arachnids are free-living. They breathe either via *tracheae (like insects) or by means of thin highly folded regions of the body wall called *lung-books*.

arachnoid membrane One of the three membranes (*meninges) that surround the brain and spinal cord of vertebrates. It lies between the *pia mater and the *dura mater. The arachnoid membrane is very delicate and carries *cerebrospinal fluid, which sustains and cushions the nervous tissue.

arbovirus One of a large heterogeneous group of RNA-containing viruses that are transmitted from animals to man through the bite of mosquitoes and ticks (i.e. arthropods, hence *ar*thropod-*bo*rne viruses). Arboviruses cause various forms of encephalitis (inflammation of the brain) and serious fevers, such as dengue and yellow fever.

archegonium The multicellular flask-shaped female sex organ of bryophytes, pteridophytes, and many gymnosperms. Such plants are described as *archegoniate* to distinguish them from algae, which do not possess archegonia. The dilated base, the *venter*, contains the oosphere (female gamete). The cells of the narrow neck liquefy to allow the male gametes to swim towards the oosphere. The archegonium is thus an adaptation to the terrestrial environment as it provides a means for the male gametes to reach the female gamete. *Compare* antheridium.

archenteron (gastrocoel) A cavity within an animal embryo at the *gastrula stage of development. All or part of the archenteron eventually forms the cavity of the gut. It is connected to the outside by an opening (the *blastopore*), which becomes either the mouth, the mouth and anus, or the anal opening of the animal.

arginine *See* amino acid.

aril An outgrowth that grows around and may completely enclose the testa (seed coat) of a seed. It develops from the placenta, funicle, or micropyle of an ovule. The aril surrounding the nutmeg seed forms the spice mace. *See also* caruncle.

arousal A level of physiological and behavioural responsiveness in an animal, which tends to vary between sleep and full alertness. It is controlled by a particular part of the brain (the *reticular activating system*) and can be detected by changes in brain electrical activity, heart rate and muscle tone, responsiveness to new stimuli, and general activity.

arteriole A small muscular blood vessel that receives blood from the arteries and carries it to the capillaries.

artery A blood vessel that carries blood away from the heart towards the other body tissues. Most arteries carry oxygenated blood (the *pulmonary artery is an exception). The large arteries branch to form smaller ones, which in turn branch into *arterioles. All arteries have muscular walls, whose contraction aids in pumping blood around the body. The accumulation of fatty deposits in the walls of the arteries leads to *atherosclerosis*, which limits and may eventually block the flow of blood. *Compare* vein.

Arthropoda A phylum of invertebrate animals comprising over one million species – the largest in the animal kingdom. Arthropods inhabit marine, freshwater, and terrestrial habitats worldwide. Characteristically, they possess an outer body layer – the *cuticle – that functions as a rigid protective exoskeleton; growth is thus possible only by periodic moults (*see* ecdysis). The arthropod body is composed of segments (*see* metameric segmentation) usually forming distinct specialized body regions, e.g. head, thorax, and abdomen. These segments may possess hardened jointed appendages, modified variously as *mouthparts, limbs, wings, reproductive organs, or sense organs. The main body cavity, containing the internal organs, is a continuous blood-filled *haemocoel*, within which lies the heart. The origins and relationships of the various groups of arthropods remain uncertain. The major classes are: *Arachnida (spiders, scorpions, mites, and ticks), *Crustacea (shrimps, barnacles, crabs, etc.), *Insecta (insects), *Myriapoda (centipedes and millipedes), and the wormlike Onychophora (e.g. *Peripatus*).

articulation The attachment of two bones, usually by means of a *joint. The thigh bone (femur), for instance, articulates with the pelvic girdle.

artificial insemination The deposition of semen, using a syringe, at the mouth of the uterus to make conception possible. It is used for selected breeding of domestic animals and in humans in some cases of impotence and infertility. It is timed to coincide with ovulation in the female.

Artiodactyla An order of hooved mammals comprising the even-toed ungulates, in which the third and fourth digits are equally developed and bear the weight of the body. The order includes cattle and other ruminants (*see* Ruminantia), camels, hippopotamuses, and pigs. All except the latter are herbivorous, having an elongated gut and teeth with enamel ridges for grinding tough grasses. *Compare* Perissodactyla.

Ascomycetes A class of *fungi that includes many yeasts. Sexual reproduction is by means of *ascospores*, eight of which are characteristically produced within a spherical or cylindrical cell, the *ascus*. The asci are usually grouped together in an *ascocarp*.

ascorbic acid *See* vitamin C.

asexual reproduction Reproduction in which new individuals are produced from a single parent without the formation of gametes. It occurs chiefly in lower animals, microorganisms, and plants. In lower animals the chief methods are *fission (e.g. in protozoans), *fragmentation (e.g. in some aquatic annelid worms), and *budding (e.g. in coelenterates). The principal methods of asexual reproduction in plants are by *vegetative propagation (e.g. bulbs, corms, tubers) and by the formation of sexual *spores. Spore formation occurs in mosses, ferns, and other plants showing alternation of generations, as a dormant stage between sporophyte and gametophyte, and in some algae and fungi, to produce replicas of the plant. *Compare* sexual reproduction.

asparagine *See* amino acid.

aspartic acid *See* amino acid.

assimilation The utilization by a living organism of absorbed food materials in the processes of growth, reproduction, or repair.

association An ecological unit in which two or more species occur in closer proximity to one another than would be expected on the basis of chance. Early plant ecologists recognized associations of fixed composition on the basis of the *dominant species present (e.g. a coniferous forest association). Associations now tend to be detected by using more objective statistical sampling methods. *See also* consociation.

aster A starlike arrangement of small fibres radiating from a *centriole. Asters become conspicuous in animal cells at the ends of the spindle when cell division starts.

astigmatism A lens defect in which when rays in one plane are in focus those in another plane are not. The eye can suffer from astigmatism, usually when the cornea is not spherical. It is corrected by using an anastigmatic lens, which has different radii of curvature in the vertical and horizontal planes.

atlas The first *cervical vertebra, a ringlike bone that joins the skull to the vertebral column in terrestrial vertebrates. In advanced vertebrates articulation between the skull and atlas permits nodding movements of the head. *See also* axis.

ATP (adenosine triphosphate) A nucleotide that is of fundamental importance as a carrier of chemical energy in all living organisms. It consists of adenine linked to D-ribose (i.e. adenosine); the D-ribose component bears three phosphate groups, linearly linked together by covalent bonds. These bonds can undergo hydrolysis to yield either a molecule of *ADP* (*adenosine diphosphate*) and inorganic phosphate or a molecule of *AMP* (*adenosine monophosphate*) and pyrophosphate. Both these reactions yield a large amount of energy (about 30.6 kJ mol^{-1}) that is used to bring about such biological processes as muscle contraction, the active transport of ions and molecules across cell membranes, and the synthesis of biomolecules. The reactions bringing about these processes often involve the enzyme-catalysed transfer of the phosphate group to intermediate substrates. Most ATP-mediated reactions require Mg^{2+} ions as *cofactors.

ATP is regenerated by the rephosphorylation of AMP and ADP using the chemical energy obtained from the oxidation of food. This takes place during *glycolysis and the *Krebs cycle but, most significantly, is also a result of the reduction–oxidation reactions of the *electron transport chain, which ultimately reduces molecular oxygen to water (oxidative phosphorylation).

atrium 1. (*or* **auricle**) A chamber of the *heart that receives blood from the veins and forces it by powerful muscular contraction into the *ventricle(s). Fish have a single atrium but all other vertebrates have two. **2.** Any of various cavities or chambers in animals, such as the chamber surrounding the gill slits of *Amphioxus* and other invertebrate chordates.

atrophy The degeneration or withering of an organ or part of the body.

atropine A poisonous crystalline alkaloid, $C_{17}H_{23}NO_3$. It can be extracted from deadly nightshade and other solanaceous plants and is used in medicine to treat colic, to reduce secretions, and to dilate the pupil of the eye.

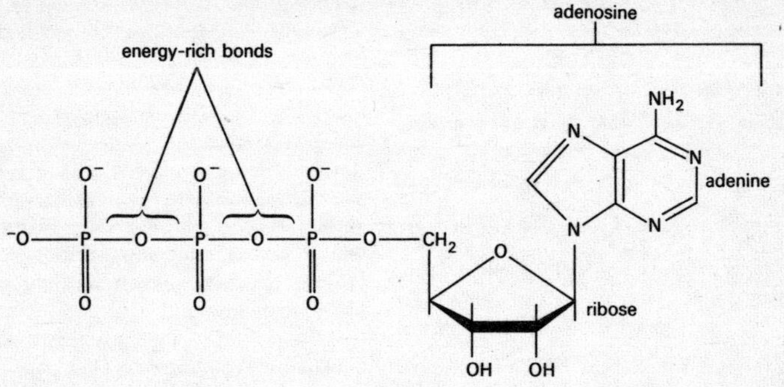

ATP

attenuation 1. (in medicine) A process of reducing the disease-producing ability of a microorganism. It can be achieved by chemical treatment, heating, drying, irradiation, by growing the organism under adverse conditions, or by serial passage through another organism. Attenuated bacteria or viruses are used for some *vaccines. **2.** (in mycology) The conversion by yeasts of carbohydrates to alcohol, as in brewing and wine and spirit production.

atto- Symbol *a*. A prefix used in the metric system to denote 10^{-18}. For example, 10^{-18} second = 1 attosecond (as).

audibility The limits of audibility of the human ear are between about 20 hertz (a low rumble) and 20 000 hertz (a shrill whistle). With increased age the upper limit falls quite considerably.

audiometer An instrument that generates a sound of known frequency and intensity in order to measure an individual's hearing ability.

auricle 1. *See* atrium. **2.** *See* pinna.

Australopithecus A genus of fossil primates that lived 1.3–4 million years ago. They walked erect and had teeth resembling those of modern man, but the brain capacity was less than half that of a human. Various finds have been made, chiefly in East and South Africa (hence the name, which means 'southern ape'). Australopithecus and related genera are known as *australopithecines. See also Homo.*

autoclave A strong steel vessel used for carrying out chemical reactions, sterilizations, etc., at high temperature and pressure.

autogamy 1. A type of reproduction that occurs in single isolated individuals of ciliate protozoan animals of the genus *Paramecium.* The nucleus divides into two genetically identical haploid nuclei, which then fuse to form a diploid zygote. The onset of autogamy is associated with changing environmental conditions and may be necessary to maintain cell vitality. **2.** Self-fertilization in plants. *See* fertilization.

autograft *See* graft.

autoimmunity A disorder of the body's defence mechanism in which an *immune response is elicited against its own tissues. Rheumatoid arthritis is basically an autoimmune disease.

autolysis The process of self-destruction of a cell, cell organelle, or tissue. It occurs by the action of enzymes within or released by *lysosomes. *See also* lysis.

autonomic nervous system (ANS) The part of the vertebrate *peripheral nervous system that supplies stimulation via motor nerves to the smooth and cardiac muscles (the involuntary muscles) and to the glands of the body. It is divided into the *parasympathetic and the *sympathetic nervous systems, which tend to work antagonistically on the same organs. The activity of the ANS is controlled principally by the *medulla oblongata and *hypothalamus of the brain.

autoradiography An experimental technique in which a radioactive specimen is placed in contact with (or close to) a photographic plate, so as to produce a record of the distribution of radioactivity in the specimen. The film is darkened by the ionizing radiation from radioactive parts of the sample. Autoradiography is used to study the distribution of particular substances in living tissues and cells. A radioactive isotope of the substance is introduced into the organism or tissue, which is killed, sectioned, and examined after enough time has elapsed for the isotope to be incorporated into the substance.

autosome Any of the chromosomes in a cell other than the *sex chromosomes.

autotomy The shedding by an animal of part of its body followed by the regeneration of the lost part. Autotomy is achieved by the contraction of muscles at specialized regions in the body. It serves as a protective mechanism if the animal is damaged or attacked (e.g. tail loss in certain reptiles) and is common as a method of asexual reproduction in polychaete worms, in which both new head and tail regions may be regenerated.

autotrophism A type of nutrition in which organisms synthesize the organic materials they require from inorganic sources. Chief sources of carbon and nitrogen are carbon dioxide and nitrates, respectively. All green plants are autotrophic and use light as a source of energy for the synthesis, i.e. they are *photoautotrophic* (*see* photosynthesis). Some bacteria are also photoautotrophic; others are *chemoautotrophic*, using energy derived from chemical processes. *Compare* heterotrophism.

auxanometer Any mechanical instrument or measuring device used to study the growth or movement of plant organs. One type of auxanometer consists of a recording device that translates any increase in stem height into movement of a needle across a scale.

auxin Any of a group of plant *growth substances responsible for such processes as the promotion of growth by cell enlargement, the maintenance of *apical dominance, and the initiation of root formation in cuttings. Naturally occurring auxins, such as *indoleacetic acid* (*IAA*), are synthesized in the shoot tips. Synthetic auxins, such as *2,4-D* and *2,4,5-T*, are used as weedkillers.

Aves The birds: a class of bipedal vertebrates with *feathers, wings, and a beak. They evolved from reptilian ancestors in the Jurassic period (190–136 million years ago) and modern birds still have scaly legs, like reptiles. Birds are warm-blooded (*see* homoiothermy). The skin is dry and loose and has no sweat glands, so cooling is effected by panting. Their efficient lungs and four-chambered heart (which completely separates oxygenated and deoxygenated blood) ensure a good supply of oxygen to the tissues. Birds can therefore sustain a high body temperature and level of activity necessary for flight. The breastbone bears a keel for the attachment of flight muscles. The skeleton is very light; many of the bones are tubular, having internal struts to provide strength and air sacs to reduce weight and provide extra oxygen in flight. Their feathers are vital for flight, streamlining the body, and insulation against heat loss.

Many birds show a high degree of social behaviour in forming large flocks and pair bonding for nesting, egg incubation, and rearing young. Fertilization is internal and the female lays hard-shelled eggs. *See also* Ratitae.

axenic culture A *culture medium in which only one type of microorganism is growing. Such cultures are widely used in microbiology to determine the basic growth requirements or degree of inhibition by antibiotics or other chemicals of a particular species.

axil The angle between a branch or leaf and the stem it grows from. *Axillary* (or *lateral*) buds develop in the axil of a leaf. The presence of axillary buds distinguishes a leaf from a leaflet.

axis The second *cervical vertebra, which articulates with the *atlas (the first cervical vertebra, which articulates with the skull). The articulation between the axis and atlas in reptiles, birds, and mammals permits side-to-side movement of the head. The body of the axis is elongated to form a peg (the *odontoid process*), which extends into the ring of the atlas and acts as a pivot on which the atlas (and skull) can turn.

axon The long threadlike part of a nerve cell (*neurone). It carries the nerve impulse (in the form of an *action potential) away from the *cell body of a neurone towards either an effector organ or the brain. *See also* nerve fibre.

B

Bacillariophyta A division of *algae comprising the diatoms. These marine or freshwater unicellular algae have cell walls composed of pectin impregnated with silica and consisting of two halves, one overlapping the other. Diatoms are found in huge numbers in plankton and are important in the food chains of seas and rivers. Past deposition has resulted in diatomaceous earths (kieselguhr) and the oil reserves of these species have contributed to oil deposits.

bacillus Any rod-shaped bacterium. Generally, bacilli are large, Gram-positive, spore-bearing, and have a tendency to form chains and produce a *capsule. Some are motile, bearing flagella. They are ubiquitous in soil and air and many are responsible for food spoilage. The group also includes *Bacillus anthracis*, which causes anthrax.

backbone *See* vertebral column.

back cross (test cross) A mating made to identify hidden *recessive alleles. If an organism displays a *dominant characteristic, it may possess two dominant alleles (i.e. it is homozygous) or a dominant and a recessive allele for that characteristic (i.e. it is heterozygous). To find out which is the case, the organism is crossed with one displaying the recessive characteristic. If all the offspring show the dominant characteristic then the organism is homozygous, but if half show the recessive characteristic, then the organism is heterozygous.

bacteria A diverse group of ubiquitous microorganisms all of which consist of only a single *cell that lacks a distinct nuclear membrane and has a *cell wall of a unique composition. Bacteria are usually classified by means of *Gram's stain, whether or not they require oxygen (*see* aerobic respiration;

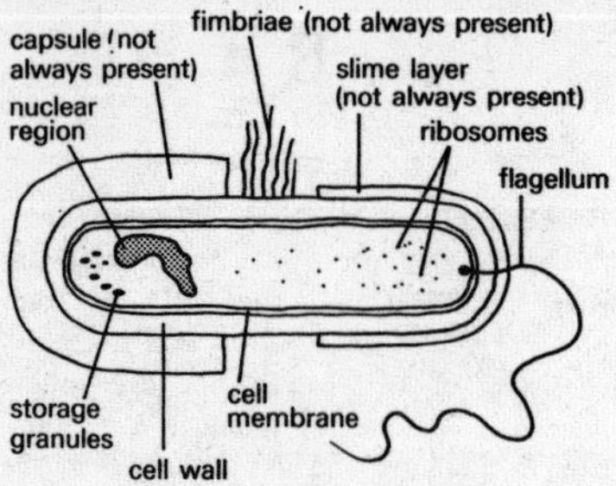

A generalized bacterial cell

anaerobic respiration), and on the basis of shape. A bacterial cell may be spherical (*see* coccus), rodlike (*see* bacillus), spiral (*see* spirillum), comma-shaped (*see* vibrio), corkscrew-shaped (*see* spirochaete), or filamentous, resembling a fungal cell. The majority of bacteria range in size from 0.5 to 5 μm. Many are motile, bearing flagella, possess an outer slimy *capsule, and produce resistant spores (*see* endospore). In general bacteria reproduce only asexually, by simple division of cells, but a few groups undergo a form of sexual reproduction (*see* conjugation). Bacteria are largely responsible for decay and decomposition of organic matter, producing a cycling of such chemicals as carbon (*see* carbon cycle), oxygen, nitrogen (*see* nitrogen cycle), and sulphur. A few bacteria obtain their food by means of *photosynthesis, some are saprophytes, and others are parasites, causing disease in man and other animals, plants, and other microorganisms. The symptoms of bacterial infections are produced by *toxins.

bacteriocidal Capable of killing bacteria. Common bacteriocides are some *antibiotics, *antiseptics, and *disinfectants. *Compare* bacteriostatic.

bacteriophage (phage) A virus that is parasitic within bacteria. Each phage is specific for only one type of bacterium. Phage infect, quickly multiply within, and destroy their host cells. They are used experimentally to identify bacteria, to control manufacturing processes (such as cheese production) that depend on bacteria, and, because they can alter the genetic make-up of bacterial cells, they are important tools in *genetic engineering.

bacteriostatic Capable of inhibiting or slowing down the growth and reproduction of bacteria. Some *antibiotics are bacteriostatic. *Compare* bacteriocidal.

baleen *See* whalebone.

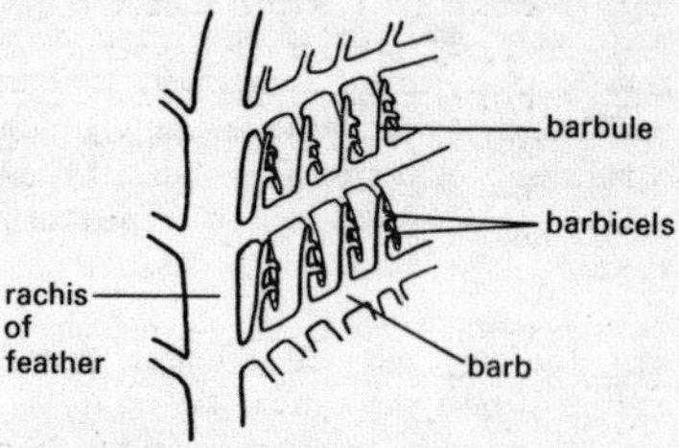

Interlocking barbs of a contour feather

barb 1. (in zoology) Any one of the stiff filaments forming a row on each side of the longitudinal shaft of a feather. Together the barbs form the expanded part (*vane*) of the feather. *See* barbule. **2.** (in botany) A hooked hair.

barbule Any of the minute filaments forming a row on each side of the *barb of a feather. In a *contour feather adjacent barbules interlock by means of hooks (*barbicels*) and grooves, forming a firm vane. Down feathers have no barbicels.

Barfoed's test A biochemical test to detect monosaccharide (reducing) sugars in solution, devised by the Swedish physician C. T. Barfoed (1815–99). *Barfoed's reagent*, a mixture of ethanoic (acetic) acid and copper(II) acetate, is added to the test solution and boiled. If any reducing sugars are present a red precipitate of copper(II) oxide is formed. The reaction will be negative in the presence of disaccharide sugars as they are weaker reducing agents.

bark The protective layer of mostly dead cells that covers the outside of woody stems and roots. It includes the living and dead tissues external to the xylem, including the phloem and periderm. The term can be used more specifically to describe the periderm together with other tissues isolated by the activity of the *cork cambium. In some species, such as birch, there is one persistent cork cambium but in the older stems of certain other species a second cork cambium becomes active beneath the periderm and further periderm layers are formed every few years. The result is a composite tissue called *rhytidome*, composed of cork, dead cortex, and dead phloem cells.

baroreceptor A *receptor that responds to changes in pressure. The *carotid sinus in the carotid artery contains baroreceptors that respond to changes in arterial pressure and are therefore involved in the regulation of blood pressure and heart beat.

Barr body A structure consisting of a condensed X chromosome (*see* sex chromosome) that is found in nondividing nuclei of female mammals. The presence of a Barr body is used to confirm the sex of athletes in sex determination tests. It is named after the Canadian anatomist M. L. Barr (1908–), who identified it in 1949.

basal ganglia Small masses of nervous tissue within the brain that connect the *cerebrum with other parts of the nervous system. They are involved with the subconscious regulation of voluntary movements.

basal metabolic rate (BMR) The rate of energy metabolism required to maintain an animal at rest. BMR is measured in terms of heat production per unit time: it indicates the energy consumed in order to sustain such vital functions as heartbeat, breathing, nervous activity, active transport, and secretion. Different tissues have different metabolic rates (e.g. the BMR of brain tissue is much greater than that of bone tissue) and therefore the tissue composition of an animal determines its overall BMR. For any comparable group of animals (such as mammals) BMR is proportional to body weight according to the allometric equation (*see* allometric growth); small animals tend to have a higher metabolic rate per unit weight than large ones.

base (in biochemistry) *See* nitrogenous base.

base pairing The chemical linking of two complementary nitrogenous bases in *DNA and in certain types of *RNA molecules. Of the four such bases in DNA, adenine pairs with thymine and cytosine with guanine. In RNA, thymine is replaced by uracil. Base pairing is responsible for holding together the two strands of a DNA molecule to form a double helix and for faithful reproduction and reading of the *genetic code. The links between bases take the form of *hydrogen bonds.

basic stains *See* staining.

Basidiomycetes A class of *fungi in which sexual reproduction is by means of *basidiospores* (spores produced externally on a club-shaped or cylindrical cell, the *basidium*). Basidia are often grouped together forming fruiting structures, such as mushrooms, puffballs, and bracket fungi. Exceptions are the *rusts and *smuts, which do not produce obvious fruiting bodies.

bast An old name for *phloem.

bats *See* Chiroptera.

becquerel Symbol Bq. The SI unit of activity (*see* radiation units). The unit is named after the discoverer of radioactivity, A. H. Becquerel (1852–1908).

bees *See* Hymenoptera.

beetles *See* Coleoptera.

beet sugar *See* sucrose.

behavioural genetics The branch of genetics concerned with determining the relative importance of the genetic constitution of animals as compared to environmental factors in influencing animal behaviour.

Benedict's test A biochemical test to detect reducing sugars in solution, devised by the US chemist S. R. Benedict (1884–1936). *Benedict's reagent* – a mixture of copper(II) sulphate and a filtered mixture of hydrated sodium citrate and hydrated sodium carbonate – is added to the test solution and boiled. A high concentration of reducing sugars induces the formation of a red precipitate; a lower concentration produces a yellow precipitate. Benedict's test is a more sensitive alternative to *Fehling's test.

benthos Flora and fauna occurring on the bottom of a sea or lake. Benthic organisms may crawl, burrow, or remain attached to a substrate. *Compare* pelagic.

berry A fleshy fruit formed from either one carpel or from several fused together and containing many seeds. The fruit wall may have two or three layers but the inner layer is never hard and stony (as in some drupes). Examples of berries are grapes and tomatoes. A berry, such as a cucumber, that develops a hard outer rind is called a *pepo.* One that is segmented and has a leathery rind, such as a citrus fruit, is called a *hesperidium.* The rind contains oil glands and is lined by the white mesocarp, commonly called *pith.*

bicuspid valve *See* mitral valve.

biennial A plant that requires two growing seasons to complete its life cycle. During the first year it builds up food reserves, which are used during the second year in the production of flowers and seeds. Examples are carrot and parsnip.

bilateral symmetry A type of arrangement of the parts and organs of an animal in which the body can be divided into two halves that are mirror images of each other along one plane only (usually passing through the midline at right angles to the dorsal and ventral surfaces). Bilaterally symmetrical animals are characterized by a type of movement in which one end of the body always leads. In botany this type of symmetry is usually called *zygomorphy* when applicable to flowers (e.g. foxglove and antirrhinum flowers are zygomorphic). *Compare* radial symmetry.

bile (gall) A bitter-tasting greenish-yellow alkaline fluid produced by the *liver, stored in the *gall bladder, and secreted into the *duodenum of vertebrates. It assists the digestion and absorption of fats by the action of *bile salts,* which chemically reduce fatty substances and decrease the surface tension of fat droplets so that they are broken down and emulsified. Bile may also stimulate gut muscle contraction (*peristalsis). Bile also contains the pigments *bilirubin* and *biliverdin,* which are produced by the breakdown of the blood pigment *haemoglobin.

bile duct The tube through which bile passes from the *liver or (when present) the *gall bladder to the duodenum.

bilirubin *See* bile.

biliverdin *See* bile.

binocular vision The ability, found only in animals with forward-facing eyes, to produce a focused image of the same object simultaneously on the retinas of both eyes.

This permits three-dimensional vision and contributes to distance judgment.

binomial nomenclature The system of naming organisms using a two-part Latinized (or scientific) name that was devised by the Swedish botanist Linnaeus (Carl Linné; 1707–78); it is also known as the *Linnaean system.* The first part is the generic name (*see* genus), the second is the specific or trivial name (*see* species). The Latin name is usually printed in italics, starting with a capital letter. For example, in the scientific name of the common frog, *Rana temporaria*, *Rana* is the generic name and *temporaria* the species name. The name of the species may be followed by an abbreviated form of the name of its discoverer; for example, the common daisy is *Bellis perennis* L. (for Linnaeus). There are several International Codes of Taxonomic Nomenclature that lay down the rules for naming organisms. *See also* classification; taxonomy.

bioassay (biological assay) A controlled experiment for the quantitative estimation of a substance by measuring its effect in a living organism. For example, the amount of the plant hormone auxin can be estimated by observing its effect on the curvature of oat coleoptiles – the concentration of the hormone is proportional to the curvature of the coleoptile.

biochemical oxygen demand (BOD) The amount of oxygen taken up by microorganisms that decompose organic waste matter in water. It is therefore used as a measure of the amount of certain types of organic pollutant in water. BOD is calculated by keeping a sample of water containing a known amount of oxygen for five days at 20°C. The oxygen content is measured again after this time. A high BOD indicates the presence of a large number of microorganisms, which suggests a high level of pollution.

biochemical taxonomy *See* taxonomy.

biochemistry The study of the chemistry of living organisms, especially the structure and function of their chemical components (principally proteins, carbohydrates, lipids, and nucleic acids). Biochemistry has advanced rapidly with the development, from the mid-20th century, of such techniques as chromatography, X-ray diffraction, radioisotopic labelling, and electron microscopy. Using these techniques to separate and analyse biologically important molecules, the steps of the metabolic pathways in which they are involved (e.g. *glycolysis and the *Krebs cycle) have been determined. This has provided some knowledge of how organisms obtain and store energy, how they manufacture and degrade their biomolecules, how they sense and respond to their environment, and how all this information is carried and expressed by their genetic material. Biochemistry forms an important part of many other disciplines, especially physiology, nutrition, and genetics, and its discoveries have made a profound impact in medicine, agriculture, industry, and many other areas of human activity.

biodegradable *See* pollution.

bioenergetics The study of the flow and the transformations of energy that occur in living organisms. Typically, the amount of energy that an organism takes in (from food or sunlight) is measured and divided into the amount used for growth of new tissues; that lost through death, wastes, and (in plants) transpiration; and that lost to the environment as heat (through respiration).

bioengineering **1.** The use of artificial tissues, organs, and organ components to replace parts of the body that are damaged, lost, or malfunctioning, e.g. artificial limbs, heart valves, and heart pacemakers. **2.** The application of engineering knowledge to medicine and zoology.

biogenesis The principle that a living organism can only arise from other living organisms similar to itself (i.e. that like gives rise to like) and can never originate from nonliving material. *Compare* spontaneous generation.

biological clock The mechanism, presumed to exist within many animals and plants, that produces regular periodic changes in behaviour or physiology. Biological clocks underlie many of the *circadian, tidal, and annual rhythms seen in organisms (e.g. hibernation in animals). They continue to run even when conditions are kept artificially constant, but eventually drift out of step with the natural environment without the specific signals that normally keep them synchronized.

biological control The control of pests by biological (rather than chemical) means. This may be achieved, for example, by breeding disease-resistant crops or by introducing a natural enemy of the pest, such as a predator or a parasite. This technique, which may offer substantial advantages over the use of pesticides or herbicides, has been employed successfully on a number of occasions. One notable example was the control of the prickly pear cactus (*Opuntia*) in Australia by introducing the cactus moth (*Cactoblastis cactorum*), whose caterpillars feed on the plant's growing shoots. Insect pests have been controlled by releasing large numbers of males of the pest species that have been sterilized by radiation: infertile matings subsequently cause a decline in the pest population.

biology The study of living organisms, which includes their structure (gross and microscopical), functioning, origin and evolution, classification, interrelationships, and distribution.

bioluminescence The emission of light without heat by living organisms. The phenomenon occurs in glow-worms and fireflies, bacteria and fungi, and in many deep-sea fish (among others); in animals it may serve as a means of protection (e.g. by disguising the shape of a fish) or species recognition or it may provide mating signals. The light is produced during the oxidation of a compound called *luciferin* (the composition of which varies according to the species), the reaction being catalysed by an enzyme, *luciferase*. Bioluminescence may be continuous (e.g. in bacteria) or intermittent (e.g. in fireflies).

biomass The total mass of all the organisms of a given type and/or in a given area; for example, the world biomass of trees, or the biomass of elephants in the Serengeti National Park.

biome A large naturally occurring assemblage of plant and animal species that are of the same general type, being adapted to the particular conditions in which they occur. Examples of biomes are tundra and tropical rain forest.

biophysics The study of the physical aspects of biology, including the application of physical laws and the techniques of physics to study biological phenomena.

biopoiesis The development of living matter from complex organic molecules that are themselves nonliving but self-replicating. It is the process by which life is assumed to have begun. *See* origin of life.

biorhythm A roughly periodic change in the behaviour or physiology of an organism that is generated and maintained by a *biological clock. A well-known example is the circadian rhythm occurring in many animals and plants. The term biorhythm has unfortunately also become associated with a pseudoscientific cult postulating that man's physical, emotional, and intellectual behaviour is influenced by three biological cycles with periodicities of exactly 23 days, 28 days, and 33 days respectively. There is no reliable evidence to support this theory.

biosphere The whole of the region of the earth's surface, the sea, and the air that is inhabited by living organisms.

biosynthesis The production of molecules by a living cell, which is the essential feature of *anabolism.

biosystematics *See* systematics.

biotechnology The development of techniques for the application of biological processes to the production of materials of use in medicine and industry. For example, the production of antibiotics, cheese, and wine rely on the activity of various fungi and bacteria. *Genetic engineering can modify bacterial cells to synthesize completely new substances, e.g. hormones.

biotic environment The part of an organism's environment that consists of other living organisms. These may affect it in many ways; for example, as competitors, predators, parasites, prey, or symbionts. In time, the distribution and abundance of the organism will be affected by its interrelationships with the biotic environment.

biotin A vitamin in the *vitamin B complex. It is the *coenzyme for various enzymes that catalyse the incorporation of carbon dioxide into various compounds. Adequate amounts are normally produced by the intestinal bacteria in animals although deficiency can be induced by consuming large amounts of raw egg white. This contains a protein, avidin, that specifically binds biotin, preventing its absorption from the gut. Other sources of biotin include cereals, vegetables, milk, and liver.

birds *See* Aves.

bisexual (in biology) *See* hermaphrodite.

biuret test A biochemical test to detect proteins in solution, named after the substance *biuret* ($H_2NCONHCONH_2$), which is formed when urea is heated. Sodium hydroxide is mixed with the test solution and drops of 1% copper(II) sulphate solution are then added slowly. A positive result is indicated by a violet ring, caused by the reaction of *peptide bonds in the proteins or peptides. Such a result will not occur in the presence of free amino acids.

bivalent (in genetics) *See* pairing.

bivalves *See* Lamellibranchia.

bladder 1. (in anatomy) **a.** A hollow muscular organ in most vertebrates, also known as the *urinary bladder*, in which urine is stored before being discharged. In mammals urine is conveyed from the *kidneys to the bladder by the *ureters and is discharged to the outside through the *urethra. **b.** Any of various other saclike organs in animals for the storage of liquid or gas. *See* gall bladder; swim bladder. **2.** (in botany) **a.** A modified submerged leaf of certain aquatic insectivorous plants, such as the bladderwort (*Utricularia*). It forms a hollow with a single opening that is sealed by a valve to trap small aquatic invertebrates after they have been sucked in. **b.** An air-filled cavity in the thallus of certain seaweeds, such as the bladderwrack (*Fucus vesiculosus*).

blastocoel *See* blastula.

blastocyst *See* blastula; implantation.

blastula The first stage of development of an animal embryo, formed as a result of *cleavage of a fertilized egg. This stage generally resembles a hollow ball with the dividing cells (*blastomeres*) of the embryo forming a layer (*blastoderm*) around a central cavity (*blastocoel*). Insect eggs have no blastula. In vertebrates the blastula forms a disc (*blastodisc*) on the surface of the yolk. In mammals the blastula stage is known as a *blastocyst*. *See also* gastrula.

blending inheritance The early theory that assumed that hereditary substances from parents merge together in their offspring. Mendel showed that this does not occur (*see* Mendel's laws). In breeding experiments an appearance of blending may result from co-dominant alleles (*see* co-dominance) and *polygenes but close study shows that the alleles retain their identity through successive generations. *Compare* particulate inheritance.

blind spot The portion of the retina at which blood vessels and nerve fibres enter the optic nerve. There are no rods or cones in this area, so no visual image can be transmitted from it.

blood A fluid body tissue that acts as a transport medium within an animal. It is contained within a blood *vascular system and in vertebrates is circulated by means of contractions of the *heart. Oxygen and food are carried to tissues, and carbon dioxide and chemical (nitrogenous) waste are transported from tissues to excretory organs for disposal (*excretion). In addition blood carries *hormones and also acts as a defence system. Blood consists of a liquid (*see* blood plasma) containing blood cells (*see* erythrocyte; leucocyte) and *platelets.

blood–brain barrier The mechanism that controls the passage of substances from the blood to the cerebrospinal fluid bathing the brain and spinal cord. It takes the form of a semipermeable lipid membrane permitting the passage of solutions but excluding particles and large molecules. This barrier provides the central nervous system with a constant environment, while not interfering with the transport of essential metabolites.

blood cell (blood corpuscle) Any of the cells that are normally found in the blood plasma. These include red cells (*see* erythrocyte) and white cells (*see* leucocyte).

blood clotting (blood coagulation) The production of a mass of semisolid material at the site of an injury that closes the wound, helping to prevent further blood loss and bacterial invasion. The clot is formed

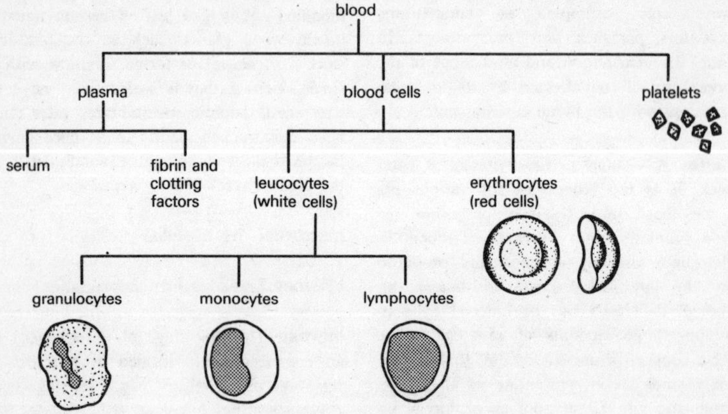

Composition of mammalian blood

by the action of *clotting factors and *platelets, which produce a network of *fibrin fibres in which blood cells become entangled. This is at first jelly-like but gradually dries to form a crust or scab.

blood groups The many types into which an individual's blood may be classified, based on the presence or absence of certain proteins (*antigens) on the surface of the red blood cells. Blood of one group contains *antibodies in the serum that react against the antigens on the cells of other groups. *Incompatibility* between groups results in clumping of cells (*agglutination), so knowledge of blood groups is important for blood transfusions. In man, the two most important blood group systems are the *ABO system and the system involving the *rhesus factor.

blood plasma The liquid part of the *blood (i.e. excluding blood cells). It consists of water containing a large number of dissolved substances, including proteins, salts (especially sodium and potassium chlorides and bicarbonates), food materials (glucose, amino acids, fats), hormones, vitamins, and excretory materials. *See also* blood serum; lymph.

blood pressure The pressure exerted by the flow of blood through the major arteries of the body. This pressure is greatest during the contraction of the ventricles of the heart (*systolic pressure*; *see* systole), which forces blood into the arterial system. Pressure falls to its lowest level when the heart is filling with blood (*diastolic pressure*; *see* diastole). Blood pressure is measured in millimetres of mercury using an instrument called a *sphygmomanometer*. Normal blood pressure for a young average adult human is in the region of 120/80 mmHg (the higher number is the systolic blood pressure; the lower number the diastolic blood pressure), but individual variations are common. Abnormally high blood pressure (*hypertension*) may be associated with disease or it may occur without an apparent cause.

blood serum Blood plasma from which the fibrin and clotting factors have been removed by centrifugation or vigorous stirring, so that it cannot clot. Serum containing a specific antibody or antitoxin may be used in the treatment or prevention of certain infections. Such serum is generally derived from a nonhuman mammal (e.g. a horse).

blood vascular system The tissues and organs of an animal that transport blood through the body. In vertebrates it consists of the heart and blood vessels. *See* vascular system.

blood vessel A tubular structure through which the blood of an animal flows. *See* artery; arteriole; capillary; venule; vein.

blue-green algae *See* Cyanophyta.

BOD *See* biochemical oxygen demand.

body cavity The internal cavity of the body of an animal, which is present in most invertebrates and all vertebrates and contains the major organs. The body cavity of vertebrates and many invertebrates is the *coelom. In vertebrates the body cavity is divided by a transverse septum just posterior to the heart into the abdominal and thoracic cavities (*see* abdomen; thorax). In mammals the septum is the *diaphragm.

bone The hard connective tissue of which the *skeleton of most vertebrates is formed. It comprises a matrix of *collagen fibres (30%) impregnated with bone salts (70%), mostly calcium phosphate, in which are embedded bone cells (*osteoblasts* and *osteocytes*), which secrete the matrix. Bone generally replaces embryonic *cartilage and is of two sorts – compact bone and spongy bone. The outer *compact bone* is formed as concentric layers (*lamellae*) that surround

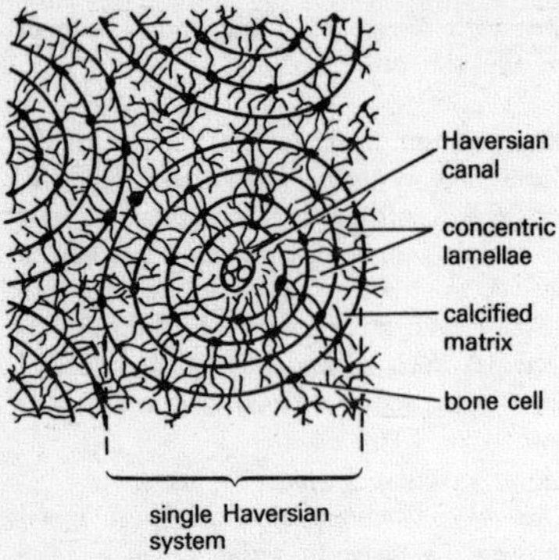

Structure of compact bone

small holes (*Haversian canals). The inner *spongy bone* is chemically similar but forms a network of bony bars. The spaces between the bars may contain bone marrow or (in birds) air for lightness. *See also* cartilage bone; membrane bone.

bone marrow A soft tissue contained within the central cavity and internal spaces of a bone. At birth and in young animals the marrow of all bones is concerned with the formation of blood cells: it contains *myeloid tissue and is known as *red marrow*. In mature animals the marrow of the long bones ceases producing blood cells and is replaced by fat, being known as *yellow marrow*.

bony fishes *See* Osteichthyes.

borax carmine A red dye, used in optical microscopy, that stains nuclei and cytoplasm pink. It is frequently used to stain large pieces of animal tissue.

botany The scientific study of plants, including their anatomy, morphology, physiology, biochemistry, taxonomy, cytology, genetics, ecology, evolution, and geographical distribution.

Bowman's capsule The cup-shaped end of a kidney *nephron. It is named after its discoverer, the British physician Sir William Bowman (1816–92).

bract A modified leaf with a flower or inflorescence in its axil. Bracts are often brightly coloured and may be mistaken for the petals of a flower. For example the showy 'flowers' of *Poinsettia* and *Bougainvillea* are composed of bracts; the true flowers are comparatively inconspicuous. *See also* involucre.

bracteole A reduced leaf that arises from the stalk of an individual flower.

bradykinin *See* kinin.

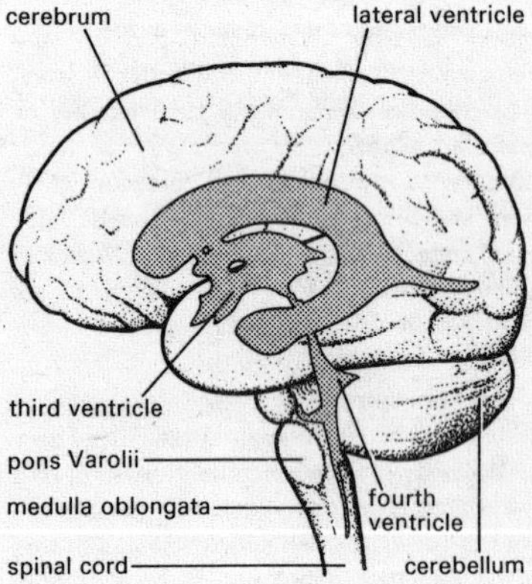

The human brain

brain 1. The enlarged anterior part of the vertebrate central nervous system, which is encased within the cranium of the skull. Continuous with the spinal cord, the brain is surrounded by three membranes (*see* meninges) and bathed in cerebrospinal fluid, which fills internal cavities (*ventricles). It functions as the main coordinating centre for nervous activity, receiving information (in the form of nerve impulses) from sense organs, interpreting it, and transmitting 'instructions' to muscles and other *effectors. It is also the seat of intelligence and memory.

The embryonic vertebrate brain is in three sections (*see* forebrain; hindbrain; midbrain), which become further differentiated during development into specialized regions. The main parts of the adult human brain are a highly developed *cerebrum in the form of two cerebral hemispheres, a *cerebellum, *medulla oblongata, and *hypothalamus. **2.** A concentration of nerve *ganglia at the anterior end of an invertebrate animal.

brain death The permanent absence of vital functions of the brain, which is marked by cessation of breathing and other reflexes controlled by the *brainstem and by a zero reading on an *electroencephalogram. Organs may be removed for transplantation when brain death is established, which may not necessarily be associated with permanent absence of heart beat.

brainstem The part of the brain comprising the *medulla oblongata, the *midbrain, and the *pons. It resembles and is continuous with the spinal cord. The midbrain controls and integrates reflex activities (such as respiration) that originate in higher centres of the brain via a network of nerve pathways (the *reticular formation*).

breathing *See* respiratory movement.

breed A domesticated *variety of an animal or, rarely, a cultivated variety of plant. Cultivated plants are more often simply called varieties or, more correctly, *cultivars. Examples of animal breeds are Friesian cattle and Shetland sheepdogs.

bronchiole A fine respiratory tube in the lungs of reptiles, birds, and mammals. It is formed by the subdivision of a *bronchus and in reptiles and mammals it terminates in a number of *alveoli.

bronchus (bronchial tube) One of the major air tubes in the *lung. The *trachea divides into two main bronchi, one for each lung, which split into smaller bronchi and then into *bronchioles. The walls of the bronchi are stiffened by rings of cartilage.

brown algae *See* Phaeophyta.

brown fat A darker coloured region of *adipose tissue found in newborn and hibernating animals (in which it may also be called the *hibernating gland*). Compared to normal white *fat, deposits of brown fat are more richly supplied with blood vessels and have a higher proportion of unsaturated fatty acids. They can also be more rapidly converted to heat energy, especially during arousal from hibernation and during cold stress in young animals. Since the deposits are strategically placed near major blood vessels, the heat they generate warms the blood returning to the heart.

It has been said that some types of obesity in man may be linked to a lack of brown fat in affected individuals: in normal individuals, an intake of excess food is converted into brown fat and rapidly metabolized.

Brownian movement The continuous random movement of microscopic solid particles (of about 1 micrometre in diameter) when suspended in a fluid medium. First observed by the botanist Robert Brown (1773–1858) in 1827 when studying pollen particles, it was originally thought to be the manifestation of some vital force. It was later recognized to be a consequence of bombardment of the particles by the continually moving molecules of the liquid. The smaller the particles the more extensive is the motion. It can be observed in the particles of a colloidal solution and in the protoplasm of dead cells.

Bryophyta A division of simple plants possessing no vascular tissue and rudimentary rootlike organs (rhizoids). It contains the classes *Musci (mosses) and *Hepaticae (liverworts). Bryophytes grow in a variety of damp habitats, from fresh water to rock surfaces. Some use other plants for support. Bryophytes show a marked *alternation of generations between gamete-bearing forms (gametophytes) and spore-bearing forms (sporophytes), the latter being dependent on the former for water and nutrients. The leaves, stems, and roots of the gametophyte generation are not equivalent to those of higher (vascular) plants since the whole structure is *haploid.

Bryozoa *See* Polyzoa.

buccal cavity The mouth cavity: the beginning of the *alimentary canal, which leads to the pharynx and (in vertebrates) to the oesophagus. In mammals it contains the tongue and teeth, which assist in the mechanical breakdown of food, and the openings of the *salivary gland ducts.

bud 1. (in botany) A condensed immature shoot with a short stem bearing small folded or rolled leaves. The outer leaves of a bud are often scalelike and protect the delicate inner leaves. A *terminal* (or *apical*) *bud* exists at the tip of a stem or branch while *axillary* (or *lateral*) *buds* develop in the *axils of leaves. However, in certain circumstances buds can be produced anywhere on the surface of a plant. Some buds remain dormant, but may become active if the terminal bud is removed. It is common gardening practice to remove the terminal buds of some shoots to induce the development of lateral shoots from axillary buds. *See also* apical dominance. **2.** (in biology) An outgrowth from a parent organism that breaks away and develops into a new individual in the process of *budding.

budding 1. (in biology) A method of asexual reproduction in which a new individual is derived from an outgrowth (*bud*) that becomes detached from the body of the parent. In animals the process is also called *gemmation*; it is common in coelenterates (e.g. *Hydra*) and also occurs in some sponges and other invertebrates. Among plants, budding is characteristic of the yeasts and other unicellular fungi. **2.** (in horticulture) A method of grafting in which a bud of the scion is inserted onto the stock, usually beneath the bark.

buffer A solution that resists change in pH when an acid or alkali is added or when the solution is diluted. Acidic buffers consist of a weak acid with a salt of the acid. The salt provides the negative ion A^-, which is the conjugate base of the acid HA. An example is carbonic acid and sodium hydrogencarbonate. Basic buffers have a weak base and a salt of the base (to provide the conjugate acid). An example is ammonia solution with ammonium chloride. In the hydrogencarbonate buffer, for example, molecules H_2CO_3 and ions HCO_3^- are present. When acid is added most of the extra protons are removed by the base:

$$HCO_3^- + H^+ \rightarrow H_2CO_3$$

When base is added, most of the extra hydroxide ions are removed by reaction with undissociated acid:

$$OH^- + H_2CO_3 \rightarrow HCO_3^- + H_2O$$

Thus, the addition of acid or base changes the pH very little.

Natural buffers occur in living organisms, where the biochemical reactions are very sensitive to change in pH (*see* acid–base balance). The main natural buffers are H_2CO_3/HCO_3^- and $H_2PO_4^-/HPO_4^{2-}$. Buffer solutions are also used in the laboratory (e.g. to keep microscopical preparations at their original pHs in order to prevent the formation of artefacts), in medicine (e.g. in intravenous injections), in agriculture, and in many industrial processes (e.g. dyeing, fermentation processes, and the food industry).

bugs *See* Hemiptera.

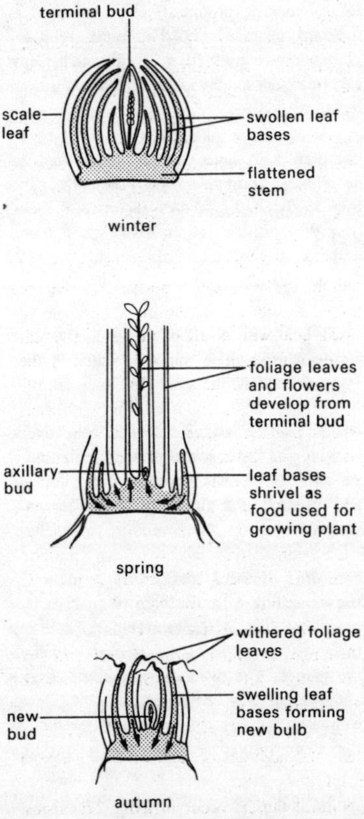

Development of a bulb

bulb An underground plant organ that enables a plant to survive from one growing season to the next. It is a modified shoot with a short flattened stem. A terminal bud develops at the centre of its upper surface, surrounded by swollen leaf bases that contain food stored from the previous growing season. Papery brown scale leaves cover the outside of the bulb. The stored food is used in the growing season when the terminal bud produces foliage leaves and flowers. The new leaves photosynthesize and some of the manufactured food passes into the leaf bases forming a new bulb. If more than one bud develops, then additional bulbs form, resulting in vegetative propagation. Examples of bulb-forming plants are daffodil, onion, and tulip. *Compare* corm.

bulbil A small bulblike organ that may develop in place of a flower, from an axillary bud, or at the base of a stem in certain plants. If it becomes detached it develops into a new plant.

bulla The rounded hollow projection of bone from the skull that encloses the *middle ear in mammals.

butanedioic acid *See* succinic acid.

butterflies *See* Lepidoptera.

buttress root *See* prop root.

C

caecum A pouch in the alimentary canal of vertebrates between the *small intestine and *colon. The caecum (and its *appendix) is large and highly developed in herbivorous animals (e.g. rabbits and cows), in which it contains a large population of bacteria essential for the breakdown of cellulose. In humans the caecum is a *vestigial organ and is poorly developed.

Cainozoic *See* Cenozoic.

calciferol *See* vitamin D.

calcitonin (thyrocalcitonin) A hormone, produced by the mammalian *thyroid gland, that lowers the concentration of calcium (and phosphate) in the blood. It operates in opposition to parathyroid hormone (*see* parathyroid glands).

calcium Symbol Ca. A soft grey metallic element that is an *essential element for living organisms, being required for normal growth and development. In animals it is an important constituent of bones and teeth and is present in the blood, being required for muscle contraction and other metabolic processes. In plants it is a constituent (in the form of calcium pectate) of the *middle lamella.

calcium cyclamate An organic calcium salt, $(C_6H_{11}NHSO_3)Ca.2H_2O$, formerly used as a sweetening agent. Its use was banned in the UK in 1969 as it was suspected of causing cancer.

callus 1. (in botany) A protective tissue, consisting of parenchyma cells, that develops over a cut or damaged plant surface. Callus tissue can also be induced to form in cell cultures by hormone treatment. **2.** (in pathology) A thick hard area of skin that commonly forms on the palms of the hands and soles of the feet as a result of continuous pressure or friction. **3.** (in physiology) Hard tissue formed round bone ends following a fracture, which is gradually converted to new bone.

calorie The quantity of heat required to raise the temperature of 1 gram of water by 1°C (1 K). The calorie, a c.g.s. unit, is now largely replaced by the *joule, an *SI unit. 1 calorie = 4.186 8 joules.

Calorie (kilogram calorie; kilocalorie) 1000 calories. This unit is still in limited use in estimating the energy value of foods, but is obsolescent.

calorific value The heat per unit mass produced by complete combustion of a given substance. Calorific values are used to express the energy values of fuels; usually these are expressed in megajoules per kilogram ($MJ\,kg^{-1}$). They are also used to measure the energy content of foodstuffs; i.e. the energy produced when the food is oxidized in the body. The units here are kilojoules per gram ($kJ\,g^{-1}$), although Calories (kilocalories) are often still used in non-technical contexts.

calyptra 1. A layer of cells that covers the developing sporophyte of bryophytes and pteridophytes. In mosses it forms a hood over the *capsule and in liverworts it forms a sheath at the base of the capsule stalk. **2.** *See* root cap.

calyptrogen The region within the root *apical meristem that divides to produce the *root cap (calyptra).

calyx The *sepals of a flower, collectively, forming the outer whorl of the *perianth. It encloses the petals, stamens, and carpels and protects the flower in bud. *See also* pappus.

cambium (lateral meristem) A plant tissue consisting of actively dividing cells (*see* meristem) that is responsible for increasing the girth of the plant, i.e. it causes secondary growth. The two most important cambia are the *vascular cambium* and the *cork cambium. The vascular cambium occurs in the stem and root; it divides to produce secondary *xylem and secondary *phloem (new food- and water-conducting tissues). *Compare* apical meristem.

Cambrian The earliest geological period of the Palaeozoic era. It is estimated to have begun about 570 million years ago and lasted for some 100 million years. During this period marine animals with mineralized shells made their first appearance and Cambrian rocks are the first to contain an abundance of fossils. Cambrian fossils are all of marine animals; they include *trilobites, which dominated the Cambrian seas, echinoderms, brachiopods, molluscs, and primitive *graptolites (from the mid Cambrian). Trace *fossils also provide evidence for a variety of worms.

Canada balsam A yellow-tinted resin used for mounting specimens in optical microscopy. It has similar optical properties to glass.

cancer Any disorder of cell growth that results in invasion and destruction of surrounding healthy tissue by the abnormal cells. Cancer cells arise from normal cells whose nature is permanently changed. They multiply more rapidly than healthy body cells and do not seem subject to normal control by nerves and hormones. They may spread via the bloodstream or lymphatic system to other parts of the body, where they produce further tissue damage (*metastases*). *Malignant tumours* are a form of cancer and *leukaemia* is cancer of white blood cells. Probable causative agents (*carcinogens*) include various chemicals, skin irritants (such as tobacco smoke), silica and asbestos particles, and *oncogenic viruses. Hereditary factors and stress may also play a role.

cane sugar *See* sucrose.

canine tooth A sharp conical *tooth in mammals that is large and highly developed in carnivores (e.g. dogs) for tearing meat. There are two canines in each jaw, each situated between the second *incisor and the first *premolar. In some animals (e.g. herbivores, such as giraffes and rabbits) canine teeth are absent.

capillarity *See* surface tension.

capillary The narrowest type of blood vessel in the vertebrate circulatory system. Capillaries conduct blood from *arterioles to all living cells: their walls are only one cell layer thick, so that oxygen and nutrients can pass through them into the surrounding tissues. Capillaries also transport waste material (e.g. urea and carbon dioxide) to venules for ultimate excretion. Capillaries can be constricted or dilated, according to local tissue requirements.

capitulum A type of flowering shoot (*see* racemose inflorescence) characteristic of plants of the family Compositae, e.g. daisy and dandelion. The tip of the shoot is flattened and bears many small stalkless flowers (*florets*) surrounded by an involucre (ring) of bracts. This arrangement gives the appearance of a single flower.

capsid The protein coat of a *virus. The chemical nature of the capsid is important in stimulating the body's *immune response against the invading virus.

capsule 1. (in botany) **a.** A dry fruit that releases its seeds when ripe; it is formed from several fused carpels and contains many seeds. The seeds may be dispersed through pores (as in the poppy), through a lid (as in plantain), or by the splitting and separation of the individual carpels (as in the crocus). Various other forms of capsules include the *silicula and *siliqua. **b.** The part of the sporophyte of mosses and liverworts in which the haploid spores are produced. It is borne on a long stalk (*seta*) and sheds its spores when mature (*see* peristome). **2.** (in microbiology) A thick gelatinous layer completely surrounding the cell wall of certain bacteria. It appears to have a protective function, making ingestion of the bacterial cell by *phagocytes more difficult and preventing desiccation. **3.** (in animal anatomy) **a.** The membranous or fibrous envelope that surrounds certain organs, e.g. the kidneys, spleen, and lymph nodes. **b.** The ligamentous sheath of connective tissue that surrounds various skeletal joints.

carapace 1. The dorsal part of the *exoskeleton of some crustaceans (e.g. crabs), which spreads like a shield over several segments of the head and thorax. **2.** The domed dorsal part of the shell of tortoises and turtles, formed of bony plates fused with the ribs and vertebrae and covered by a horny epidermal layer. The ventral part of the shell (*plastron*) is similar but flatter.

carbamide *See* urea.

carbohydrate One of a group of organic compounds based on the general formula $C_x(H_2O)_y$. The simplest carbohydrates are the *sugars (saccharides), including glucose and sucrose. *Polysaccharides are carbohydrates of much greater molecular weight and complexity; examples are starch, glycogen, and cellulose. Carbohydrates perform many vital roles in living organisms. Sugars, notably glucose, and their derivatives are essential intermediates in the conversion of food to energy. Starch and other polysaccharides serve as energy stores in plants, particularly in seeds, tubers, etc., which provide a major energy source for animals, including man. Cellulose, lignin, and others form the supporting cell walls and woody tissue of plants. Chitin is a structural polysaccharide found in the body shells of many invertebrate animals. Carbohydrates also occur in the surface coat of animal cells and in bacterial cell walls.

carbon Symbol C. A nonmetallic element that occurs in all organic compounds and is therefore basic to the structure of all living organisms. It is an *essential element for plants and animals, being ultimately derived from atmospheric carbon dioxide assimilated by plants during photosynthesis (*see* carbon cycle). The ubiquitous nature of carbon in living organisms is due to its unique ability to form stable covalent bonds with other carbon atoms and also with hydrogen, oxygen, nitrogen, and sulphur atoms, resulting in the formation of a variety of compounds containing chains and rings of carbon atoms.

There are two stable isotopes of carbon (proton numbers 12 and 13) and four radioactive ones (10, 11, 14, 15). Carbon-14 is used in *carbon dating.

carbon cycle One of the major cycles of chemical elements in the environment. Carbon (as carbon dioxide) is taken up from the atmosphere and incorporated into the tissues of plants in *photosynthesis. It may then pass into the bodies of animals as the plants are eaten (*see* food chain). During the respiration of plants, animals, and organisms that bring about decomposition, carbon dioxide is returned to the atmosphere. The combustion of fossil fuels (e.g. coal and peat, which are of organic origin) also releases carbon dioxide into the atmosphere.

carbon dating (radiocarbon dating) A method of estimating the ages of archaeological specimens of biological origin. As a result of cosmic radiation a small number of atmospheric nitrogen nuclei are continuously being transformed by neutron bombardment into radioactive nuclei of carbon-14. Some of these radiocarbon atoms find their way into living trees and other plants in the form of carbon dioxide, as a result of *photosynthesis. When the tree is cut down photosynthesis stops and the ratio of radiocarbon atoms to stable carbon atoms begins to fall as the radiocarbon decays. The ratio $^{14}C/^{12}C$ in the specimen can be measured and enables the time that has elapsed since the tree was cut down to be calculated. The method has been shown to give consistent results for specimens up to some 40 000 years old, though its accuracy depends upon assumptions concerning the past intensity of the cosmic radiation. The technique was developed by Willard F. Libby (1908-80) and his coworkers in 1946-47.

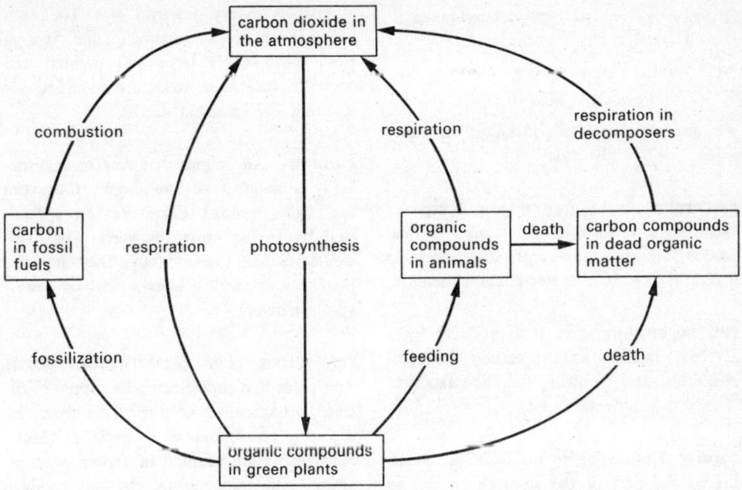

The carbon cycle in nature

carbon dioxide A colourless odourless gas, CO_2. It occurs in the atmosphere (0.03% by volume) but has a short residence time in this phase as it is both consumed by plants during *photosynthesis and produced by *respiration and by combustion.

The level of carbon dioxide in the atmosphere has been the subject of much environmental controversy as it is argued that extensive burning of fossil fuels will increase the overall CO_2 concentration and then, by the *greenhouse effect, increase atmospheric temperatures and cause climatic modification. The full significance of all the factors remains to be established.

Carboniferous A geological period in the Palaeozoic era. It began about 345 million years ago, following the Devonian period, and extended until the beginning of the Permian period, about 280 million years ago. In Europe the period is divided into the Lower and Upper Carboniferous, which roughly correspond to the Mississippian and Pennsylvanian periods, respectively, of North America. During the Lower Carboniferous a marine transgression occurred and the characteristic rock of this division – the Carboniferous limestone – was laid down in the shallow seas. Fauna included foraminiferans, corals, bryozoans, brachiopods, blastoids, and other invertebrates. The Upper Carboniferous saw the deposition of the millstone grit, a mixture of shale and sandstone formed in deltaic conditions, followed by the coal measures, alternating beds of coal, sandstone, shale, and clay. The coal was formed from the vast swamp forests composed of seed ferns, lycopsids, and other plants. During the period fishes continued to diversify and amphibians became more common.

carboxyl group The organic group –CO.OH, present in *carboxylic acids.

carboxylic acids Organic compounds containing the group –CO.OH (the *carboxyl group*; i.e. a carbonyl group attached to a hydroxyl group). Many long-chain carboxylic acids occur naturally as esters in fats and

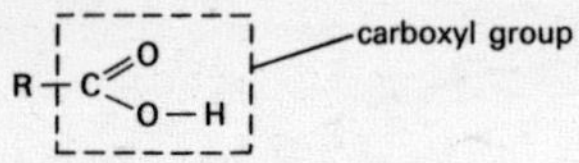

Carboxylic acid structure

oils and are therefore also known as *fatty acids. *See also* glycerides.

carcerulus A dry fruit that is a type of *schizocarp. It consists of a number of one-seeded fragments (*mericarps*) that adhere to a central axis. It is characteristic of mallow.

carcinogen Any agent that produces *cancer, e.g. tobacco smoke, certain industrial chemicals, and *ionizing radiation (such as X-rays and ultraviolet rays).

cardiac **1.** Relating to the heart. **2.** Relating to the part of the stomach nearest to the oesophagus.

cardiac muscle A specialized form of *muscle that is peculiar to the vertebrate heart. The muscle fibres are branched and interlock, and the muscle itself shows spontaneous contraction and does not need nervous stimulation (*see* pacemaker). The vagus nerve to the heart can, however, affect the rate of contraction.

carnassial teeth Molar and premolar teeth modified for shearing flesh by having cusps with sharp cutting edges. They are typical of animals of the order *Carnivora (e.g. tigers, wolves), in which they are the first molars in the lower jaw and the last premolars in the upper.

Carnivora An order of mainly flesh-eating mammals that includes the dogs, wolves, bears, badgers, weasels, and cats. Carnivores typically have very keen sight, smell, and hearing. The hinge joint between the lower jaw and skull is very tight, allowing no lateral movement of the lower jaw. This – together with the arrangement of jaw muscles – enables a very powerful bite. The teeth are specialized for stabbing and tearing flesh: canines are large and pointed and some of the cheek teeth are modified for shearing (*see* carnassial teeth).

carnivore An animal that eats meat, especially a member of the order *Carnivora (e.g. tigers, wolves). Carnivores are specialized by having strong powerful jaws and well-developed canine teeth. They may be predators or carrion eaters. *Compare* herbivore; omnivore.

carnivorous plant (insectivorous plant) Any plant that supplements its supply of nitrates in conditions of nitrate deficiency by digesting small animals, especially insects. Such plants are adapted in various ways to attract and trap the insects and produce proteolytic enzymes to digest them. Venus' fly trap (*Dionaea*), for example, has spiny-margined hinged leaves that snap shut on an alighting insect. Sundews (*Drosera*) trap and digest insects by means of glandular leaves that secrete a sticky substance, and pitcher plants (families Nepenthaceae and Sarraceniaceae) have leaves modified as pitchers into which insects fall, drowning in the water and digestive enzymes at the bottom.

carotene A member of a class of *carotenoid pigments. Examples are β-carotene and lycopene, which colour carrot roots and ripe tomato fruits respectively. α- and β-carotene yield vitamin A when they are broken down during animal digestion.

carotenoid Any of a group of yellow, orange, red, or brown plant pigments chemically related to terpenes. Carotenoids are responsible for the characteristic colour of many plant organs, such as ripe tomatoes, carrots, and autumn leaves. They also function in the light reactions of *photosynthesis. *See* carotene; xanthophyll.

carotid artery The major artery that supplies blood to the head. A pair of *common carotid arteries* arise from the aorta (on the left) and the innominate artery (on the right) and run up the neck; each branches into an *external* and an *internal carotid artery*, which supply the head.

carotid body One of a pair of tissue masses adjacent to the *carotid sinus. Each contains receptors that are sensitive to oxygen and pH levels (acidity) in the blood. High levels of carbon dioxide in the blood lower the pH (i.e. increase the acidity). By responding to fluctuations in pH, the carotid body coordinates reflex changes in respiration rate.

carotid sinus An enlarged region of the *carotid artery at its major branching point in the neck. Its walls contain many receptors that are sensitive to changes in pressure and it regulates blood pressure by initiating reflex changes in heart rate and dilation of blood vessels.

carpal (carpal bone) One of the bones that form the wrist (*see* carpus) in terrestrial vertebrates.

carpel The female reproductive organ of a flower. Typically it consists of a *stigma, *style, and *ovary. It is thought to have evolved by the fusion of the two edges of a flattened megasporophyll (*see* sporophyll). Each flower may have one carpel (*monocarpellary*) or many (*polycarpellary*), either free (*apocarpous*) or fused together (*syncarpous*). *See also* pistil.

carpus The wrist (or corresponding part of the forelimb) in terrestrial vertebrates, consisting of a number of small bones (*carpals*). The number of carpal bones varies with the species. The rabbit, for example, has two rows of carpals, the first (proximal) row containing three bones and the second (distal) row five. In humans there are also eight carpals. This large number of bones enables flexibility at the wrist joint, between the hand and forelimb. *See also* pentadactyl limb.

carrier 1. (in medicine) An individual who harbours a particular disease-causing microorganism without ill-effects and who can transmit the microorganism to others. *Compare* vector. **2.** (in genetics) An individual with an *allele for some defective condition that is masked by a normal *dominant allele. Such individuals therefore do not suffer from the condition themselves but they may pass on the defective allele to their offspring. In humans, women may be carriers of such conditions as red–green colour blindness and haemophilia, the alleles for which are carried on the X chromosomes (*see* sex linkage).

cartilage (gristle) A firm flexible connective tissue that forms the adult skeleton of cartilaginous fish (e.g. sharks). It comprises a matrix consisting chiefly of a mucopolysaccharide called *chondroitin sulphate* secreted by cells (*chondroblasts*) that become embedded in the matrix as *chondrocytes*. In other vertebrates cartilage forms the skeleton of the embryo, being largely replaced by *bone in mature animals. It persists in areas of high wear (e.g. bone ends, joints, and intervertebral discs) and in the nose and pinna of the ear.

cartilage bone (replacing bone) *Bone that is formed by replacing the cartilage of an embryo skeleton. The process, called *ossification*, is brought about by the cells (osteoblasts) that secrete bone. *Compare* membrane bone.

cartilaginous fishes *See* Elasmobranchii.

caruncle A small outgrowth from the testa of a seed that develops from the placenta, funicle, or micropyle. Examples include the warty outgrowth from the castor-oil seed and the tuft of hairs on the testa of the seed of willowherb. *See also* aril.

caryopsis A dry single-seeded indehiscent fruit that differs from an *achene in that the fruit wall is fused to the testa of the seed. It is the grain of cereals and grasses.

casein One of a group of phosphate-containing proteins (phosphoproteins) found in milk. Caseins are easily digested by the enzymes of young mammals and represent a major source of phosphorus. *See* rennin.

caste A division found in social insects, such as the *Hymenoptera (ants, bees, wasps) and the Isoptera (termites), in which the individuals are structurally and physiologically specialized to perform a particular function. For example, in honeybees there are queens (fertile females), workers (sterile females), and drones (males). There are several different castes of workers (all sterile females) among ants.

catabolism The metabolic breakdown of large molecules in living organisms to smaller ones, with the release of energy. Respiration is an example of a catabolic series of reactions. *See* metabolism. *Compare* anabolism.

catalysis The process of changing the rate of a chemical reaction by use of a *catalyst.

catalyst A substance that increases the rate of a chemical reaction without itself undergoing any permanent chemical change (*see also* inhibition). The catalyst provides an alternative pathway by which the reaction can proceed, in which the activation energy is lower. It thus increases the rate at which the reaction comes to equilibrium, although it does not alter the position of the equilibrium. *Enzymes are the catalysts in biochemical reactions; they are highly specific in the type of reaction they catalyse.

catecholamine Any of a class of amines (including *dopamine, *adrenaline, and *noradrenaline) that function as *neurotransmitters and/or hormones.

category *See* rank.

catkin A type of flowering shoot (*see* racemose inflorescence) in which the axis, which is often long, bears many small stalkless unisexual flowers. Usually the male catkins hang down from the stem; the female catkins are shorter and often erect. Examples include birch and hazel. Most plants with catkins are adapted for wind pollination, the male flowers producing large quantities of pollen; willows are an exception, having nectar-secreting flowers and being pollinated by insects.

CAT scanner (computerized axial tomography scanner) *See* tomography.

caudal vertebrae The bones (*see* vertebra) of the tail, which articulate with the *sacral vertebrae. The number of caudal vertebrae varies with the species. Rabbits, for example, have 15 caudal vertebrae, while in man these vertebrae are fused to form a single bone, the *coccyx.

cell The structural and functional unit of all living organisms. Cell size varies, but most cells are microscopic (average diameter 0.01–0.1 mm). Cells may exist as independent units of life, as in bacteria and protozoans, or they may form colonies or tissues, as in all higher plants and animals. Each cell consists of a mass of protein material (*see* protoplasm) that is differentiated into a jelly-like substance (*see* cytoplasm) and a *nucleus, which contains DNA. The protoplasm is bounded by a *cell membrane, which in plant and bacterial cells is surrounded by a *cell wall. There are two main types of cell. *Prokaryotic cells* (bacteria and blue-green algae) are the more primitive. The nuclear material is not bounded

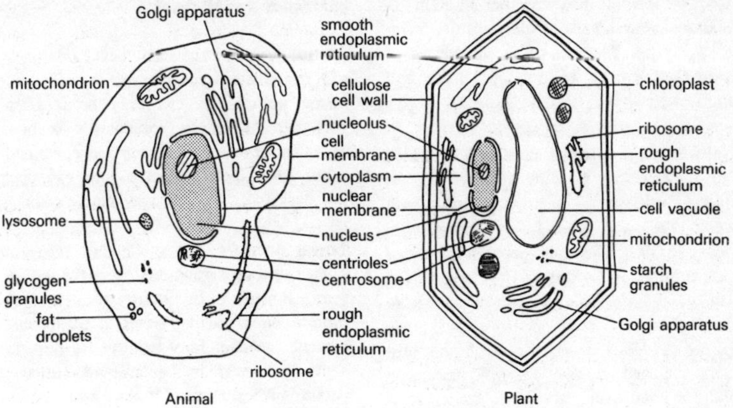

Generalized eukaryotic cells

by a membrane and chemicals involved in cell metabolism are associated with the cell membrane. Reproduction is asexual and involves simple cell cleavage. In *eukaryotic cells* the nucleus is bounded by a nuclear membrane and the cytoplasm is divided by membranes into a system of interconnected cavities and separate compartments (*organelles*), e.g. *mitochondria, *endoplasmic reticulum, *Golgi apparatus, *lysosomes, and *ribosomes. Reproduction can be either asexual (*see* mitosis) or sexual (*see* meiosis). Plants and animals consist of eukaryotic cells but plant cells possess *chloroplasts and other *plastids and bear a rigid cellulose cell wall.

cell body (perikaryon) The part of a *neurone that contains the nucleus. The cell processes that are involved in the transmission and reception of nervous impulses (the axon and the dendrites respectively) develop as extensions from the cell body.

cell division The formation of two daughter cells from a single mother cell. The nucleus divides first and this is followed by the formation of a cell membrane between the daughter nuclei. In *mitosis the daughter nuclei are identical to the original nucleus; *meiosis results in daughter nuclei each with half the number of chromosomes in the mother cell nucleus.

cell membrane (plasmalemma; plasma membrane) The semipermeable *membrane forming the outer limit of a *cell. It consists mostly of protein and lipid. The membrane regulates the flow of materials in and out of the cell and plays a role in the production or assembly of the *cell wall or *capsule (when this is present).

cellulose A polysaccharide that consists of a long unbranched chain of glucose units. It is the main constituent of the cell walls of all higher plants, many algae, and some fungi and is responsible for providing the rigidity of the cell wall. The fibrous nature of extracted cellulose has led to its use in the textile industry for the production of cotton, artificial silk, etc.

cell wall A rigid outer layer that surrounds the cell membrane of plant and bacterial (but not animal) cells. It protects and/or gives shape to a cell, and in herbaceous plants provides mechanical support for the plant body. Most plant cell walls are composed of the polysaccharide *cellulose and may be strengthened by the addition of *lignin. The cell walls of fungi consist mainly of *chitin. Bacterial cell walls consist of complex polymers of polysaccharides and amino acids.

Celsius scale A temperature scale in which the fixed points are the temperatures at standard pressure of ice in equilibrium with water (0°C) and water in equilibrium with steam (100°C). The scale, between these two temperatures, is divided in 100 degrees. The degree Celsius (°C) is equal in magnitude to the *kelvin. This scale was formerly known as the *centigrade scale*; the name was officially changed in 1948 to avoid confusion with a hundredth part of a grade. It is named after the Swedish astronomer Anders Celsius (1701–44), who devised the inverted form of this scale (ice point 100°, steam point 0°) in 1742.

cement (cementum) A thin layer of bony material that fixes teeth to the jaw. It covers the dentine of the root of a *tooth, below the level of the gum, and is attached to the *periodontal membrane lining the tooth socket in the jawbone.

Cenozoic (Cainozoic; Kainozoic) The geological era that began about 65 million years ago and extends to the present. It followed the *Mesozoic era and is subdivided into the *Tertiary and *Quaternary periods. The Cenozoic is often known as the *Age of Mammals* as these evolved to become an abundant, diverse, and dominant group. Birds and flowering plants also flourished. The era saw the formation of the major mountain ranges of the Himalayas and the Alps.

centi- Symbol c. A prefix used in the metric system to denote one hundredth. For example, 0.01 metre = 1 centimetre (cm).

centigrade scale *See* Celsius scale.

centipedes *See* Myriapoda.

Central Dogma The basic belief originally held by molecular geneticists, that flow of genetic information can only occur from *DNA to *RNA to proteins. It is now known, however, that information contained within RNA molecules of viruses can also flow back to DNA. *See also* genetic code.

central nervous system (CNS) The part of the nervous system that coordinates all neural functions. In invertebrates it may comprise simply a few *nerve cords and associated *ganglia. In vertebrates it consists of the *brain and the *spinal cord. The vertebrate CNS contains *reflex arcs, which produce automatic and rapid responses to particular stimuli.

centre (in neurology) A part of the nervous system, consisting of a group of nerve cells, that coordinates a particular process. An example is the respiratory centre in the vertebrate brainstem, which controls breathing movements. The stimulation of a centre will initiate the process, while destruction of the centre will prevent or impair it.

centrifuge A device in which solid or liquid particles of different densities are separated by rotating them in a tube in a horizontal circle. The denser particles tend to move along the length of the tube to a greater radius of rotation, displacing the lighter particles to the other end.

centriole A cylindrical structure in a cell that has a role in cell division. Centrioles are not normally found in plant cells but are commonly present in pairs in animal cells. During cell division they move to opposite sides of the nucleus to form the ends of the *spindle. The spindle fibres develop from them.

centromere (kinomere; spindle attachment) The part of a *chromosome that contains no genes. It usually appears as a constriction when chromosomes contract during cell division (*see* meiosis; mitosis). The position of the centromere is a distinguishing feature of individual chromosomes and it attaches them to the *spindle fibres. *See also* kinetochore.

centrum *See* vertebra.

cephalization The tendency among animal groups for the major sense organs, mouth, and brain to be grouped together at the front (anterior) end of the body. These are usually contained in a specialized cephalic region – the head.

Cephalopoda The most advanced class of molluscs, containing the squids, cuttlefishes, octopuses, and the extinct *ammonites. Cephalopods have a highly concentrated central nervous system within a protective cartilaginous case. The eye has a well-developed retina and is comparable to that of vertebrates. All cephalopods are predacious carnivores capable of swimming by jet propulsion; they have highly mobile tentacles for catching and holding prey.

cephalothorax The fused head and thorax of crustaceans and arachnids (spiders, scorpions), which is connected to the abdomen.

cerebellum The part of the vertebrate brain concerned with the coordination and regulation of muscle activity and the maintenance of muscle tone and balance. In mammals it consists of two connected hemispheres, composed of a core of white matter and a much-folded outer layer of grey matter, and is situated above the medulla oblongata and partly beneath the cerebrum.

cerebral cortex (pallium) The layer of *grey matter that forms the outer layer of the hemispheres of the *cerebrum in many vertebrates. It is most highly developed in mammals. The cortex is responsible for the control and integration of voluntary movement and the senses of vision, hearing, touch, etc.; it also contains centres concerned with memory, language, thought, and intellect.

cerebral hemisphere Either of the two halves of the vertebrate *cerebrum.

cerebrospinal fluid (CSF) The fluid, similar in composition to *lymph, that bathes the central nervous system of vertebrates. It is secreted by the *choroid plexus into the *ventricles of the brain, filling these and other cavities in the brain and spinal cord, and is reabsorbed by veins on the brain surface. Its function is to protect the central nervous system from mechanical injury.

cerebrum The largest part of the vertebrate brain. It consists of two *cerebral hemispheres*, which develop from the embryonic *forebrain. The hemispheres have an outer convoluted layer of grey matter – the *cerebral cortex – which contains an estimated ten billion nerve cells. Underneath this is *white matter. The two halves of the cerebrum are linked by several neural structures. The function of the cerebrum is to integrate complex sensory and neural functions. The cerebrum is thought to play a critical role in the process of learning, which involves both short-term and long-term memory.

cervical vertebrae The *vertebrae of the neck. The number of cervical vertebrae varies: for example, rabbits have 7 and humans have 12. Their main functions are to support the head and to provide articulating surfaces against which it can move relative to the backbone. *See* atlas; axis.

Cestoda A class of flatworms (*see* Platyhelminthes) comprising the tapeworms – ribbon-like parasites within the gut of vertebrates. Tapeworms are surrounded by partially digested food in the host gut so they are able to absorb nutrients through their whole body surface. The body consists of a head (*scolex*), bearing suckers and hooks for attachment, and a series of segments (*proglottides*), containing male and female reproductive systems. The life cycle of a tapeworm requires two hosts, the primary host usually being a predator of the secondary host. *Taenia soleum* has man for its primary host and the pig as its secondary host. Mature segments, containing thousands of fertilized eggs, leave the primary host with its faeces and develop into embryos and then larvae that continue the life cycle in the gut of a secondary host.

Cetacea An order of marine mammals comprising the whales, which includes what is probably the largest known animal – the blue whale (*Balaenoptera musculus*), over 30 m long and over 150 tonnes in weight. The forelimbs of whales are modified as short stabilizing flippers and the skin is very thin and almost hairless. A thick layer of blubber insulates the body against heat loss and is an important food store. Whales breathe through a dorsal blowhole, which is closed when the animal is submerged. The toothed whales (suborder Odontoceti), such as the dolphins and killer whale, are carnivorous; whalebone whales (suborder Mysticeti), such as the blue whale, feed on plankton filtered by *whalebone plates.

c.g.s. units A system of *units based on the centimetre, gram, and second. Derived from the metric system, it was badly adapted to use with thermal quantities (based on the inconsistently defined *calorie) and with electrical quantities (in which two systems, based respectively on unit permittivity and unit permeability of free space, were used). For scientific purposes c.g.s. units have now been replaced by *SI units.

chaeta A bristle, made of *chitin, occurring in annelid worms. In the earthworm they occur in small groups projecting from the skin in each segment and function in locomotion. The chaetae of polychaete worms (e.g. ragworm) are borne in larger groups on paddle-like appendages (*parapodia*).

chalaza **1.** A twisted strand of fibrous albumen in a bird's egg that is attached to the membrane at either end of the yolk and thus holds the yolk in position in the albumen. **2.** The part of a plant *ovule where the nucellus and integuments merge.

chalk A very fine-grained white rock composed of the fossilized skeletal remains of marine plankton known as *coccoliths* and consisting largely of calcium carbonate ($CaCO_3$). It is the characteristic rock of the *Cretaceous period. It should not be confused with blackboard 'chalk', which is made from calcium sulphate.

character A distinctive inherited feature of an organism. Organisms in a population may display different aspects of a particular character, e.g. the A, B, and O human blood groups (*see* ABO system) are different aspects of the blood group character.

chemical dating An absolute *dating technique that depends on measuring the chemical composition of a specimen. Chemical dating can be used when the specimen is known to undergo slow chemical change at a known rate. For instance, phosphate in buried bones is slowly replaced by fluoride ions from the ground water. Measurement of the proportion of fluorine present gives a rough estimate of the time that the bones have been in the ground. Another, more accurate, method depends on the fact that amino acids in living organisms are L- optical isomers. After death, these racemize and the age of bones can be estimated by measuring the relative amounts of D- and L-amino acids present.

chemical fossil Any of various organic compounds found in ancient geological strata that appear to be biological in origin and are assumed to indicate that life existed when the rocks were formed. The presence of chemical fossils in Precambrian strata indicates that life existed over 3000 million years ago.

chemoreceptor A *receptor that detects the presence of particular chemicals and (in multicellular organisms) transmits this information to sensory nerves. Examples include the *taste buds and the receptors in the *carotid body.

chemosystematics *See* systematics.

chemotaxis *See* taxis.

chemotaxonomy The *classification of plants and microorganisms based on similarities and differences in their natural products and the biochemical pathways involved in their manufacture. *See also* taxonomy.

chemotropism The growth or movement of a plant or plant part in response to a chemical stimulus. An example is the growth of a pollen tube down the style during fertilization in response to the presence of sugars in the style.

chiasma The point at which paired *homologous chromosomes remain in contact as they begin to separate during the first prophase of *meiosis, forming a cross shape. A number of chiasmata can usually be identified and at these points *crossing over occurs.

chimaera An organism composed of tissues that are genetically different. Chimaeras can develop if a *mutation occurs in a cell of a developing embryo. All the cells arising from it have the mutation and therefore produce tissue that is genetically different from adjacent tissue, e.g. brown patches in otherwise blue eyes in humans. *Graft hybrids are examples of plant chimaeras.

Chiroptera An order of flying mammals comprising the bats. Their membranous wings are supported by very elongated forelimbs and digits and stretch along the sides of the body to the hindlimbs and tail. Whenever bats rest they allow their body temperature to fall, hibernating in winter when food is scarce. Most bats are nocturnal; their ears are enlarged and specialized for *echolocation, which they use to hunt prey and avoid obstacles. Bats feed variously on insects, fruit, nectar, or blood.

chlorophyll The pigment responsible for the green colour of most plants. The chlorophyll molecule is the principal site of light absorption in the light reactions of *photosynthesis. It is a magnesium-containing *porphyrin, chemically related to *cytochrome and *haemoglobin.

Chlorophyta (green algae) A large division of *algae, the members of which possess chlorophylls *a* and *b*, store food reserves as starch, and have cellulose cell walls. In these respects they resemble higher plants more closely than do any of the other algal divisions. The Chlorophyta are widely distributed and diverse in form. Unicellular forms may occur singly (sometimes with *flagella for motility) or in colonies, while multicellular forms may be filamentous (e.g. *Spirogyra*) or platelike (e.g. *Ulva*).

chloroplast Any of the chlorophyll-containing organelles (*see* plastid) that are found in large numbers in those plant cells undergoing *photosynthesis. Chloroplasts are typically lens-shaped and made up of stacks of membranes (*see* granum) enclosed in a gel-like matrix. The light reactions of photosynthesis occur on the membranes while the dark reactions take place in the matrix.

chlorosis The abnormal condition in plant stems and leaves in which synthesis of the green pigment chlorophyll is inhibited, resulting in a pale yellow coloration. This may be caused by lack of light, mineral deficiency, infection (particularly by viruses), or genetic factors.

choanae (internal nares) *See* nares.

cholecalciferol *See* vitamin D.

cholecystokinin (pancreozymin) A hormone, produced by the duodenal region of the small intestine, that induces the gall bladder to contract and eject bile into the intestine and stimulates the pancreas to secrete its digestive enzymes. Cholecystokinin output is stimulated by contact with the contents of the stomach.

cholesterol A *sterol occurring widely in animal tissues and also in some higher plants and algae. It can exist as a free sterol or esterified with a long-chain fatty acid. Cholesterol is absorbed through the intestine or manufactured in the liver. It serves principally as a constituent of blood plasma lipoproteins and of the lipid–protein complexes that form cell membranes. It is also important as a precursor of various steroids, especially the bile acids, sex hormones, and adrenocorticoid hormones. The derivative 7-dehydrocholesterol is converted to vitamin D_3 by the action of sunlight on skin. Increased levels of dietary and blood cholesterol have been associated with *atherosclerosis*, a condition in which lipids accumulate on the inner walls of arteries and eventually obstruct blood flow.

choline An amino alcohol, $CH_2OHCH_2N(CH_3)_3OH$. It occurs widely in living organisms as a constituent of certain types of phospholipids – the *lecithins and sphingomyelins. It is sometimes classified as a member of the *vitamin B complex.

cholinergic Describing a nerve fibre that either releases *acetylcholine when stimulated or is itself stimulated by acetylcholine. *Compare* adrenergic.

Chondrichthyes *See* Elasmobranchii.

Chordata A phylum of animals characterized by a hollow dorsal nerve cord and, at some stage in their development, a flexible skeletal rod (the *notochord) and *gill slits opening from the pharynx. There are four subphyla, the largest of which is the *Vertebrata, in which the notochord is only present in the embryo or larva and becomes replaced by the backbone before birth or metamorphosis. The other three subphyla are often grouped together as the *Protochordata*. They are relatively primitive chordates including the lancelet (*Amphioxus*) and the sea squirts.

chorion 1. A membrane enclosing the embryo yolk sac and allantois of reptiles, birds, and mammals. *See* extraembryonic membranes. **2.** The protective shell of an insect egg, produced by the ovary. It is pierced by a small pore (*micropyle*) that allows the entry of spermatozoa for fertilization. *See also* egg membrane.

choroid A pigmented layer, rich in blood vessels, that lies between the retina and the sclerotic of the vertebrate eye. At the front of the eye the choroid is modified to form the *ciliary body and the *iris.

choroid plexus A membrane rich in blood vessels that lines the *ventricles of the brain. It is an extension of the *pia mater and secretes *cerebrospinal fluid into the ventricles; it also controls exchange of materials between the blood and cerebrospinal fluid.

chromatid A threadlike strand formed from a *chromosome during the early stages of cell division. Each chromosome divides along its length into two chromatids, which are at first held together at the centromere. They separate completely at a later stage. The DNA of the chromosome reproduces itself exactly so that each chromatid has the complete amount of DNA and becomes a daughter chromosome with exactly the same genes as the original chromosome from which it was formed.

chromatin The substance of which *chromosomes are composed. It consists of proteins (principally histones), DNA, and small amounts of RNA. In a metabolically inactive nucleus, chromatin is mainly in a condensed form, *heterochromatin*, but in an active nucleus most of the chromatin is in an expanded form, *euchromatin*. Messenger RNA molecules are formed in euchromatic regions.

chromatogram A record obtained by chromatography. The term is applied to the developed records of *paper chromatography and *thin-layer chromatography and also to the graphical record produced in *gas chromatography.

chromatography A technique for analysing or separating mixtures of gases, liquids, or dissolved substances, such as mixtures of amino acids or chlorophyll pigments. The original technique (invented by the Russian botanist Mikhail Tsvet in 1906) is a good example of *column chromatography*. A vertical glass tube is packed with an adsorbing material, such as alumina. The sample is poured into the column and continuously washed through with a solvent (a process known as *elution*). Different components of the sample are adsorbed to different extents and move down the column at different rates. In Tsvet's original application, plant pigments were used and these separated into coloured bands in passing down the column (hence the name chromatography). The usual method is to collect the liquid (the *eluate*) as it passes out from the column in fractions.

In general, all types of chromatography involve two distinct phases – the *stationary phase* (the adsorbent material in the column in the example above) and the *moving phase* (the solution in the example). The separation depends on competition for molecules of sample between the moving phase and the stationary phase. The form of column chromatography above is an example of *adsorption chromatography*, in which the sample molecules are adsorbed on the alumina. In *partition chromatography*, a liquid (e.g. water) is first absorbed by the stationary phase and the moving phase is an immiscible liquid. The separation is then by partition between the two liquids. In *ion-exchange chromatography* the process involves competition between different ions for ionic sites on the stationary phase. *Gel filtration is another chromatographic technique in which the size of the sample molecules is important.

See also gas–liquid chromatography; paper chromatography; thin-layer chromatography.

chromatophore **1.** A pigment-containing cell found in the skin of many lower vertebrates (e.g. chameleon) and in the integument of crustaceans. Concentration or dispersion of the pigment granules in the cytoplasm of the cell causes the colour of the animal to alter to match its surroundings. A common type of chromatophore is the *melanophore*, which contains the pigment *melanin. **2.** *See* chromoplast.

chromoplast (chromatophore) Any of various organelles in plant cells that contain pigments. Red, orange, and yellow chromoplasts contain carotenoid pigments, while green chromoplasts (called *chloroplasts) contain the pigment chlorophyll and provide the green coloration of most plants. *Compare* leucoplast. *See* plastid.

chromosome A threadlike structure several to many of which are found in the nucleus of plant and animal cells. Chromosomes are composed of *chromatin and carry the *genes, which determine the individual characteristics of an organism. When the nucleus is not dividing, individual chromosomes cannot be identified with a light microscope. During the first stage of nuclear division, however, the chromosomes contract and, when stained, can be clearly seen under a microscope. Each consists of two *chromatids held together at the *centromere (*see also* meiosis; mitosis). The number of chromosomes in each cell is constant for and characteristic of the species concerned. In the normal body cells of *diploid organisms the chromosomes occur in pairs (*see* homologous chromosomes); in the gamete-forming germ cells, however, the diploid number is halved and each cell contains only one member of each chromosome pair. Thus in man each body cell contains 46 chromosomes (22 matched pairs and one pair of *sex chromosomes) and each germ cell 23. Abnormalities in the number or structure of chromosomes may give rise to abnormalities in the individual; *Down's syndrome is the result of one such abnormality.

chromosome map (linkage map) A plan showing the relative positions of *genes along the length of the chromosomes of an organism. It is constructed by making crosses and observing whether certain characteristics tend to be inherited together. The closer together two allele pairs are situated on *homologous chromosomes, the less often will they be separated and rearranged as the reproductive cells are formed (*see* chiasma; crossing over). The proportion of offspring that show *recombination of the alleles concerned thus reflects their spacing and is used as a unit of length in mapping chromosomes.

chrysalis *See* pupa.

chyle A milky fluid consisting of *lymph that contains absorbed food materials (especially emulsified fats). Most chyle occurs in the lymphatic ducts (*lacteals) in the *villi of the small intestine during the absorption of fat.

chyme The semisolid and partly digested food that is discharged from the stomach into the duodenum.

ciliary body The circular band of tissue surrounding and supporting the *lens of the vertebrate eye. It contains the *ciliary muscles*, which bring about changes in the shape of the lens (*see also* accommodation). The ciliary body produces the *aqueous humour.

ciliary feeding A method of feeding used by *Amphioxus* and many aquatic invertebrates. The movement of cilia causes a current of water to be drawn towards and through the animal, and microorganisms in the water are filtered out by the cilia.

cilium A short minute hairlike structure (up to 10 μm long) present on the surface of many animal and a few plant cells, notably in certain protozoans and some types of vertebrate epithelium. Cilia usually occur in large groups. Beating of cilia can produce cell movement or create a current in fluid surrounding a cell. *Compare* flagellum.

circadian rhythm (diurnal rhythm) Any 24-hour periodicity in the behaviour or physiology of animals or plants. Examples are the sleep/activity cycle in many animals and the growth movements of plants. Circadian rhythms are generally controlled by *biological clocks.

circulatory system The heart, blood vessels, blood, lymphatic vessels, and lymph, which together serve to transport materials throughout the body. *See also* vascular system.

citric acid A white crystalline hydroxy carboxylic acid, $HOOCCH_2C(OH)(COOH)CH_2COOH$. It is present in citrus fruits and is an intermediate in the *Krebs cycle in plant and animal cells.

citric-acid cycle *See* Krebs cycle.

clade A group of organisms that share a common ancestor. *See* cladistics.

cladistics A controversial method of classification in which animals and plants are

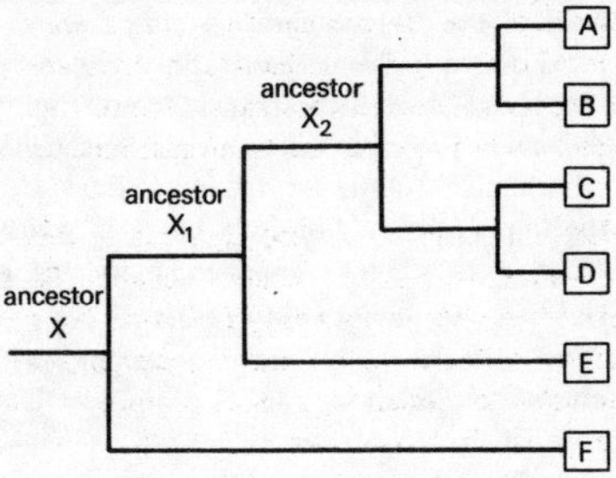

A cladogram showing the relationships of six species (A–F)

placed into taxonomic groups called *clades* when they share characteristics (known as *homologies*) that are thought to indicate common ancestry. It is based on the assumption that two new species are formed suddenly, by splitting from a common ancestor, and not by gradual evolutionary change. A diagram indicating these relationships (called a *cladogram*) therefore consists of a system of dichotomous branches: each point of branching represents divergence from a common ancestor, as shown in the diagram. Thus the species A to F form a clade as they share the common ancestor X, and species A to D form a clade of a different taxonomic rank, sharing the ancestor X_2. Species C to F do not form a clade, since the latter must include *all* the descendants of a common ancestor.

Cladistics has been used as evidence by the opponents of neo-Darwinism in the debate on the nature of evolution. Cladists argue that the major branching patterns of cladograms correspond to large-scale evolutionary events that cannot be explained by orthodox neo-Darwinism, which postulates a series of small changes occurring over a long period of time.

cladode A flattened stem or internode that resembles and functions as a leaf. It is an adaptation to reduce water loss, since it contains fewer *stomata than a leaf. An example of a plant with cladodes is asparagus.

cladogram *See* cladistics.

class A category used in the *classification of organisms that consists of similar or closely related orders. Similar classes are grouped into a phylum in the case of animals and a division in the case of plants. Examples include, among animals, Mammalia (mammals), Aves (birds), and Agnatha (jawless fish).

classification The arrangement of organisms into a series of groups based on physiological, biochemical, anatomical, or other relationships. An *artificial classification* is based on one or a few characters simply for ease of identification or for a specific purpose; for example, birds are often arranged according to habit and habitat (seabirds, songbirds, birds of prey, etc.) while fungi may be classified as edible or poisonous. Such systems do not reflect evolutionary relationships. A *natural classification* is based on resemblances and is a hierarchical arrangement. The smallest group commonly used is the *species. In plant classification, species are grouped into genera (*see* genus), the hierarchy continuing up through *tribes, *families, *orders, *classes, and *divisions to *kingdom. Animal classification follows the same system except that the *phylum replaces the division. Higher up in the hierarchy the similarities between members of a group become fewer. Present-day natural classifications try to take into account as many features as possible and in so doing aim to reflect evolutionary relationships (*see* cladistics). Natural classifications are also predictive. Thus if an organism is placed in a particular genus because it shows certain features characteristic of the genus, then it can be assumed it is very likely to possess most (if not all) of the other features of that genus. *See also* binomial nomenclature; taxonomy.

clavicle A bone that forms part of the *pectoral (shoulder) girdle, linking the *scapula (shoulder blade) to the sternum (breastbone). In man it forms the collar bone and serves as a brace for the shoulders.

cleavage (in embryology) The series of cell divisions by which a single fertilized egg cell is transformed into a multicellular body, the *blastula. Characteristically no growth occurs during cleavage, the shape of the embryo is unchanged except for the formation of a central cavity (the blastocoel), and the ratio of nuclear material (DNA) to cytoplasm increases.

climax community A relatively stable ecological *community that is achieved at the end of a *succession.

clinostat A mechanical device that rotates whole plants (usually seedlings), so removing the effect of any stimulus that normally acts in one particular direction. It is most often used to study the growth of plant organs when the influence of gravity has been removed.

clitoris An erectile rod of tissue in female mammals (and also some reptiles and birds) that is the equivalent of the male penis. It lies in front of the *urethra and *vagina.

cloaca The cavity in the pelvic region into which the terminal parts of the alimentary canal and the urinogenital ducts open in most vertebrates. Placental mammals, however, have a separate anus and urinogenital opening.

clone A group of organisms or cells that have arisen from a single individual by asexual reproduction and are therefore all genetically identical, e.g. a group of plants propagated by bulbs or cuttings.

clotting factors Substances that cause a blood clot to form at the site of an injury. Calcium ions and *platelets trigger the release of the enzyme *thromboplastin* from injured tissues. This converts *prothrombin* in the blood to its enzymically active form *thrombin*. Thrombin catalyses the formation of insoluble *fibrin from soluble *fibrinogen*; the fibrin forms a fibrous network in which blood cells become enmeshed, producing a clot. *See also* blood clotting.

clubmoss *See* Lycopsida.

cnidoblast *See* thread cell.

CNS *See* central nervous system.

CoA *See* coenzyme A.

coagulation The process in which colloidal particles come together to form larger masses. Coagulation can be brought about by adding ions to neutralize the charges stabilizing the colloid. Ions with a high charge are particularly effective (e.g. alum, containing Al^{3+}, is used in styptics to coagulate blood). Heating is another way of coagulating certain colloids (e.g. boiling an egg coagulates the albumin). *See also* blood clotting.

coal A brown or black carbonaceous deposit derived from the accumulation and alteration of ancient vegetation, which originated largely in swamps or other moist environments. Most deposits of coal were formed during the Carboniferous and Permian periods; more recent periods of coal formation occurred during the early Jurassic and Tertiary periods. As the vegetation decomposed it formed layers of peat, which were subsequently buried (for example, by marine sediments following a rise in sea level or subsidence of the land). Under the increased pressure and resulting higher temperatures the peat was transformed into coal. Two types of coal are recognized: *humic* (or *woody*) *coals*, derived from plant remains; and *sapropelic coals*, which are derived from algae, spores, and finely divided plant material.

As the processes of coalification (i.e. the transformation resulting from the high temperatures and pressures) continue, there is a progressive transformation of the deposit: the proportion of carbon relative to oxygen rises and volatiles and water are driven out. The various stages in this process are referred to as the *ranks* of the coal. In ascending order, the main ranks of coal are: *lignite* (or *brown coal*), which is soft, brown, and has a high moisture content; *subbituminous coal*, which is used chiefly by generating stations; *bituminous coal*, which is the most abundant rank of coal; *semibituminous coal*; *semianthracite coal*, which has a fixed carbon content of between 86% and 92%; and *anthracite coal*, which is hard and black with a fixed carbon content of between 92% and 98%.

coccus Any spherical bacterium. Cocci may occur singly (*monococcus*), in pairs (*diplococcus*), in groups of four or more, in cubical packets (*sarcina*), in grapelike clusters (*staphylococcus*), or in chains (*streptococcus*). Diplococci, staphylococci, and streptococci include pathogenic species, causing pneumonia, suppurative infections, and scarlet fever, respectively. They are generally nonmotile and do not form spores.

coccyx The last bone in the *vertebral column in apes and man (i.e. tailless primates). It is formed by the fusion of 3–5 *caudal vertebrae.

cochlea Part of the *inner ear of mammals, birds, and some reptiles that transforms sound waves into nerve impulses. In mammals it is coiled, resembling a snail shell. The cochlea is lined with sensitive cells that bear tiny hairs; it is filled with fluid (*endolymph*) and surrounded by fluid (*perilymph*). Sound-induced vibrations of the *fenestra ovalis are transmitted through the perilymph and endolymph and stimulate the cells that line the cochlea. These in turn stimulate nerve cells that transmit information, via the *auditory nerve, to the brain for interpretation of the sounds.

cockroaches *See* Dictyoptera.

cocoon A protective covering for eggs and/or larvae produced by many invertebrates. For example, the larvae of many insects spin a cocoon in which the pupae develop (that of the silkworm moth produces silk), and earthworms secrete a cocoon for the developing eggs.

co-dominance The condition that arises when both alleles in a *heterozygous organism are dominant and are fully expressed. For example, the human blood group AB is the result of two alleles, A and B, both being expressed. A is not dominant to B, nor vice versa. *Compare* incomplete dominance.

codon A triplet of nitrogen-containing bases within a molecule of *DNA or messenger *RNA that codes for a particular amino acid during the synthesis of proteins in a cell (*see* translation). *See also* genetic code.

coelacanth A bony fish, *Latimeria chalumnae*, that was believed to have been extinct until 1938, when the first specimen of modern times was discovered. The coelacanth belongs to the same order (Crossopterygii – lobe-finned fishes) as the ancestors of the amphibians. It lives at a depth of 200–300 m in the Indian Ocean. The coelacanth is a large fish, 1–2 m long and weighing 80 kg or more, with a three-lobed tail fin. The body is covered in rough heavy scales and the pectoral fins can be used like crutches to help movement across the sea bed. The young are born alive. Fossil coelacanths are most abundant in deposits about 400 million years old and no fossils less than 70 million years old have been found.

Coelenterata A phylum of aquatic invertebrates that includes *Hydra*, jellyfish, sea anemones, and *corals. A coelenterate's body is *diploblastic, with two cell layers of the body wall separated by *mesoglea, and shows *radial symmetry. The body cavity (*coelenteron*) is sac-shaped, with one opening acting as both mouth and anus. This opening is surrounded by tentacles bearing *thread cells. Coelenterates exist both as free-swimming *medusae (e.g. jellyfish) and as sedentary *polyps. The latter may be colonial (e.g. corals) or solitary (e.g. sea anemones and *Hydra*). In many coelenterates the life cycle alternates between these two forms (*see* alternation of generations).

coelom A fluid-filled cavity that forms the main *body cavity of vertebrate and most invertebrate animals. It is formed by the splitting of the *mesoderm. Ciliated ducts (*coelomoducts*) connect the coelom to the exterior allowing the exit of waste products and gametes; in higher animals these are specialized as oviducts, etc. The coelom is large and often subdivided in annelid worms (in which it functions as a hydrostatic skeleton) and vertebrates. In arthropods it is restricted to the cavities of the gonads and excretory organs, the body cavity being a blood-filled *haemocoel*.

coelomoduct *See* coelom.

coenocyte A mass of protoplasm containing many nuclei and enclosed by a cell wall. It is found in certain algae and fungi. *Compare* cell; plasmodium; syncytium.

coenzyme An organic nonprotein molecule that associates with an enzyme molecule in catalysing biochemical reactions. Coenzymes usually participate in the substrate–enzyme interaction by donating or accepting certain chemical groups. Many vitamins are precursors of coenzymes. *See also* cofactor.

coenzyme A (CoA) A complex organic compound that acts in conjunction with enzymes involved in various biochemical reactions, notably the oxidation of pyruvate via the *Krebs cycle and fatty-acid oxidation and synthesis. It comprises principally the B vitamin *pantothenic acid, the nucleotide *adenine, and a ribose–phosphate group.

coenzyme Q (ubiquinone) Any of a group of related quinone-derived compounds that serve as electron carriers in the *electron transport chain reactions of cellular respiration. Coenzyme Q molecules have side chains of different lengths in different types of organisms but function in similar ways.

cofactor A nonprotein component essential for the normal catalytic activity of an enzyme. Cofactors may be organic molecules (*coenzymes) or inorganic ions. They may activate the enzyme by altering its shape or they may actually participate in the chemical reaction.

colchicine An *alkaloid derived from the autumn crocus, *Colchicum autumnale*. It inhibits *spindle formation in cells, so preventing their division and inducing multiple sets of chromosomes (*polyploidy). Colchicine is used in genetics, cytology, and plant breeding research and also in cancer therapy to inhibit cell division.

cold-bloodedness *See* poikilothermy.

Coleoptera An order of insects comprising the beetles and weevils and containing about 330 000 known species – the largest order in the animal kingdom. The forewings are hardened and thickened to form *elytra*, which meet at a precise mid-dorsal line and protect the underlying pair of hindwings and abdomen. The mouthparts are generally modified for biting and in some species assume antler-like proportions. Beetles occur in a wide variety of terrestrial and aquatic habitats; many feed on decaying organic matter, some eat living vegetation, while others prey on other arthropods. A number of beetles and weevils are economically important pests of stored grain, timber, and crops. The young emerge as larvae and generally undergo metamorphosis via a pupal stage to form the adult beetle.

coleoptile A protective sheath that covers the young shoot of the embryo in plants of the grass family. It bursts open when the first leaves develop. Experiments investigating growth movements of the oat coleoptile led to the discovery of the plant growth substance indoleacetic acid (IAA).

coleorhiza A protective sheath that covers the young root of the embryo in plants of the grass family.

collagen An insoluble fibrous protein found extensively in the connective tissue of skin, tendons, and bone. The polypeptide chains of collagen (containing the amino acids glycine and proline predominantly) form triple-stranded helical coils that are bound together to form fibrils, which have great strength and limited elasticity. Collagen accounts for over 30% of the total body protein of mammals.

collenchyma *See* ground tissues.

colloids Colloids were originally defined by Thomas Graham in 1861 as substances, such as starch or gelatin, which will not diffuse through a membrane. He distinguished them from *crystalloids* (e.g. inorganic salts), which would pass through membranes. Later it was recognized that colloids were distinguished from true solutions by the presence of particles that were too small to be observed with a normal microscope yet were much larger than normal molecules. Colloids are now regarded as systems in which there are two or more phases, with one (the *dispersed phase*) distributed in the other (the *continuous phase*). Moreover, at least one of the phases has small dimensions (in the range 10^{-9}–10^{-6} m). Colloids are classified in various ways.

Sols are dispersions of small solid particles in a liquid. The particles may be macromolecules or clusters of small molecules. *Lyophobic sols* are those in which there is no affinity between the dispersed phase and the liquid. An example is silver chloride dispersed in water. In such colloids the solid particles have a surface charge, which tends to stop them coming together. Lyophobic sols are inherently unstable and in time the particles aggregate and form a precipitate. *Lyophilic sols*, on the other hand, are more like true solutions in which the solute molecules are large and have an affinity for the solvent. Starch in water is an example of such a system. *Association colloids* are systems in which the dispersed phase consists of clusters of molecules that have lyophobic and lyophilic parts. Soap in water is an association colloid (*see* micelle).

Emulsions are colloidal systems in which the dispersed and continuous phases are both liquids, e.g. oil-in-water or water-in-oil. Such systems require an emulsifying agent to stabilize the dispersed particles.

Gels are colloids in which both dispersed and continuous phases have a three-dimensional network throughout the material, so that it forms a jelly-like mass. Gelatin is a common example. One component may sometimes be removed (e.g. by heating) to leave a rigid gel (e.g. silica gel).

Other types of colloid include *aerosols* (dispersions of liquid or solid particles in a gas, as in a mist or smoke) and *foams* (dispersions of gases in liquids or solids).

colon The section of the vertebrate *large intestine that lies between the *caecum and the *rectum. Its prime function is to absorb water and minerals from indigestible food residues passing from the small intestine, which results in the formation of *faeces.

colony **1.** (in zoology) A group of animals of the same species living together and dependent upon each other. Some, such as the corals and sponges, are physically connected and function as a single unit. Others, such as insect colonies, are not physically joined but show a high level of social organization with members specialized for different functions (*see* caste). **2.** (in microbiology) A group of microorganisms, usually bacteria or yeasts, that are considered to have developed from a single parent cell. Colonies that grow on *agar plates differ in shape, colour, surface texture, and translucency and can therefore be used as a means of identification.

colostrum A liquid with a high content of nitrogen, antibodies, and vitamins that is secreted from the mammary glands before and just after giving birth. The change of secretion from colostrum to proper milk takes place gradually during the days after birth.

colour blindness Any disorder of vision in which colours are confused. The most common type is red–green colour blindness. This is due to a recessive gene carried on the X chromosome (*see* sex linkage), and therefore men are more likely to show the defect although women may be *carriers. It results in absence or malfunctioning of nerve cells sensitive to coloured light (*cones) in the retina of the eyes.

commensalism An interaction between two animal or plant species that habitually live together in which one species (the *commensal*) benefits from the association while the other is not significantly affected. For example, the burrows of many marine worms contain commensals that take advantage of the shelter provided but do not affect the worm.

community A naturally occurring assemblage of plant and animal species living within a defined area or habitat. Communities are named after one of their *dominant species (e.g. a pine community) or the major physical characteristics of the area (e.g. a freshwater pond community). Members of a community interact in various ways (e.g. through *food chains and *competition). Large communities may be divided into smaller component communities. *See* association.

companion cell A type of cell found within the *phloem of flowering plants. Each companion cell is usually closely associated with a *sieve element. Its function is uncertain, though it appears to regulate the activity of the adjacent sieve element. In gymnosperms a similar function is attributed to *albuminous cells*, which are found closely associated with gymnosperm sieve elements.

compass plant A plant that has its leaves permanently orientated in a north–south direction. Such an arrangement enables the plant to take full advantage of morning and evening sun, while avoiding the stronger midday sunlight. An example is the compass plant of the prairies (*Silphium laciniatum*).

competent Describing embryonic tissue that is capable of developing into a specialized tissue when suitably stimulated. *See* induction; evocation.

competition The interaction that occurs between two or more organisms, populations, or species that share some environmental resource when this is in short supply. Competition is an important force in evolution: plants, for example, become tall to compete for light, and animals evolve various foraging methods to compete for food. There may be a direct confrontation between competitors, as occurs between barnacles competing for space on a rock, or the numbers or fecundity of the competitors are indirectly reduced through joint dependence on limited resources. Competition occurs both between members of a species (*intraspecific competition*) and between different species (*interspecific competition*). Since competition ultimately results in the displacement by one competitor of the others, it is to the advantage of the competitors to avoid one another wherever possible. Thus in time the competitors become separated from each other geographically or ecologically, which promotes evolutionary change. Competition for mates may lead to *sexual selection.

complement A group of proteins present in blood plasma and tissue fluid that aids the body's defences following an *immune response. Following an antibody–antigen reaction, complement is activated chemically and becomes bound to the antibody–antigen complex (*complement fixation*); it becomes involved in destroying foreign cells and attracting scavenging white blood cells (*phagocytes) to the area of conflict in the body.

complemental males The small males of certain animals that live in or on the females and are usually more or less degenerate apart from the reproductive organs. They occur in certain crustaceans (e.g. some barnacles), in which the normal individuals are hermaphrodite but the complemental males have suppressed ovaries, lose their alimentary canal, and lead a semiparasitic existence in the mantle cavity of the larger partner. This may ensure that cross fertilization occurs.

composite fruit A type of fruit that develops from an inflorescence rather than from a single flower. *See* pseudocarp; sorosis; strobilus; syconus.

compound eye The eye of insects and crustaceans, which consists of numerous visual units, the *ommatidia*. Each ommatidium consists of a lens beneath which is a crystalline cone and light-sensitive cells. The eye is convex, with the apices of the cones converging onto the optic nerve. There are two types of compound eye. In *apposition eyes*, typical of diurnal insects, each ommatidium focuses rays parallel to its long axis so that each gives an image of a minute part of the visual field, producing a detailed mosaic image. In *superposition eyes*, typical of nocturnal insects, each ommatidium receives light from a larger part of the visual field and the image may overlap with those received by many neighbouring ommatidia. This produces an image that is bright but lacks sharpness of detail.

compound microscope *See* microscope.

concentration The quantity of dissolved substance per unit quantity of solvent in a solution. Concentration is measured in various ways. The amount of substance dissolved per unit volume (symbol c) has units of $\mathrm{mol\,dm^{-3}}$ or $\mathrm{mol\,l^{-1}}$. It is now called 'concentration' (formerly *molarity*). The *mass concentration* (symbol ρ) is the mass of solute per unit volume of solvent. It has units of $\mathrm{kg\,dm^{-3}}$, $\mathrm{g\,cm^{-3}}$, etc. The *molal concentration* (or *molality*; symbol m) is the amount of substance per unit mass of solvent, commonly given in units of $\mathrm{mol\,kg^{-1}}$.

conceptacle A flask-shaped cavity with a small opening (the *ostiole*) that is found in the swollen tip of certain brown algae, such as *Fucus*. It contains the sex organs.

conditional response (conditioned reflex) A learned response that develops to an initially ineffective stimulus in classical *conditioning.

conditioning A process by which animals learn about a relation between two events. In *classical* (or *Pavlovian*) *conditioning*, repeated presentations of a neutral stimulus (e.g. the sound of a bell or buzzer) are followed each time by a biologically important stimulus (such as food or electric shock), which elicits a response (e.g. salivation). Eventually the neutral stimulus presented by itself produces a response (the *conditional response*, or *conditioned reflex*) similar to that originally evoked by the biologically important stimulus. For example, Pavlov's dogs learned to salivate in response to the sound of a metronome that preceded the presentation of food. In *instrumental* (or *operant*) *conditioning* the animal is rewarded (or punished) each time it makes a particular response; this eventually causes the frequency of the response to increase (or decrease). For example, a rat will learn to press a lever in order to obtain food. *See* reinforcement.

condyle A smooth round knob of bone that fits into a socket on an adjoining bone, forming a *joint. Such a joint permits up-and-down or side-to-side movement but does not allow rotation. There are condyles where the lower jawbone (mandible) is attached to the skull, which permits chewing movements. *See also* occipital condyle.

cone 1. (in botany) A reproductive structure occurring in gymnosperms and some pteridophytes, known technically as a *strobilus*. It consists of *sporophylls bearing the spore-producing sporangia. Gymnosperms produce different male and female cones. The large woody female cones of pines, firs, and other conifers are made up of structures called *ovuliferous scales*, which bear the ovules. **2.** (in animal anatomy) A type of light-sensitive receptor cell, found in the *retinas of all diurnal vertebrates. Cones are specialized to transmit information about colour and they function best in bright light. They are not evenly distributed on the retina, being concentrated in the *fovea and absent on the margin of the retina. *Compare* rod.

congenital Present at birth. Congenital disorders of the body may be due to genetic factors, e.g. *Down's syndrome, or caused by injury or environmental factors, e.g. drugs (such as thalidomide) and chemicals (such as dioxin).

conidiospore *See* conidium.

conidium (conidiospore) A spore of certain fungi, such as mildews and moulds, that is produced by the constriction of the tip of a specialized hypha, the *conidiophore*. Chains of conidia may be cut off in this way.

Coniferales An order within the *Gymnospermae containing some extinct species and most of the present-day gymnosperms, such as the pines, firs, and spruces. They are typically evergreen trees inhabiting cool temperate regions and have leaves reduced to needles or scales. The wood of conifers, which is called *softwood* in contrast to the *hardwood* of angiosperm trees, is widely used for timber and pulp.

conjugation 1. The fusion of two reproductive cells, particularly when these are both the same size (*see* isogamy). **2.** A form of sexual reproduction seen in some algae (e.g. *Spirogyra*), some bacteria (e.g. *Escherichia coli*), and ciliate protozoans. Two individuals are united by a tube formed by outgrowths from one or both of the cells. Genetic material from one cell (designated the male) then passes through the tube to unite with that in the other (female) cell.

conjunctiva The delicate membrane that covers the cornea and lines the inside of the eyelid of a vertebrate eye. It is kept clean by secretions of the *lachrimal (tear) gland and the reflex blink mechanism.

connective tissue An animal tissue consisting of a small number of cells (e.g. *fibroblasts and *mast cells) and fibres and a large amount of intercellular material (*matrix*). It is widely distributed and has many functions, including support, packing, defence, and repair. The individual constituents vary, depending on the function of the tissue. Different types of connective tissue include *mesenchyme* in the embryo, *adipose tissue, loose *areolar tissue* for packing and support, blood, lymph, cartilage, and bone.

conservation The sensible use of the earth's natural resources in order to avoid excessive degradation and impoverishment of the environment. It should include the search for alternative food and fuel supplies when these are endangered (as by deforestation and overfishing); an awareness of the dangers of *pollution; and the maintenance and preservation of natural habitats and the creation of new ones (e.g. nature reserves and national parks).

consociation A climax plant *community that is dominated by one particular species, e.g. a pine forest. *See* dominant. *Compare* association.

consumer An organism that feeds upon those below it in a *food chain (i.e. at the preceding *trophic level). Herbivores, which feed upon green plants, are *primary consumers*; a carnivore that feeds only upon herbivores is a *secondary consumer*. *Compare* producer.

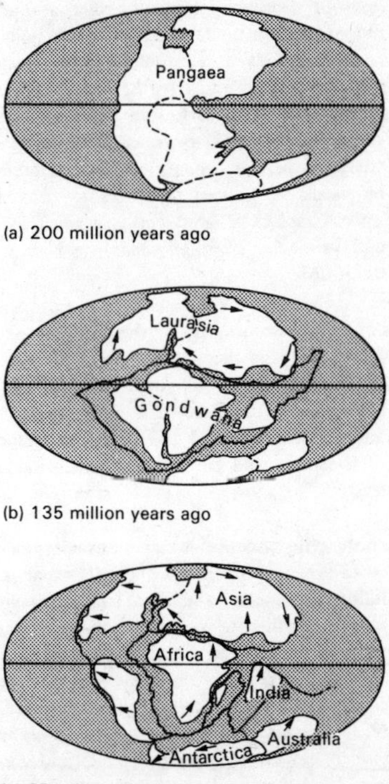

Continental drift

continental drift The theory that the earth's continents once formed a single mass and have since moved relative to each other. It was first postulated by A. Snider in 1858 and greatly developed by Alfred Wegener in 1912. He used evidence, such as the fit of South America into Africa and the distribution of rock types, flora, fauna, and geological structures, to suggest that the present distribution of the continents results from the breaking up of one or two greater land masses. The original land mass was named Pangaea and it was suggested that this broke up into the northerly Laurasia and the southerly Gondwanaland. The theory was not accepted for about 50 years by the majority of geologists but during the early 1960s, H. H. Hess's seafloor-spreading hypothesis and the subsequent development of *plate tectonics produced a mechanism to explain the drift of the continents.

continuous variation (quantitative variation) The range of differences that can be observed in many characteristics in a population. Characteristics resulting from *multifactorial inheritance show continuous variation, e.g. the wide range of foot sizes in an adult human population. *Compare* discontinuous variation.

contour feathers *Feathers that are arranged in regular rows on a bird's body, giving the body its streamlined shape. Each has a central horny shaft (the *rachis*) with a flattened *vane* on each side. Each vane is composed of two rows of filament-like *barbs, which are connected to each other by means of hooked *barbules to form a smooth surface. There is often a small second vane, the *aftershaft*, near the base of the feather.

contractile root Any of the modified adventitious roots that develop from the base of the stem of a bulb or corm. The new bulb or corm develops at a higher level in the soil than the old one. The contractile roots shorten and pull it down to a suitable level.

contractile vacuole A membrane-surrounded cavity in a cell that periodically expands, filling with water, and then suddenly contracts, expelling its contents to the cell's exterior. It is thus an organ of *osmoregulation and excretion. Contractile vacuoles are common in freshwater sponges and typical of freshwater protozoans, such as *Amoeba* (which has one spherical vacuole) and *Euglena* and *Paramecium* (in which a number of accessory vacuoles are attached to a main vacuole).

convergent evolution The development of superficially similar structures in unrelated organisms, usually because the organisms live in the same kind of environment. Examples are the wings of insects and birds and the streamlined bodies of whales and fish. *Compare* adaptive radiation.

Copepoda A subclass of crustaceans occurring in marine and freshwater habitats. Copepods are usually 0.5–2 mm long and lack both a carapace and compound eyes. Copepods are important members of plankton: some are free-living, feeding on microscopic organisms; others are parasitic. A familiar freshwater genus is *Cyclops*, so named because the members have a single median eye.

coral Any of a group of sedentary colonial marine invertebrates belonging to the class Anthozoa of the phylum *Coelenterata. A coral colony consists of individual *polyps within a protective skeleton that they secrete: this skeleton may be soft and jelly-like, horny, or stony. The horny skeleton secreted by corals of the genus *Corallium*, especially *C. rubrum*, constitutes the red, or precious, coral used as a gemstone. The skeleton of stony, or true, corals consists of almost pure calcium carbonate and forms the coral reefs common in tropical seas.

cork (phellem) A protective waterproof plant tissue produced by the *cork cambium. It develops in plants undergoing *secondary growth and replaces the epidermis. Its cells, whose walls are impregnated with *suberin, are arranged in radial rows and fit closely together except where the cork is interrupted by *lenticels. Some cork cells become air-filled while others contain deposits of lignin, tannins, and fatty acids, which give the cork a particular colour. The cork oak (*Quercus suber*) produces cork that can be used commercially.

cork cambium (phellogen) A type of *cambium arising within the outer layers of the stems of woody plants, usually as a complete ring surrounding the inner tissues. The cells of the cork cambium divide to produce an outer corky tissue (*cork or *phellem*) and an inner secondary cortex (*phelloderm*). Cork, cork cambium, and phelloderm together make up the *periderm*, an impermeable outer layer that protects the inner stem tissues if the outer tissues split as the stem girth increases with age. It thus takes over the functions of the epidermis.

corm An underground organ formed by certain plants, e.g. crocus and gladiolus, that

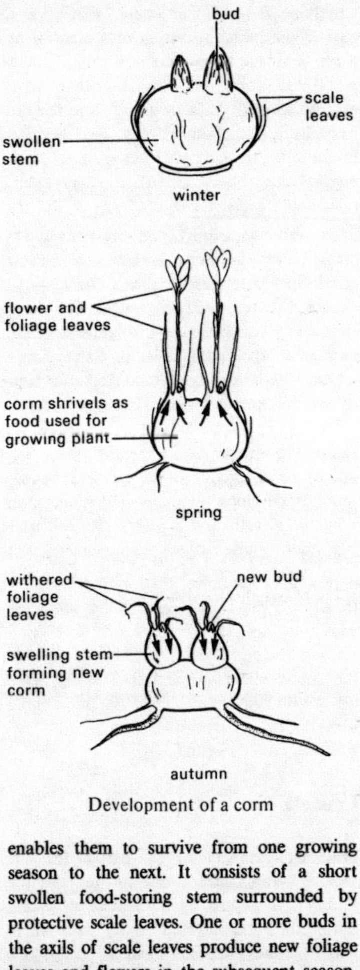

Development of a corm

enables them to survive from one growing season to the next. It consists of a short swollen food-storing stem surrounded by protective scale leaves. One or more buds in the axils of scale leaves produce new foliage leaves and flowers in the subsequent season, using up the food stored in the stem. *Compare* bulb.

cornea A transparent layer of tissue, continuous with the *sclerotic, that forms the front part of the vertebrate eye, over the iris and lens. The cornea refracts light entering the eye onto the lens, thus assisting in the focusing of images onto the *retina. *See also* astigmatism.

cornification *See* keratinization.

corolla The *petals of a flower, collectively, forming the inner whorl of the *perianth. It encircles the stamens and carpels. The form of the corolla is very variable. The petals may either be free (*polypetalous*) or united to form a tube (*gamopetalous*).

coronary vessels Two pairs of blood vessels (the coronary arteries and coronary veins) that supply the muscles of the heart itself. The coronary arteries arise from the aorta and divide into branches that encircle the heart. A blood clot in a coronary artery (*coronary thrombosis*) is one of the causes of a 'heart attack'.

corpus luteum The yellowish body that forms in the cavity of a *Graafian follicle in the ovary of a mammal after the release of the egg cell. It secretes the hormone *progesterone. Some species of sharks, reptiles, and birds have similar structures in their ovaries but the function of these is less well understood.

cortex 1. (in botany) The tissue between the epidermis and the vascular system in plant stems and roots. It is composed of *parenchyma cells and shows little or no structural differentiation. Cortex is produced by activity of the *apical meristem. *See also* endodermis. **2.** (in zoology) The outermost layer of tissue of various organs, including the adrenal glands (*adrenal cortex*), kidneys (*renal cortex*), and cerebral hemispheres (*see* cerebral cortex).

corticosteroid Any of several hormones produced by the cortex of the *adrenal glands. *Glucocorticoids* regulate the use of carbohydrates, proteins, and fats in the body and include *cortisol and *cortisone.

Mineralocorticoids regulate salt and water balance (*see* aldosterone).

corticotrophin *See* ACTH.

cortisol (hydrocortisone) A hormone (*see* corticosteroid), produced by the adrenal glands, that promotes the synthesis and storage of glucose and is therefore important in the normal response to stress, suppresses or prevents inflammation, and regulates deposition of fat in the body. It is used as treatment for various allergies and for rheumatic fever, certain skin conditions, and adrenal failure (Addison's disease).

cortisone A hormone (*see* corticosteroid), produced by the adrenal glands, that is closely related to, and has an action similar to that of, *cortisol. Cortisone is also the name for a synthetic version of cortisol.

corymb A type of flowering shoot (*see* racemose inflorescence) in which the lower flower stalks are longer than the higher ones, resulting in a flat-topped cluster of flowers. Examples are candytuft and wallflower.

cosmoid scale *See* scales.

cotyledon (seed leaf) A part of the embryo in a seed plant. The number of cotyledons is an important feature in classifying plants. Among the flowering plants, the class *Monocotyledonae have a single cotyledon and *Dicotyledonae have two. Gymnosperms have either two cotyledons, as in *Taxus* (yews) and *Cycas*, or five to ten, as in *Pinus* (pines). In seeds without an *endosperm, e.g. garden pea and broad bean, the cotyledons store food, which is used in germination. In seeds showing *epigeal germination, e.g. runner bean, they emerge above the soil surface and become the first photosynthetic leaves.

courtship Behaviour in animals that plays a part in the initial attraction of a mate or as a prelude to copulation. Courtship often takes the form of *displays that have evolved through *ritualization; some are derived from other contexts (e.g. food begging in some birds). Chemical stimuli (*see* pheromone) are also important in many mammals and insects.

As well as ensuring that the prospective mate is of the same species, the male's courtship performance allows females to choose between different males. The later stages of courtship may involve both partners in an alternating series of displays that inhibit *aggression and fear responses and ensure synchrony of sexual arousal.

coxa The first segment, attached to the thorax, of an insect's leg. *See also* femur; trochanter.

cranial nerves Ten to twelve pairs of nerves in vertebrates that emerge directly from the brain. They supply the sense organs and muscles of the head, neck, and viscera. Examples of cranial nerves include the *optic nerve (II) and the *vagus nerve (X). With the *spinal nerves, the cranial nerves form an important part of the *peripheral nervous system.

Craniata *See* Vertebrata.

cranium (brain case) The part of the vertebrate *skull that encloses and protects the brain. It is formed by the fusion of several flattened bones, which have immovable joints (sutures) between them.

creatine A compound, synthesized from the amino acids arginine, glycine, and methionine, that occurs in muscle. In the form of *creatine phosphate* (or *phosphocreatine*), it is an important reserve of energy for muscle contraction, which is released when creatine phosphate loses its phosphate and is converted to *creatinine*, which is excreted in the urine. *See also* phosphagen.

creationist A proponent of the theory of *special creation.

cremocarp A dry fruit that is a type of *schizocarp formed from two one-seeded carpels. The carpels remain separate and form indehiscent *mericarps* that are attached to a central supporting strand (*carpophore*) for some time before dispersal. It is characteristic of the Umbelliferae (carrot family).

Cretaceous The final geological period of the Mesozoic era. It extended from about 136 million years ago, following the Jurassic, to about 65 million years ago, when it was succeeded by the Palaeocene epoch. The name of the period is derived from *creta* (Latin: chalk) and the Cretaceous was characterized by the deposition of large amounts of *chalk in western Europe. The Cretaceous was the time of greatest flooding in the Mesozoic. Angiosperm plants made their first appearance on land and in the early Cretaceous Mesozoic reptiles reached their peak. At the end of the period there was a widespread extinction of the dinosaurs, flying reptiles, and ammonites, the cause of which is not definitely known, although it may be partly related to changing environmental conditions at the time.

crista 1. *See* semicircular canals. **2.** *See* mitochondrion.

critical group A large group of related organisms that, although variations exist between them, cannot be divided into smaller groups of equivalent taxonomic rank to the parent group.

Cromagnon man The earliest form of modern man (*Homo sapiens*), which is believed to have appeared in Europe about 35 000 years ago and possibly earlier in Africa and Asia. Fossils indicate that these hominids were taller and more delicate than *Neanderthal man, which they replaced. They used intricately worked tools of stone and bone and left the famous cave drawings at Lascaux in the Dordogne. The name is derived from the site at Cromagnon, France, where the first fossils were found in 1868.

crop An enlarged portion of the anterior section of the alimentary canal in some animals, in which food may be stored and/or undergo preliminary digestion. The term is most commonly applied to the thin-walled sac in birds between the oesophagus and the *proventriculus. In female pigeons the crop contains glands that secrete *crop milk*, used to feed nestlings.

cross 1. A mating between two selected individuals. Controlled crosses are made for many reasons, e.g. to investigate the inheritance of a particular characteristic or to improve a livestock or crop variety. *See also* back cross; reciprocal cross. **2.** An organism resulting from such a mating.

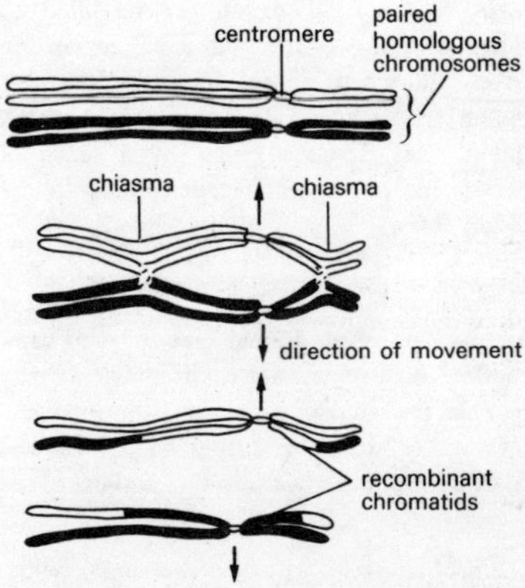

Crossing over at two chiasmata in a pair of homologous chromosomes

crossing over An exchange of portions of chromatids between *homologous chromosomes. As the chromosomes begin to move apart at the end of the first prophase of *meiosis, they remain in contact at a number of points (*see* chiasma). At these points the chromatids break and rejoin in such a way that sections are exchanged (see illustration). Crossing over thus alters the pattern of genes in the chromosomes. *See* recombination.

Crustacea A class of arthropods containing over 35 000 species distributed worldwide, mainly in freshwater and marine habitats, where they constitute a major component of plankton. Crustaceans include shrimps, crabs, and lobsters (*see* Decapoda), barnacles, water fleas (*see Daphnia*), copepods (*see* Copepoda), and the terrestrial woodlice. The segmented body usually has a distinct head (bearing *compound eyes, two pairs of *antennae, and various mouthparts), thorax, and abdomen, and is protected by a shell-like carapace. Each body segment may bear a pair of branched (*biramous*) appendages used for locomotion, as gills, and for filtering food particles from the water. Appendages in the head region are modified to form jaws and in the abdominal region are often reduced or absent. Typically, the eggs hatch to produce a free-swimming *nauplius* larva. This develops either by a series of moults or undergoes metamorphosis to the adult form.

cryptic coloration The type of colouring or marking of an animal that helps to camouflage it in its natural environment. It may enable the animal to blend with its background or, like the stripes of zebras and tigers, help to break up the outline of its body.

CSF *See* cerebrospinal fluid.

cultivar A *variety of a plant that has been developed by cultivation as a result of agricultural or horticultural practices. The term is derived from *culti*vated *var*iety.

culture medium A nutrient material, either solid or liquid, used to support the growth and reproduction of microorganisms or to maintain tissue or organ cultures. *See also* agar.

cupule **1.** A hard or membranous cup-shaped structure formed from bracts and enclosing various fruits, such as the hazelnut and acorn. **2.** A structure in club mosses (*Lycopodium* species) that protects the gemma (resting bud) during its development. It is composed of six leaflike structures. **3.** The bright red tissue around the seed of yew (*Taxus*), forming the yew 'berry'.

cusp **1.** A sharp raised protuberance on the surface of a *molar tooth. The cusps of opposing molars (i.e. on opposite jaws) are complementary to each other, which increases the efficiency of grinding food during chewing. **2.** A flap forming part of a *valve.

cuticle **1.** (in botany) The continuous waxy layer that covers the aerial parts of a plant. Composed of *cutin, it is secreted by the *epidermis and its primary function is to prevent water loss. **2.** (in zoology) A layer of horny noncellular material covering, and secreted by, the epidermis of many invertebrates. It is usually made of a collagen-like protein or of chitin and its main function is protection. In arthropods it is also strong enough to act as a skeleton (*see* exoskeleton) and in insects it reduces water loss. Growth is allowed by moulting of the cuticle (*see* ecdysis).

cuticularization The secretion by the outer (epidermal) layer of cells of plants and many invertebrates of substances that then harden to form a *cuticle.

cutin A mixture of waxy substances that impregnates the cell wall and covers the outer surface of mature epidermal plant cells, forming a *cuticle. The deposition of cutin (*cutinization*) reduces water loss by the plant and helps prevent the entry of pathogens. *See also* suberin.

cutinization The deposition of *cutin in plant cell walls, principally in the outermost layers of leaves and young stems.

cutis *See* dermis.

cutting A part of a plant, such as a bud, leaf, or a portion of a root or shoot, that, when detached from the plant and inserted in soil, can take root and give rise to a new daughter plant. Taking or striking cuttings is a horticultural method for propagating plants. *See also* vegetative propagation.

cyanocobalamin *See* vitamin B complex.

Cyanophyta (blue-green algae) A division containing very simple *algae that resemble photosynthetic bacteria in their internal organization (*see* prokaryote). Cyanophytes are unicellular but sometimes become joined in colonies or filaments by a sheath of mucilage. They occur in all aquatic habitats. A few species fix atmospheric nitrogen and thus contribute to soil fertility (*see* nitrogen fixation). Others exhibit *symbiosis (*see* Lichenes).

Cycadales An order within the *Gymnospermae containing many extinct species and a few of the present-day gymnosperms. Cycads inhabit tropical and subtropical regions, sometimes growing to a height of 20 m. The stem bears a crown of fernlike leaves. These species are among the most primitive of living seed plants.

cyclamates Salts of the acid $C_6H_{11}.NH.SO_3H$, where $C_6H_{11}-$ is a cyclohexyl group. Sodium and calcium cyclamates were formerly used as sweetening agents in soft drinks, etc, until their use was banned when they were suspected of causing cancer.

cyclic AMP A derivative of *ATP that is widespread in animal cells as an intermediate messenger in many biochemical reactions induced by hormones. Upon reaching their target cells, the hormones activate the enzyme that catalyses cyclic AMP production. Cyclic AMP ultimately activates the enzymes of the reaction induced by the hormone concerned. Cyclic AMP is also involved in controlling gene expression and cell division, in immune responses, and in nervous transmission.

Cyclostomata *See* Agnatha.

cyme *See* cymose inflorescence.

cymose inflorescence (cyme; definite inflorescence) A type of flowering shoot (*see* inflorescence) in which the first-formed flower develops from the growing region at the top of the flower stalk. Thus no new flower buds can be produced at the tip and other flowers are produced from lateral buds beneath. In a *monochasial cyme* (or *monochasium*), the development of the flower at the tip is followed by a new flower axis growing from a single lateral bud. Subsequent new flowers may develop from the same side of the lateral shoots, as in the buttercup, or alternately on opposite sides, as in forget-me-not. In a *dichasial cyme* (or *dichasium*), the development of the flower at the apex is followed by two new flower axes developing from buds opposite one another, as in plants of the family Caryophyllaceae (such as stitchwort). *Compare* racemose inflorescence.

cypsela A dry single-seeded fruit that does not split open during seed dispersal and is formed from a double ovary in which only one ovule develops into a seed. It is similar to an *achene and characteristic of members of the family Compositae, such as the dandelion. *See also* pappus.

cysteine *See* amino acid.

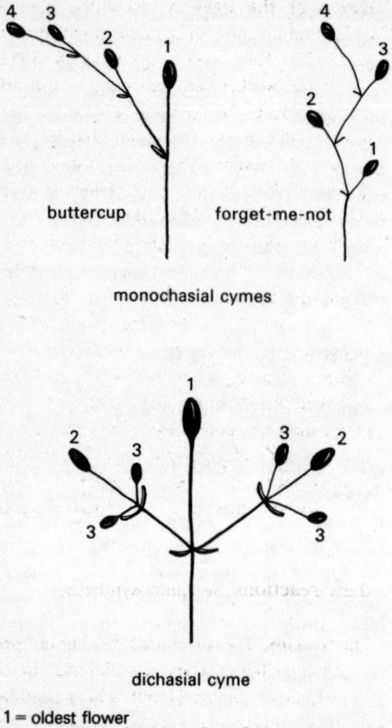

Types of cymose inflorescence

cystine A molecule resulting from the oxidation reaction between the sulphydryl (–SH) groups of two cysteine molecules (*see* amino acid). This often occurs between adjacent cysteine residues in polypeptides. The resultant *disulphide bonds* (–S–S–) are important in stabilizing the structure of protein molecules.

cytidine A nucleoside comprising one cytosine molecule linked to a D-ribose sugar molecule. The derived nucleotides, cytidine mono-, di-, and triphosphate (CMP, CDP, and CTP respectively) participate in various biochemical reactions, notably in phospholipid synthesis.

cytochrome Any of a group of proteins, each with an iron-containing *haem group, that form part of the *electron transport chain in mitochondria and chloroplasts. Electrons are transferred by reversible changes in the iron atom between the reduced Fe(II) and oxidized Fe(III) states.

cytogenetics The study of inheritance in relation to the structure and function of cells. For example, the results of breeding experiments can be explained in terms of the behaviour of chromosomes during the formation of the reproductive cells.

cytokinin (kinin) Any of a group of plant *growth substances chemically related to the purine adenine. Cytokinins stimulate cell division in the presence of *auxin and have also been found to delay senescence, overcome *apical dominance, and promote cell expansion. Zeatin is a naturally occurring cytokinin.

cytology The study of the structure and function of cells. The development of the light and electron microscopes has enabled the detailed structure of the nucleus (including the chromosomes) and other organelles to be elucidated. Microscopic examination of cells, either live or as stained sections on a slide, is also used in the detection and diagnosis of various diseases, especially *cancer.

cytolysis The breakdown of cells, usually as a result of destruction or dissolution of their outer membranes. Certain drugs (*cytotoxic drugs*) have this effect and are used in the treatment of some forms of cancer.

cytomegalovirus A virus belonging to the herpes group (*see* herpesvirus). In man it normally causes symptoms that are milder than the common cold, but it can produce more serious symptoms in those whose *immune response is disturbed (e.g. cancer patients). Infection in pregnant women may cause congenital handicap in their children.

cytoplasm The jelly-like material surrounding the nucleus of a *cell. It can be differentiated into dense outer *ectoplasm*, which is concerned primarily with cell movement, and less dense *endoplasm*, which contains most of the cell's structures.

cytoplasmic inheritance The inheritance of genes contained in the cytoplasm of a cell, rather than the nucleus. Only a very small number of genes are inherited in this way. The phenomenon occurs because certain organelles, the *mitochondria and (in plants) the *chloroplasts, contain their own genes and can reproduce independently. The female reproductive cell (the egg) has a large amount of cytoplasm containing many such organelles, which are consequently incorporated into the cytoplasm of all the cells of the embryo. The male reproductive cells (sperm or pollen), however, consist almost solely of a nucleus. Cytoplasmic organelles are thus not inherited from the male parent. In plants, male sterility can be inherited via the cytoplasm. The inheritance of any such factors does not follow Mendelian laws.

cytosine A *pyrimidine derivative. It is one of the principal component bases of *nucleotides and the nucleic acids *DNA and *RNA.

cytotaxonomy *See* taxonomy.

D

2,4-D 2,4-dichlorophenoxyacetic acid: a synthetic *auxin frequently used as a weedkiller of broad-leaved weeds.

dance of the bees A celebrated example of communication in animals, first investigated by Karl von Frisch (1886–1982). Honeybee workers on returning to the hive after a successful foraging expedition perform a 'dance' on the comb that contains coded information about the distance and direction of the food source. Other workers, sensing vibrations from the dance, follow the instructions to find the food source.

Daphnia A genus of crustaceans belonging to the suborder Cladocera (water fleas). *Daphnia* species have a transparent carapace and a protruding head with a pair of highly branched antennae for swimming and a single median compound eye. The five pairs of thoracic appendages form an efficient filter-feeding mechanism. Reproduction can take place without mating, i.e. by *parthenogenesis.

dark reactions *See* photosynthesis.

Darwinism The theory of *evolution proposed by Charles Darwin (1809–82) in *On the Origin of Species* (1859), which postulated that present-day species have evolved from simpler ancestral types by the process of *natural selection acting on the variability found within populations. *On the Origin of Species* caused a furore when it was first published because it suggested that species are not immutable nor were they specially created – a view directly opposed to the doctrine of *special creation. However the wealth of evidence presented by Darwin gradually convinced most people and the only major unresolved problem was to explain how the variations in populations arose and were maintained from one generation to the next. This became clear with the rediscovery of Mendel's work on classical genetics in the 1900s and led to the present theory of *neo-Darwinism.

Darwin's finches (Galapagos finches) The 14 species of finch, unique to the

Galapagos Islands, that Charles Darwin studied during his journey on HMS *Beagle*. Each is adapted to exploit a different food source. They are not found on the mainland because competition there for these food sources from other birds is fiercer. Darwin believed all the Galapagos finches to be descendants of a few that strayed from the mainland, and this provided important evidence for his theory of evolution. *See also* adaptive radiation.

dating techniques Methods of estimating the age of rocks, palaeontological specimens, archaeological sites, etc. *Relative dating techniques* date specimens in relation to one another; for example, stratigraphy is used to establish the succession of fossils. *Absolute* (or *chronometric*) *techniques* give an absolute estimate of the age and fall into two main groups. The first depends on the existence of something that develops at a seasonally varying rate, as in *dendrochronology and *varve dating. The other uses some measurable change that occurs at a known rate, as in *chemical dating, *radioactive* (or *radiometric*) *dating* (*see* carbon dating; fission-track dating; potassium–argon dating; rubidium–strontium dating; uranium–lead dating), and *thermoluminescence.

day-neutral plant A plant in which flowering can occur irrespective of the day length. Examples are cucumber and maize. *See* photoperiodism. *Compare* long-day plant; short-day plant.

DDT Dichlorodiphenyltrichloroethane; a colourless organic crystalline compound, $(ClC_6H_4)_2CH(CCl_3)$. DDT is the best known of a number of chlorine-containing pesticides used extensively in agriculture in the 1940s and 1950s. The compound is stable, accumulates in the soil, and concentrates in fatty tissue, reaching dangerous levels in carnivores high in the food chain. Restrictions are now placed on the use of DDT and similar pesticides.

deca- Symbol da. A prefix used in the metric system to denote ten times. For example, 10 coulombs = 1 decacoulomb (daC).

Decapoda An order of crustaceans distributed worldwide, mainly in marine habitats. Decapods comprise swimming forms (shrimps and prawns) and crawling forms (crabs, lobsters, and crayfish). All are characterized by five pairs of walking legs, the first pair of which are highly modified in crawling forms to form powerful grasping pincers. The carapace is fused with the thorax and head forming a *cephalothorax. The antennae are especially long in shrimps and prawns, which also possess several pairs of well-developed swimming appendages (*pleopods*) posterior to the walking legs. Following fertilization by the male, females usually carry the eggs until they hatch. The larvae undergo several transformations before attaining adult form.

deci- Symbol d. A prefix used in the metric system to denote one tenth. For example, 0.1 coulomb = 1 decicoulomb (dC).

decibel A unit used to compare two power levels, usually applied to sound or electrical signals. Although the decibel is one tenth of a *bel*, it is the decibel, not the bel, that is invariably used. Two power levels P and P_0 differ by n decibels when $n = 10\log_{10}P/P_0$. If P is the level of sound intensity to be measured, P_0 is a reference level, usually the intensity of a note of the same frequency at the threshold of audibility.

The logarithmic scale is convenient as human audibility has a range of 1 (just audible) to 10^{12} (just causing pain) and one decibel, representing an increase of some 26%, is about the smallest change the ear can detect.

deciduous Describing plants in which all the leaves are shed at the end of each growing season, usually the autumn in temperate regions or at the beginning of a dry season in the tropics. This seasonal leaf fall helps the plant retain water that would otherwise be lost by transpiration from the leaves. Examples of deciduous plants are rose and horse chestnut. *Compare* evergreen.

deciduous teeth (milk teeth) The first of two sets of teeth of a mammal. These teeth are smaller than those that replace them (the *permanent teeth) and fewer in number, since there are no deciduous *molars. *See also* diphyodont.

decomposer An organism that obtains energy from the chemical breakdown of dead organisms or animal or plant wastes. Examples are earthworms and many bacteria and fungi. Many decomposers (e.g. nitrifying bacteria) are specialized to break down organic materials that are difficult for other organisms to digest. Decomposers fulfil a vital role in the *ecosystem, returning the constituents of organic matter to the environment in inorganic form so that they can again be assimilated by plants. *See also* carbon cycle; nitrogen cycle.

deficiency disease Any disease caused by an inadequate intake of an essential nutrient in the diet, primarily vitamins, minerals, and amino acids. Examples are scurvy (lack of vitamin C), rickets (lack of vitamin D), and iron-deficiency anaemia.

definite inflorescence *See* cymose inflorescence.

degeneration **1.** Changes in cells, tissues, or organs due to disease, etc., that result in an impairment or loss of function and possibly death and breakdown of the affected part. **2.** The reduction in size or complete loss of organs during evolution. The human appendix has undergone this process and performs no function in man. Degeneration of external organs may cause animals to appear to be more primitive than they really are; for example, early zoologists believed whales were fish rather than mammals because of the degeneration of their limbs. *See also* vestigial organ.

deglutition (swallowing) A reflex action initiated by the presence of food in the pharynx. During deglutition, the soft *palate is raised, which prevents food from entering the nasal cavity; the *epiglottis closes, which blocks the entrance to the windpipe; and the oesophagus starts to contract (*see* peristalsis), which ensures that food is conveyed to the stomach.

dehiscence The spontaneous and often violent opening of a fruit, seed pod, or anther to release and disperse the seeds or pollen. Examples are the splitting of laburnum pods and primrose capsules.

dehydrogenase Any enzyme that catalyses the removal of hydrogen atoms in biological reactions. Dehydrogenases occur in many biochemical pathways but are particularly important in driving the *electron-transport-chain reactions of cell respiration. They work in conjunction with the hydrogen-accepting coenzymes *NAD and *FAD.

deme A group of organisms in the same *taxon. The term is used with various prefixes that denote how the group differs from other groups. For example, an *ecodeme* occurs in a particular ecological habitat, *cytodemes* differ from each other cytologically, and *genodemes* differ genetically.

denature To produce a structural change in a protein or nucleic acid that results in the reduction or loss of its biological properties. Denaturation is caused by heat, chemicals, and extremes of pH. The differences between raw and boiled eggs are largely a result of denaturation.

dendrite A slender branching process of the cell body of a *neurone. It forms connections (*see* synapse) with the axons of other neurones and transmits nerve impulses from these to the cell body.

dendrochronology An absolute *dating technique using the *growth rings of trees. It depends on the fact that trees in the same locality show a characteristic pattern of growth rings resulting from climatic conditions. Thus it is possible to assign a definite date for each growth ring in living trees, and to use the ring patterns to date fossil trees or specimens of wood (e.g. used for buildings or objects on archaeological sites) with lifespans that overlap those of living trees. The bristlecone pine (*Pinus aristata*), which lives for up to 5000 years, has been used to date specimens over 8000 years old. Fossil specimens accurately dated by dendrochronology have been used to make corrections to the *carbon-dating technique. Dendrochronology is also helpful in studying past climatic conditions. Analysis of trace elements in sections of rings can also provide information on past atmospheric pollution.

denitrification A chemical process in which nitrates in the soil are reduced to molecular nitrogen, which is released into the atmosphere. This process is effected by the bacterium *Pseudomonas denitrificans*, which uses nitrates as a source of energy for other chemical reactions in a manner similar to respiration in other organisms. *Compare* nitrification. *See* nitrogen cycle.

dental formula A representation of the dentition of an animal. A dental formula consists of eight numbers, four above and four below a horizontal line. The numbers represent (from left to right) the numbers of incisors, canines, premolars, and molars in either half of the upper and lower jaws. The total number of teeth in both jaws is therefore obtained by adding up all the numbers in the dental formula and multiplying by 2. Representative dental formulas are shown in the illustration.

$$\frac{2\quad 1\quad 2\quad 3}{1\quad 1\quad 2\quad 3}\quad \text{man (32 teeth)}$$

$$\frac{2\quad 0\quad 3\quad 3}{1\quad 0\quad 2\quad 3}\quad \text{rabbit (28 teeth)}$$

$$\frac{3\quad 1\quad 4\quad 2}{3\quad 1\quad 4\quad 3}\quad \text{bear (42 teeth)}$$

Representative dental formulas

denticle (placoid scale) *See* scales.

dentine The bony material that forms the bulk of a *tooth. Dentine is similar in composition to bone but is perforated with many tiny canals for nerve fibres, blood capillaries, and processes of the dentine-forming cells (*odontoblasts). Ivory, the material that forms elephant tusks, is made of dentine.

dentition The type, number, and arrangement of teeth in a species. This can be represented concisely by a *dental formula. *See also* diphyodont; monophyodont; polyphyodont; heterodont; homodont.

deoxyribonucleic acid *See* DNA.

depolarization A reduction in the difference of electrical potential that exists across the cell membrane of a nerve or muscle cell. Depolarization of a nerve-cell membrane occurs during the passage of an *action potential along the axon when the nerve is transmitting an impulse.

dermal bone *See* membrane bone.

Dermaptera An order of insects comprising the earwigs. Earwigs typically have long thin cylindrical bodies with biting mouthparts and a stout pair of curved forceps at the tip of the abdomen, used for catching prey and in courtship. Some species have a single pair of wings, which at rest are folded back over the abdomen like a fan; others are wingless. Most earwigs are nocturnal and omnivorous.

dermis (corium; cutis) The thicker and innermost layer of the *skin of vertebrates, the other layer being the *epidermis. The dermis consists of fibrous connective tissue in which are embedded blood vessels, sensory nerve endings, and (in mammals) hair follicles, sebaceous glands, and sweat ducts. Beneath the dermis lies the *subcutaneous tissue.

desiccator A container for drying substances or for keeping them free from moisture. Simple laboratory desiccators are glass vessels containing a drying agent, such as silica gel. They can be evacuated through a tap in the lid.

desmids A group of unicellular algae belonging to the *Chlorophyta. Like *Spirogyra*, they have an elaborate chloroplast. The cells of desmids are characteristically split into two halves joined by a narrow neck, each half being a mirror image of the other. The outer wall of the cell is patterned with various protuberances. Desmids are an important component of the plankton.

desorption The removal of adsorbed atoms, molecules, or ions from a surface.

determined Describing embryonic tissue at a stage when it can develop only as a certain kind of tissue (rather than as any kind)

Devonian A geological period in the Palaeozoic era that extended from the end of the Silurian (about 395 million years ago) to the beginning of the Carboniferous (about 345 million years ago). It was named by Adam Sedgwick (1785–1873) and Roderick Murchison (1792–1871) in 1839. The Devonian is divided into seven stages based on invertebrate fossil remains, such as corals, brachiopods, ammonoids, and crinoids, found in marine deposits. There were also extensive continental deposits consisting of conglomerates, red silts, and sandstones, forming the Old Red Sandstone facies. Fossils in the Old Red Sandstone include fishes and the earliest land plants. Graptolites became extinct early in the Devonian and the trilobites declined.

dextrose *See* glucose.

***d*-form** *See* optical activity.

diabetes *See* insulin; vasopressin.

diakinesis The period at the end of the first prophase of *meiosis when the separation of *homologous chromosomes is almost complete and *crossing over has occurred.

dialysis A method by which large molecules (such as starch or protein) and small molecules (such as glucose or amino acids) in solution may be separated by selective diffusion through a semipermeable membrane. For example, if a mixed solution of starch and glucose is placed in a closed container made of a semipermeable substance (such as Cellophane), which is then immersed in a beaker of water, the smaller glucose molecules will pass through the membrane into the water while the starch molecules remain behind. The cell membranes of living organisms are semipermeable, and dialysis takes place naturally in the kidneys for the excretion of nitrogenous waste. An artificial kidney (*dialyser*) utilizes the principle of dialysis by taking over the functions of diseased kidneys.

diapause A period of suspended development or growth occurring in some insects during which metabolism is greatly decreased. It is often triggered by seasonal changes and regulated by an inborn rhythm and enables the insect to survive unfavourable environmental conditions so that its offspring may be produced in more favourable ones.

diaphragm The muscular membrane that divides the thorax (chest) from the abdomen in mammals. It plays an essential role in breathing (*see* respiratory movement), being depressed during inhalation and raised during exhalation.

diaphysis The shaft of a limb bone, which in immature animals is separated from the ends of the bone (*see* epiphysis) by cartilage.

diastase *See* amylase.

diastole The phase of a heart beat that occurs between two contractions of the heart, during which the heart muscles relax and the ventricles fill with blood. *Compare* systole. *See* blood pressure.

diatoms *See* Bacillariophyta.

dichasium *See* cymose inflorescence.

2,4-dichlorophenoxyacetic acid *See* 2,4-D.

dichogamy The condition in which the male and female reproductive organs of a flower mature at different times, thereby ensuring that self-fertilization does not occur. *Compare* homogamy. *See also* protandry; protogyny.

dichotomous Describing the type of branching in plants that results when the growing point (apical bud) divides into two equal growing points, which in turn divide in a similar manner after a period of growth, and so on. Dichotomous branching is common is ferns and mosses.

Dicotyledonae One of the two classes or subclasses of plants within the *Angiospermae, distinguished by having two seed leaves (*cotyledons) within the seed. The dicotyledons usually have leaf veins in the form of a net, a ring of vascular bundles in the stem, and flower parts in fours or fives or multiples of these. Dicotyledons include many food plants (e.g. potatoes, peas, beans), ornamentals (e.g. roses, ivies, honeysuckles), and hardwood trees (e.g. oaks, limes, beeches). *Compare* Monocotyledonae.

Dictyoptera An order of insects comprising the cockroaches (suborder Blattaria) and the mantids (suborder Mantodea), occurring mainly in tropical regions. Cockroaches are oval and flattened in shape; some have a single well-developed pair of wings, folded back over the abdomen at rest, while in others the wings may be reduced or absent. They are usually found in forest litter, feeding on dead organic matter, but some species, e.g. the American cockroach (*Periplaneta americana*), are major household pests, scavenging on starchy foods, fruits, etc. In most species the females produce capsules (*oothecae*) containing 16–40 eggs. These are either deposited or carried by the female during incubation.

differentiation The changes from simple to more complex forms undergone by developing tissues and organs so that they become specialized for particular functions. Differentiation occurs during embryonic development, *regeneration, and (in plants) meristematic activity (*see* meristem).

digestion The breakdown by a living organism of ingested food material into chemically simpler forms that can be readily absorbed and assimilated by the body. This process requires the action of digestive enzymes and may take place extracellularly (i.e. in the *alimentary canal), as is the case in most animals; or intracellularly (e.g. by engulfing phagocytic cells), as occurs in protozoans and coelenterates.

digit A finger or toe. In the basic limb structure of terrestrial vertebrates there are five digits (*see* pentadactyl limb). This number is retained in man and other primates, but in some other species the number of digits is reduced. Frogs, for example, have four fingers and five toes, and in ungulate (hooved) mammals, the digits are reduced and their tips are enclosed in horn, forming hooves.

digitigrade Describing the gait of most fast-running animals, such as dogs and cats, in which only the toes are on the ground and the rest of the foot is raised off the ground. *Compare* plantigrade; unguligrade.

dilation *See* vasodilation.

dimethylbenzenes (xylenes) Three compounds with the formula $(CH_3)_2C_6H_4$, each having two methyl groups substituted on the benzene ring. 1,2-dimethylbenzene is *o*-xylene, etc. A mixture of the isomers is obtained from petroleum and is used as a clearing agent in preparing specimens for optical microscopy.

dimorphism The existence of two distinctly different types of individual within a species. An obvious example is *sexual dimorphism* in certain animals, in which the two sexes differ in colouring, size, etc. Dimorphism also occurs in some lower plants, such as mosses and ferns, that show an *alternation of generations.

dinosaur An extinct terrestrial reptile belonging to a group that constituted the dominant land animals of the Jurassic and Cretaceous periods, 190–65 million years ago. There were two orders. The Ornithischia were typically quadrupedal herbivores, many with heavily armoured bodies, and included *Stegosaurus*, *Triceratops*, and *Iguanodon*. They were all characterized by birdlike pelvic girdles. The Saurischia included many bipedal carnivorous forms, such as *Tyrannosaurus* (the largest known carnivore), and some quadrupedal herbivorous forms, such as *Apatosaurus* (*Brontosaurus*) and *Diplodocus*. They all had lizard-like pelvic girdles. Many of the herbivorous dinosaurs were amphibious or semiaquatic.

dioecious Describing plant species that have male and female flowers on separate plants. Examples of dioecious plants are willows. *Compare* monoecious.

diphyodont Describing a type of dentition that is characterized by two successive sets of teeth: the *deciduous (milk) teeth, which are followed by the *permanent (adult) teeth. Mammals have a diphyodont dentition. *Compare* monophyodont; polyphyodont.

diploblastic Describing an animal with a body wall composed only of two layers, *ectoderm and *endoderm, sometimes with a noncellular *mesoglea between them. The Coelenterata are diploblastic; all other Metazoa are *triploblastic.

diploid Describing a nucleus, cell, or organism with twice the *haploid number of chromosomes characteristic of the species. The diploid number is designated as $2n$. Two sets of chromosomes are present, one set being derived from the female parent and the other from the male. In animals, all the cells except the reproductive cells are diploid.

diplotene The period in the first prophase of *meiosis when paired *homologous chromosomes begin to move apart. They remain attached at a number of points (*see* chiasma).

Dipnoi A subclass of bony fishes that contains the lungfishes, which have lungs and breathe air. They are found in Africa, Australia, and South America, where they live in freshwater lakes and marshes that tend to become stagnant or even dry up in summer. They survive in these conditions by burrowing into the mud, leaving a small hole for breathing air, and entering a state of *aestivation, in which they can remain for six months or more. The Dipnoi date from the Devonian era (395–345 million years ago) and share many features with the modern *Amphibia.

Diptera An order of insects comprising the true, or two-winged, flies. Flies possess only one pair of wings – the forewings; the hindwings are modified to form small clublike *halteres* that function as balancing organs. Typically fluid feeders, flies have mouthparts adapted for piercing and sucking or for lapping; the diet includes nectar, sap, decaying organic matter, and blood. Some species prey on insects; others are parasitic. Dipteran larvae (*maggots*) are typically wormlike with an inconspicuous head. They undergo metamorphosis via a pupal stage to the adult form. Many flies or their larvae are serious pests, either by feeding on crops (e.g. fruit flies) or as vectors of disease organisms (e.g. the house fly (*Musca domestica*) and certain mosquitoes).

disaccharide A sugar consisting of two linked *monosaccharide molecules. For example, sucrose comprises one glucose molecule and one fructose molecule bonded together.

discontinuous variation (qualitative variation) Clearly defined differences in a characteristic that can be observed in a population. Characteristics that are determined by different *alleles at a single locus show discontinuous variation, e.g. garden peas are either wrinkled or smooth. *Compare* continuous variation.

disinfectant Any substance that kills or inhibits the growth of disease-producing microorganisms and is in general toxic to human tissues. Disinfectants include cresol, bleaching powder, and phenol. They are used to cleanse surgical apparatus, sickrooms, and household drains and if sufficiently diluted can be used as *antiseptics.

displacement activity An activity shown by an animal that appears to be irrelevant to its situation. Displacement activities are frequently observed when there is conflict between opposing tendencies. For example, birds in aggressive situations, in which there are simultaneous tendencies to attack and to flee, may preen their feathers as a displacement activity.

display behaviour Stereotyped movement or posture that serves to influence the behaviour of another animal. Many displays in *courtship and *aggression are conspicuous and characteristic of the species; special markings or parts of the body may be prominently exhibited (for example, the male peacock spreads its tail in courtship). Other displays are cryptic and make it harder for a predator to recognize the displaying animal as potential prey. For example, geometer moth caterpillars, which look like twigs, hold themselves on plant stems with one end sticking into the air.

distal Describing the part of an organ that is farthest from the organ's point of attachment to the rest of the body. For example, hands and feet are at the distal ends of arms and legs, respectively. *Compare* proximal.

diurnal Daily; denoting an event that happens once every 24 hours.

diurnal rhythm *See* circadian rhythm.

divergent evolution *See* adaptive radiation.

diverticulum A saclike or tubular outgrowth from a tubular or hollow internal organ. Diverticula may occur as normal structures (e.g. the *caecum and *appendix in the alimentary canal) or abnormally, from a weakened area of the organ.

division A category used in the *classification of plants that consists of one or several similar classes. Division names end in -phyta. Examples are the Bryophyta (mosses and liverworts) and the Tracheophyta (vascular plants). The divisions are grouped in the kingdom Plantae. In animal classification, the term *phylum is substituted for division.

dizygotic twins *See* fraternal twins.

***dl*-form** *See* optical activity.

DNA (deoxyribonucleic acid) The genetic material of most living organisms, which is a major constituent of the *chromosomes within the cell nucleus and plays a central role in the determination of hereditary characteristics by controlling *protein synthesis in cells (*see also* genetic code). DNA is a nucleic acid composed of two chains of *nucleotides in which the sugar is *deoxyribose* and the bases are *adenine, *cytosine, *guanine, and *thymine (*compare*

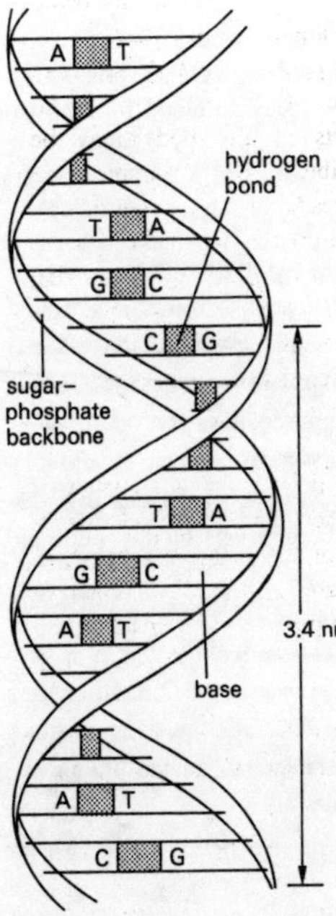

Double helical structure of DNA

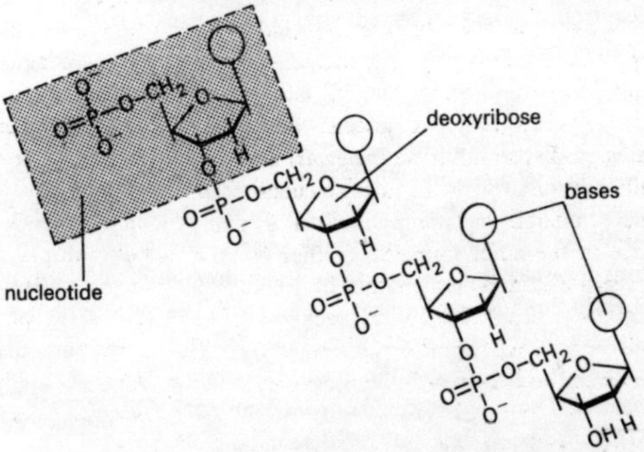

Detail of molecular structure of sugar–phosphate backbone. Each deoxyribose unit is attached to a phosphate group and a base, forming a nucleotide

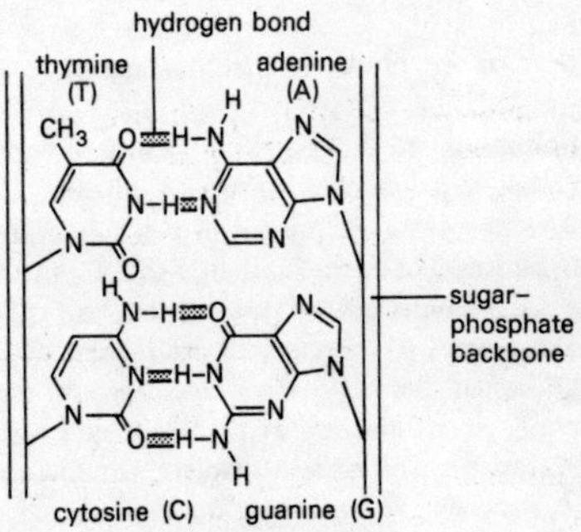

The four bases of DNA, showing the hydrogen bonding between base pairs

Molecular structure of DNA

RNA). The two chains are wound round each other and linked together by hydrogen bonds between specific complementary bases (*see* base pairing) to form a spiral ladder-shaped molecule (*double helix*).

When the cell divides, its DNA also replicates in such a way that each of the two daughter molecules is identical to the parent molecule. The hydrogen bonds between the complementary bases on the two strands of the parent molecule break and the strands unwind. Using as building bricks nucleotides present in the nucleus, each strand directs the synthesis of a new one complementary to itself. Replication is initiated, controlled, and stopped by means of polymerase enzymes.

DNA hybridization A method of determining the similarity of DNA from different sources. Single strands of DNA from two sources, e.g. different bacterial species, are put together and the extent to which double hybrid strands are formed is estimated. The greater the tendency to form these hybrid molecules, the greater the extent of complementary base sequences, i.e. gene similarity. The method is one way of determining the genetic relationships of species.

dodecanoic acid *See* lauric acid.

dominance hierarchy *See* dominant.

dominant 1. (in genetics) Describing the *allele that functions when two different alleles of a gene are present in the cells of an organism. For example, the height of garden peas is controlled by two alleles, 'tall' (T) and 'dwarf' (t). When both are present (Tt), i.e. when the cells are *heterozygous, the plant is tall since T is dominant and t is *recessive. *See also* incomplete dominance. **2.** (in ecology) Describing the most conspicuously abundant and characteristic species in a *community. The term is usually used of a plant species in plant ecology; for example, pine trees in a pine forest. **3.** (in animal behaviour) Describing an animal that is allowed priority in access to food, mates, etc., by others of its species because of its success in previous aggressive encounters. Less dominant animals frequently show *appeasement behaviour towards a more dominant individual, so overt *aggression is minimized. In a stable group there may be a linear *dominance hierarchy* or *peck order* (so called because it was first observed in domestic fowl), with each animal being subservient to those above it in the hierarchy and taking precedence over those below it.

donor An individual whose tissues or organs are transferred to another (the *recipient*). Donors may provide blood for transfusion or a kidney or heart for transplantation.

dopa (dihydroxyphenylalanine) A derivative of the amino acid tyrosine. It is found in particularly high levels in the adrenal glands and is a precursor in the synthesis of *dopamine, *noradrenaline, and *adrenaline. The laevorotatory form, *L-dopa*, is administered in the treatment of Parkinson's disease, in which brain levels of dopamine are reduced.

dopamine A *catecholamine that is a precursor in the synthesis of *noradrenaline and *adrenaline. It is also believed to function as a neurotransmitter in the brain.

dormancy An inactive period in the life of an animal or plant during which growth slows or completely ceases. Physiological changes associated with dormancy help the organism survive adverse environmental conditions. Annual plants survive the winter as dormant seeds while many perennial plants survive as dormant tubers, rhizomes, or bulbs. *Hibernation and *aestivation in animals help them survive extremes of cold and heat, respectively.

dorsal Describing the surface of a plant or animal that is farthest from the ground or other support, i.e. the upper surface. In vertebrates, the dorsal surface is that down which the backbone runs. Thus in upright (bipedal) mammals, such as man and kangaroos, it is the backward-directed (*posterior) surface. *Compare* ventral.

dose A measure of the extent to which matter has been exposed to *ionizing radiation. The *absorbed dose* is the energy per unit mass absorbed by matter as a result of such exposure. The SI unit is the gray, although it is often measured in rads (1 rad = 0.01 gray; *see* radiation units). The *maximum permissible dose* is the recommended upper limit of absorbed dose that a person or organ should receive in a specified period according to the International Commission on Radiological Protection.

double helix *See* DNA.

double recessive An organism with two *recessive alleles for a particular characteristic.

down feathers (plumules) Small soft feathers that cover and insulate the whole body of a bird. In nestlings they are the only feathers; in adults they lie between and beneath the *contour feathers. Down feathers have a fluffy appearance as their *barbs are not joined together to form a smooth vane.

Down's syndrome (mongolism) A congenital form of mental retardation due to a chromosome defect in which there are three copies of chromosome no. 21 instead of the usual two. The affected individual has a short broad face and slanted eyes (as in the Mongolian races), short fingers, and weak muscles. Down's syndrome can be detected before birth by *amniocentesis. It is named after the British physician John Down (1828–96), who first studied the incidence of the disorder.

dragonflies *See* Odonata.

drupe (pyrenocarp) A fleshy fruit that develops from either one or several fused carpels and contains one or many seeds. The seeds are enclosed by the hard protective endocarp (*see* pericarp) of the fruit. Thus the stone of a peach is the endocarp containing the seed. Plums, cherries, coconuts, and almonds are other examples of one-seeded drupes; holly and elder fruits are examples of many-seeded drupes. *See also* etaerio.

Dryopithecus A genus of extinct apes, fossils of which have been found in India and dated as 30–40 million years old. Fossils of *Dryopithecus* and of the similar genus *Proconsul* are often referred to as *dryopithecines*. Dryopithecines are believed to be either the ancestors of the fossil hominids or ancestral to both the human and ape lines.

ductless gland *See* endocrine gland.

duodenum The first section of the *small intestine of vertebrates. It is the site where food from the stomach is subjected to the action of bile (from the bile duct) and pancreatic enzymes (from the pancreatic duct) as well as the enzymes secreted by digestive glands in the duodenum itself (*see* succus entericus), which are required in the breakdown of proteins, carbohydrates, and fats. By neutralizing the acidic secretions of the stomach, the duodenum provides an alkaline environment necessary for the action of the intestinal enzymes.

dura mater The outermost and toughest of the three membranes (*meninges) that surround the central nervous system in vertebrates. It lies adjacent to the skull and its purpose is to protect the delicate inner meninges (the *arachnoid membrane and the *pia mater).

duramen *See* heartwood.

E

ear The sense organ in vertebrates that is specialized for the detection of sound and the maintenance of balance. It can be divided into the *outer ear and *middle ear, which collect and transmit sound waves, and the *inner ear, which contains the organs of balance and (except in fish) hearing. The term ear is often used for the *pinna of the mammalian outer ear.

eardrum *See* tympanum.

ear ossicles Three small bones – the *incus* (*anvil*), *malleus* (*hammer*), and *stapes* (*stirrup*) – that lie in the mammalian *middle ear, forming a bridge between the tympanum (eardrum) and the fenestra ovalis. The function of the ossicles is to transmit (and amplify) vibrations of the tympanum across the middle ear to the fenestra ovalis, which transfers them to the *inner ear. Muscles of the middle ear constrict the movement of the ossicles. This serves to safeguard the ear from damage caused by excessively loud noise.

earwigs *See* Dermaptera.

ecdysis (moulting) 1. The periodic loss of the outer cuticle of arthropods. It starts with the reabsorption of some materials in the inner part of the old cuticle and the formation of a new soft cuticle. The remains of the old cuticle then split; the animal emerges and absorbs water or swallows air and increases in size while the new cuticle is still soft. This cuticle is then hardened with chitin and lime salts. In insects and crustaceans ecdysis is controlled by the hormone *ecdysone. **2.** The periodic shedding of the

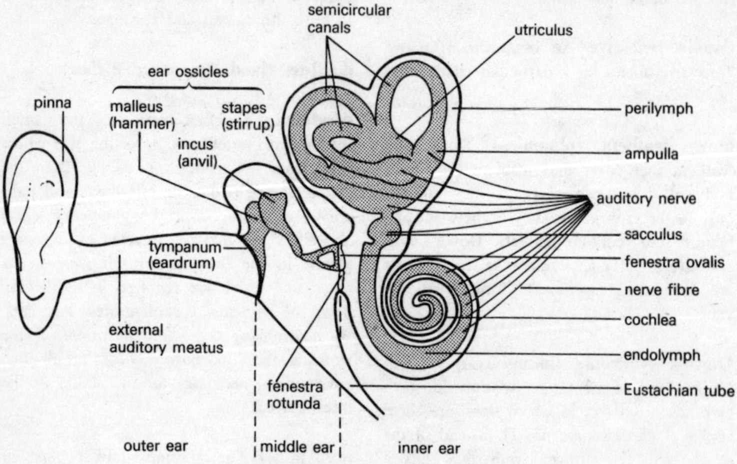

Structure of the mammalian ear

outer layer of the epidermis of reptiles (except crocodiles) to allow growth to occur.

ecdysone A steroid hormone, produced by insects and crustaceans, that stimulates moulting (*see* ecdysis) and metamorphosis. It acts on specific gene loci, stimulating the synthesis of proteins involved in these bodily changes.

ECG *See* electrocardiogram.

Echinodermata A phylum of marine invertebrates that includes the sea urchins, starfish, brittlestars, and sea cucumbers. Echinoderms have an exoskeleton (*test*) of calcareous plates embedded in the skin. In many species (e.g. sea urchins) spines protrude from the test. A system of water-filled canals (the *water vascular system*) provides hydraulic power for thousands of *tube feet*: saclike protrusions of the body wall used for locomotion, feeding, and respiration. Echinoderms have a long history: fossils of primitive echinoderms are known from rocks 500 million years old.

echolocation A method used by some animals (such as bats, dolphins, and certain birds) to detect objects in the dark. The animal emits a series of high-pitched sounds that echo back from the object and are detected by the ear or some other sensory receptor. From the direction of the echo and from the time between emission and reception of the sounds the object is located, often very accurately.

ecological niche The status or role of an organism in its environment. An organism's niche is defined by the types of food it consumes, its predators, temperature tolerances, etc. Two species cannot coexist stably if they occupy identical niches.

ecology The study of the interrelationships between organisms and their natural environment, both living and nonliving. For this purpose, ecologists study organisms in the context of the *populations and *communities in which they can be grouped and the *ecosystems of which they form a part. The study of ecological interactions provides important information on the nature and mechanisms of evolutionary change. Advances made in ecology over the last 25 years have led to increased concern about the effects of man's activities on the environment (notably the effects of *pollution), which has resulted in a greater awareness of the importance of *conservation.

ecosystem A biological *community and the physical environment associated with it. Nutrients pass between the different organisms in an ecosystem in definite pathways; for example, nutrients in the soil are taken up by plants, which are then eaten by herbivores, which in turn may be eaten by carnivores (*see* food chain). Organisms are classified on the basis of their position in an ecosystem into various *trophic levels. Nutrients and energy move round ecosystems in loops or cycles (in the case above, for example, nutrients are returned to the soil via animal wastes and decomposition). *See* carbon cycle; nitrogen cycle.

ectoderm The external layer of cells of the *gastrula, which will develop into the skin epidermis and the nervous system in the adult. *See also* germ layers.

ectoparasite A parasite that lives on the outside of its host's body. *See* parasitism.

ectoplasm *See* cytoplasm.

edaphic factor A factor relating to the physical or chemical composition of the soil found in a particular area. For example, very alkaline soil may be an edaphic factor limiting the variety of plants growing in a region.

EDTA Ethylenediaminetetracetic acid,

$(HOOCCH_2)_2N(CH_2)_2N(CH_2COOH)_2$
A compound that acts as a chelating agent, reversibly binding with iron, magnesium, and other metal ions. It is used in certain culture media bound with iron, which it slowly releases into the medium.

EEG *See* electroencephalogram.

effector A cell or organ that produces a physiological response when stimulated by a nerve impulse. Examples include muscles and glands.

efferent Carrying (nerve impulses, blood, etc.) away from the centre of a body or organ towards peripheral regions. The term is usually applied to types of nerve fibres or blood vessels. *Compare* afferent.

effusion The flow of a gas through a small aperture. The relative rates at which gases effuse, under the same conditions, is approximately inversely proportional to the square roots of their densities.

egg 1. The fertilized ovum (*zygote) in egg-laying animals, e.g. birds and insects, after it emerges from the body. The egg is covered by *egg membranes that protect it from environmental damage, such as drying. **2.** (*or* **egg cell)** The mature female reproductive cell in animals and plants. *See* oosphere; ovum.

egg membrane The layer of material that covers an animal egg cell. *Primary membranes* develop in the ovary and cover the egg surface in addition to the normal cell membrane. The primary membrane is called the *vitelline membrane* in insects, molluscs, birds, and amphibians, the *chorion* in tunicates and fish, and the *zona pellucida* in mammals. Insects have a second thicker membrane, also called the chorion. *Secondary membranes* are secreted by the oviducts and parts of the genital system while the egg is passing to the outside. They include the jelly coat of frogs' eggs and the albumen and shell of birds' eggs.

Elasmobranchii (Chondrichthyes) A class of vertebrates comprising the fishes with cartilaginous skeletons. The majority belong to the order Selachi (skates, rays, and sharks). Most cartilaginous fishes are marine carnivores with powerful jaws. Unlike bony fishes, they have no swim bladder, and therefore avoid sinking only by constant swimming with the aid of an asymmetrical (*heterocercal*) tail. There is no operculum covering the gill slits, the first of which is modified as a *spiracle. Fertilization is internal so the few eggs produced are consequently yolky, large, and well-protected. Some elasmobranchs show viviparous development of the young (*see* viviparity).

electric organ An organ occurring on the body or tail of certain fish, such as the electric ray (*Torpedo*) and electric eel (*Electrophorus electricus*). It gives an electric shock when touched and is used either to stun prey or predators or, in some species, to maintain a weak electric field in the surrounding water that is used in navigation. The organ is composed of modified muscle cells (*electroplate cells*), nervous stimulation of which greatly increases the potential difference across the cell. The electroplates are in series so a high overall voltage can be achieved.

electrocardiogram (ECG) A tracing or graph of the electrical activity of the heart. Recordings are made from electrodes fastened over the heart and usually on both arms and a leg. Changes in the normal pattern of an ECG may indicate heart irregularities or disease.

electroencephalogram (EEG) A tracing or graph of the electrical activity of the brain. Electrodes taped to the scalp record electrical waves from different parts of the brain. The pattern of an EEG reflects an individual's level of consciousness and can be used to detect such disorders as epilepsy, tumours, or brain damage. *See also* brain death.

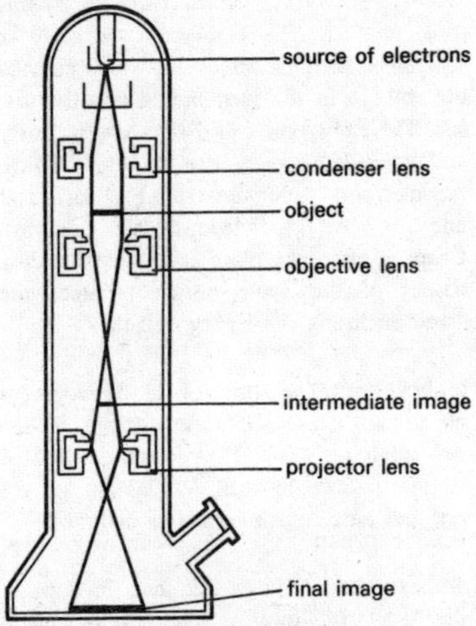

Principle of transmission electron microscope

electron microscope A form of microscope that uses a beam of electrons instead of a beam of light (as in the optical microscope) to form a large image of a very small object, such as a cell organelle, a virus, or a DNA molecule. In optical microscopes the resolution is limited by the wavelength of the light. High-energy electrons, however, can be associated with a considerably shorter wavelength than light; for example, electrons accelerated to an energy of 10^5 electronvolts have a wavelength of 0.04 nanometre, enabling a resolution of 0.2–0.5 nm to be achieved. The *transmission electron microscope* has an electron beam, sharply focused by electron lenses (coils producing a magnetic field or electrodes between which an electric field is created), passing through a very thin metallized specimen (less than 50 nanometres thick) onto a fluorescent screen, where a visual image is formed. This image can be photographed. The *scanning electron microscope* can be used with thicker specimens and forms a perspective image, although the resolution and magnification are lower. It is used particularly for examining surface features of small objects, such as pollen grains. In this type of instrument a beam of primary electrons scans the specimen and those that are reflected, together with any secondary electrons emitted, are collected. This current is used to modulate a separate electron beam in a TV monitor, which scans the screen at the same frequency, consequently building up a picture of the specimen. The resolution is limited to about 10–20 nm. *See also* field-emission microscope; field-ionization microscope.

electron transport chain (respiratory chain) A sequence of biochemical reduction–oxidation reactions that forms the final stage of *aerobic respiration. It results in the transfer of electrons or hydrogen atoms derived from the *Krebs cycle to molecular oxygen, with the formation of water. At the same time it conserves energy from food or light in the form of *ATP. The chain comprises a series of electron carriers that undergo reversible reduction–oxidation reactions, accepting electrons and then donating them to the next carrier in the chain. In the mitochondria, NADH and $FADH_2$, generated by the Krebs cycle, transfer their electrons to a chain comprising flavin mononucleotide (FMN), *coenzyme Q, and a series of *cytochromes. This process is coupled to the formation of ATP at three sites along the chain (*see* oxidative phosphorylation). The ATP is then carried across the mitochondrial membrane in exchange for ADP. An electron transport chain also occurs in the light reaction of *photosynthesis.

electrophoresis (cataphoresis) A technique for the analysis and separation of colloids, based on the movement of charged colloidal particles in an electric field. There are various experimental methods. In one the sample is placed in a U-tube and a buffer solution added to each arm, so that there are sharp boundaries between buffer and sample. An electrode is placed in each arm, a voltage applied, and the motion of the boundaries under the influence of the field is observed. The rate of migration of the particles depends on the field, the charge on the particles, and on other factors, such as the size and shape of the particles. More simply, electrophoresis can be carried out using an adsorbent, such as a strip of filter paper, soaked in a buffer with two electrodes making contact. The sample is placed between the electrodes and a voltage applied. Different components of the mixture migrate at different rates, so the sample separates into zones. The components can be identified by the rate at which they move. This technique has also been called *electrochromatography*.

Electrophoresis is used extensively in studying mixtures of proteins, nucleic acids, carbohydrates, enzymes, etc. In clinical medicine it is used for determining the protein content of body fluids.

elytra The thickened horny forewings of the *Coleoptera (beetles), which cover and protect the membranous hind wings when the insect is at rest.

emasculation The removal of the anthers of a flower in order to prevent self-pollination or the undesirable pollination of neighbouring plants.

Embden–Meyerhof pathway *See* glycolysis.

embryo 1. An animal in the earliest stages of its development, from the time when the fertilized ovum starts to divide (*see* cleavage), while it is contained within the egg or reproductive organs of the mother, until hatching or birth. A human embryo is called a *fetus after the first eight weeks of pregnancy. **2.** The structure in bryophytes, pteridophytes, and seed plants that develops from the zygote prior to germination. In seed plants the zygote is situated in the *embryo sac of the ovule. It divides by mitosis to form the embryonic cell and a structure called the *suspensor*, which embeds the embryo in the surrounding nutritive tissue. The embryonic cell divides continuously and eventually gives rise to the *radicle (young root), *plumule (young shoot), and one or two *cotyledons (seed leaves). Changes also take place in the surrounding tissues of the ovule, which becomes the *seed enclosing the embryo plant.

embryology The study of the development of animals from the fertilized egg to the new adult organism. It is sometimes limited to the period between fertilization of the egg and hatching or birth (*see* embryo).

embryo sac A large cell that develops in the *ovule of flowering plants. It is equivalent to the female *gametophyte of lower plants, although it is very much reduced. Typically, it contains eight nuclei formed by division of the original female gamete. One, the *oosphere (egg nucleus), is fertilized by a male nucleus and becomes the *embryo. Two of the remaining nuclei fuse with a second male nucleus to form a triploid nucleus that gives rise to the *endosperm.

emulsion A *colloid in which small particles of one liquid are dispersed in another liquid. Usually emulsions involve a dispersion of water in an oil or a dispersion of oil in water, and are stabilized by an *emulsifier*. Commonly emulsifiers are substances, such as *detergents, that have lyophobic and lyophilic parts in their molecules.

enamel The material that forms a covering over the crown of a *tooth (i.e. the part that projects above the gum). Enamel is smooth, white, and extremely hard, being rich in minerals containing calcium. It is

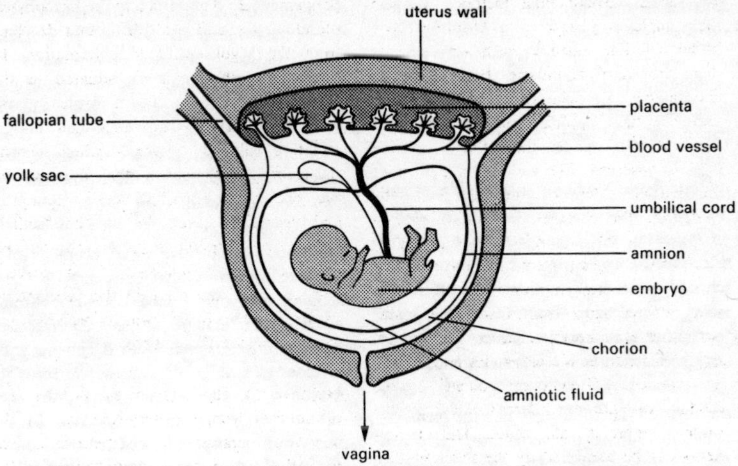

A developing human embryo

produced by certain cells (*ameloblasts*) of the oral epithelium and protects the underlying dentine of the tooth. Enamel may also be found in the placoid *scales of certain fish, which demonstrates the common developmental origin of scales and teeth.

encephalin (enkephalin) A peptide *neurotransmitter found principally in the brain. Two encephalins have so far been isolated. They bind to opiate receptors in the brain and their release is thought to control levels of pain and other sensations. Encephalins have also been found in spinal cord nerve cells and in gut epithelial cells. *Compare* endorphin.

endocarp *See* pericarp.

endocrine gland (ductless gland) Any gland in an animal that manufactures *hormones and secretes them directly into the bloodstream to act at distant sites in the body (known as *target organs* or *cells*). Endocrine glands tend to control slow long-term activities in the body, such as growth and sexual development. In mammals they include the *pituitary, *adrenal, *thyroid, and *parathyroid glands, the *ovary and *testis, the *placenta, and part of the pancreas (*see* islets of Langerhans). The activity of endocrine glands is controlled by negative feedback, i.e. a rise in output of hormone inhibits a further increase in its production, either directly or indirectly via the target organ or cell. *See also* neuroendocrine system. *Compare* exocrine gland.

endocrinology The study of the structure and functions of the *endocrine glands and of the *hormones they produce.

endoderm (entoderm) The internal layer of cells of the *gastrula, which will develop into the alimentary canal (gut) and digestive glands of the adult. *See also* germ layers.

endodermis The innermost layer of the root *cortex of a plant, lying immediately outside the vascular tissue. Various modifications of the endodermal cell walls indicate that they regulate the passage of materials both into and out of the vascular system.

An endodermis may also be seen in the stems of some plants.

endogamy The fusion of reproductive cells from closely related parents, i.e. *inbreeding. *Compare* exogamy.

endogenous Describing a substance, stimulus, organ, etc., that originates from within an organism. For example, growth rhythms not directed by environmental stimuli are termed endogenous rhythms. Lateral roots, which always grow from inside the main root rather than from its surface, are said to arise endogenously. *Compare* exogenous.

endolymph The fluid that fills the membranous labyrinth of the vertebrate *inner ear. *See* cochlea; semicircular canals. *Compare* perilymph.

endoparasite A parasite that lives inside its host's body. *See* parasitism.

endoplasm *See* cytoplasm.

endoplasmic reticulum (ER) A system of membranes within the cytoplasm of plant and animal *cells. It forms a link between the cell and nuclear membranes and is the site of protein synthesis. It is also concerned with the transport of proteins and lipids within the cell. *Rough ER* has *ribosomes attached to its surface; in *smooth ER* ribosomes are absent.

end organ The structure at the end of a peripheral nerve. Examples of end organs are the muscle *end plate at the end of a motor neurone and the *receptor at the end of a sensory neurone.

endorphin A peptide neurotransmitter found in the pituitary gland and hypothalamus that has pain-relieving properties similar to the *encephalins.

endoskeleton A supporting framework that lies entirely within the body of an animal, such as the bony *skeleton of vertebrates or the spicules of a sponge. The function of an endoskeleton is to support the body and in vertebrates it also protects the organs and provides a system of levers on which the muscles can act to produce movement. *Compare* exoskeleton.

endosperm A nutritive tissue, characteristic of flowering plants, that surrounds the developing embryo in a seed. It develops from nuclei in the *embryo sac and its cells are triploid. In *endospermic* seeds it remains and increases in size; in *nonendospermic* seeds it disappears as the food is absorbed by the embryo, particularly the *cotyledons. Many plants with endospermic seeds, such as cereals and oil crops, are cultivated for the rich food reserves in the endosperm.

endospore The resting stage of certain bacteria, formed in response to adverse conditions. The bacterial cell becomes enclosed in a tough resistant spore coat. On return to favourable conditions the spore changes back to the normal vegetative form of the organism.

endothelium A single layer of thin plate-like cells that line the inner surfaces of blood and lymph vessels and the heart. Endothelium is derived from the *mesoderm. *Compare* epithelium; mesothelium.

endotoxin *See* toxin.

end plate The area of muscle cell membrane that lies immediately beneath a motor nerve ending. Release of a *neurotransmitter at the end plate induces contraction of the muscle fibre.

enthalpy Symbol H. A thermodynamic property of a system defined by $H = U + pV$, where H is the enthalpy, U is the internal energy of the system, p its pressure, and

V its volume. In a chemical reaction carried out in the atmosphere the pressure remains constant and the enthalpy of reaction, ΔH, is equal to $\Delta U + p\Delta V$. For an exothermic reaction ΔH is taken to be negative.

entoderm *See* endoderm.

entomology The study of insects.

entomophily Pollination of a flower in which the pollen is carried on an insect. Entomophilous flowers are usually brightly coloured and scented and often secrete nectar. In some species (e.g. primulas) there are structural differences between the flowers to ensure that cross-pollination occurs. Other examples of entomophilous flowers are orchids and antirrhinums. *Compare* anemophily; hydrophily.

entropy Symbol S. A measure of the unavailability of a system's energy to do work; an increase in entropy is accompanied by a decrease in energy availability. When a system undergoes a reversible change the entropy (S) changes by an amount equal to the energy (Q) absorbed by the system divided by the thermodynamic temperature (T) at which the energy is absorbed, i.e. $\Delta S = \Delta Q/T$. However, all real processes are to a certain extent irreversible changes and in any closed system an irreversible change is always accompanied by an increase in entropy.

In a wider sense entropy can be interpreted as a measure of a system's disorder; the higher the entropy the greater the disorder. As any real change to a closed system tends towards higher entropy, and therefore higher disorder, it follows that the entropy of the universe (if it can be considered a closed system) is increasing and its available energy is decreasing. This increase in the entropy of the universe is one way of stating the second law of thermodynamics.

environment (in ecology) The physical, chemical, and biological conditions of the region in which an organism lives. *See also* ecology; ecosystem.

enzyme A protein that acts as a *catalyst in biochemical reactions. Each enzyme is specific to a particular reaction or group of similar reactions. Many require the association of certain nonprotein *cofactors in order to function. The molecule undergoing reaction (the *substrate*) binds to a specific *active site on the enzyme molecule to form a short-lived intermediate: this greatly increases (by a factor of up to 10^{20}) the rate at which the reaction proceeds to form the product. Enzyme activity is influenced by substrate concentration and by temperature and pH, which must lie within a certain range. Other molecules may compete for the active site, causing *inhibition of the enzyme or even irreversible destruction of its catalytic properties.

Enzyme production is governed by a cell's genes. Enzyme activity is further controlled by pH changes, alterations in the concentrations of essential cofactors, feedback inhibition by the products of the reaction, and activation by another enzyme, either from a less active form or an inactive precursor (*zymogen). Such changes may themselves be under the control of hormones or the nervous system.

The names of most enzymes end in *-ase*, which is added to the names of the substrates on which they act. Thus *lactase* is the enzyme that acts to break down lactose.

Eocene The second geological epoch of the *Tertiary period. It extended from the end of the Palaeocene epoch, about 54 million years ago, to the beginning of the Oligocene epoch, about 38 million years ago. The term was first proposed by Sir Charles Lyell (1797–1875) in 1833. In some classifications of geological time the *Palaeocene is included as part of the Eocene. Mammals were dominant in the Eocene: rodents, artiodactyls, carnivores, perissodactyls (including early horses), and whales were among the groups to make their first appearance.

eosin One of a series of acidic dyes, used in optical microscopy, that colours cytoplasm pink and cellulose red. It is frequently used as a counterstain with *haematoxylin for colouring tissue smears and sections of animal tissue.

ephemeral **1.** (in botany) An *annual plant that completes its life cycle in considerably less than one growing season. A number of generations can therefore occur in one year. Many troublesome weeds, such as groundsel and willowherb, are ephemerals. Certain desert plants are also ephemerals, completing their life cycles in a short period following rain. **2.** (in zoology) A short-lived animal, such as a mayfly.

epicalyx A ring of bracts below a flower that resembles a calyx. It is seen, for example, in the strawberry flower.

epicarp *See* pericarp.

epicotyl The region of a seedling stem above the stalks of the seed leaves (*cotyledons) of an embryo plant. It grows rapidly in seeds showing *hypogeal germination and lifts the stem above the soil surface. *Compare* hypocotyl.

epidermis **1.** (in zoology) The outermost layer of cells of the body of an animal. In invertebrates the epidermis is normally only one cell thick and is covered by an impermeable *cuticle. In vertebrates the epidermis is the thinner of the two layers of *skin (*compare* dermis). It consists of a basal layer of actively dividing cells (*see* Malpighian layer), covered by layers of cells that become impregnated with keratin (*see* keratinization). The outermost layers of epidermal cells (the *stratum corneum) form a water-resistant protective layer. The epidermis may bear a variety of specialized structures (e.g. *feathers, *hairs). **2.** (in botany) The outermost layer of cells covering a plant. It is overlaid by a *cuticle and its functions are principally to protect the plant from injury and to reduce water loss. Some epidermal cells are modified to form guard cells (*see* stoma) or hairs of various types (*see* piliferous layer). In woody plants the functions of the shoot epidermis are taken over by the periderm tissues (*see* cork cambium) and in mature roots the epidermis is sloughed off and replaced by the *hypodermis.

epididymis A long coiled tube in which spermatozoa are stored in vertebrates. In reptiles, birds, and mammals it is attached at one end to the *testis and opens into the sperm duct (*vas deferens) at the other.

epigamic Serving to attract a mate. Epigamic characters include the bright plumage of some male birds.

epigeal Describing seed germination in which the seed leaves (cotyledons) emerge from the ground and function as true leaves. Examples of epigeal germination are seen in sycamore and sunflower. *Compare* hypogeal.

epiglottis A flexible flap of cartilage in mammals that is attached to the wall of the pharynx near the base of the tongue. During swallowing (*see* deglutition) it covers the *glottis (the opening to the respiratory tract), so preventing food from entering the trachea (windpipe).

epinephrine *See* adrenaline.

epiphysis The terminal section of a growing bone (especially a long limb bone). It is separated from the bone shaft (*diaphysis*) by cartilage. New bone is produced on the side of the cartilage facing the diaphysis, while new cartilage is produced on the other side of the cartilage disc. When the bone reaches adult length the epiphysis merges with the diaphysis.

epiphyte A plant that grows upon another plant but is neither parasitic on it nor rooted in the ground. Epiphytes include many mosses and lichens and some tropical orchids.

epithelium A tissue in vertebrates consisting of closely packed cells in a sheet with little intercellular material. It forms a membrane over the outer surfaces of the body and walls of the internal cavities (coeloms). It also forms glands and parts of sense organs. Its functions are protective, absorptive, secretory, and sensory. The types of cell vary, giving rise to *squamous*, *cuboidal*, *columnar*, and *ciliated epithelia*. Epithelium is derived from *ectoderm and *endoderm. *Compare* endothelium; mesothelium.

Epstein–Barr virus *See* herpesvirus.

erepsin An obsolete name for a mixture of protein-degrading enzymes secreted by the small intestine.

ergocalciferol *See* vitamin D.

ergonomics The study of the engineering aspects of the relationship between workers and their working environment.

ergosterol A *sterol occurring in fungi, bacteria, algae, and higher plants. It is converted into vitamin D_2 by the action of ultraviolet light.

erythroblast Any of the cells in the *myeloid tissue of red bone marrow that develop into erythrocytes (red blood cells). Erythroblasts have a nucleus and are at first colourless, but fill with *haemoglobin as they develop. In mammals the nucleus disappears.

erythrocyte (red blood cell) The most numerous type of blood cell, which contains the red pigment *haemoglobin and is responsible for oxygen transport. Mammalian erythrocytes are disc-shaped and lack a nucleus; those of other vertebrates are oval and nucleated. In man the number of erythrocytes in the blood varies between 4.5 and 5.5 million per cubic millimetre. They survive for about four months and are then destroyed in the spleen and liver. *See also* erythroblast. *Compare* leucocyte.

essential amino acid An *amino acid that an organism is unable to synthesize in sufficient quantities. It must therefore be present in the diet. In man the essential amino acids are arginine, histidine, lysine, threonine, methionine, isoleucine, leucine, valine, phenylalanine, and tryptophan. These are required for protein synthesis and deficiency leads to retarded growth and other symptoms. Most of the amino acids required by man are also essential for all other multicellular animals and for most protozoans.

essential element Any of a number of elements required by living organisms to ensure normal growth, development, and maintenance. Apart from the elements found in organic compounds (i.e. carbon, hydrogen, oxygen, and nitrogen), plants, animals, and microorganisms all require a range of elements in inorganic forms in varying amounts, depending on the type of organism. The *major elements*, present in tissues in relatively large amounts (greater than 0.005%), are calcium, phosphorus, potassium, sodium, chlorine, sulphur, and magnesium. The *trace elements* occur at much lower concentrations and thus requirements are much less. The most important are iron, manganese, zinc, copper, iodine, cobalt, selenium, molybdenum, chromium, and silicon. Each element may fulfil one or more of a variety of metabolic roles. Sodium, potassium, and chloride ions are the chief electrolytic components of cells and body fluids and thus determine their electrical and osmotic status. Calcium, phosphorus, and magnesium are all present in bone. Calcium is also essential to nerve and muscle activity, while phosphorus is a key constituent of the chemical energy carriers (e.g. *ATP) and the nucleic acids. Sulphur is needed primarily for amino acid synthesis (in plants and microorganisms). The trace elements may serve as *cofactors or as constituents of complex molecules, e.g. iron in haem and cobalt in vitamin B_{12}.

essential fatty acids *Fatty acids that must normally be present in the diet of certain animals, including man. Essential fatty acids, which include *linoleic and *linolenic acids, all possess double bonds at the same two positions along their hydrocarbon chain and so can act as precursors of *prostaglandins. Deficiency of essential fatty acids can cause dermatosis, weight loss, irregular oestrus, etc. An adult human requires 2–10 g linoleic acid or its equivalent per day.

essential oil A natural oil with a distinctive scent secreted by the glands of certain aromatic plants. *Terpenes are the main constituents. Essential oils are extracted from plants by steam distillation, extraction with cold neutral fats or solvents (e.g. alcohol), or pressing and used in perfumes, flavourings, and medicine. Examples are citrus oils, flower oils (e.g. rose, jasmine), and oil of cloves.

etaerio A cluster of fruits formed from the unfused carpels of a single flower. For example, the anemone has an etaerio of *achenes, larkspur an etaerio of *follicles, and blackberry an etaerio of *drupes.

ethanedioic acid *See* oxalic acid.

ethanoic acid *See* acetic acid.

ethene *See* ethylene.

ethology The study of the biology of *animal behaviour. Central to the ethologist's approach is the principle that animal behaviour (like physical characteristics) is subject to evolution through natural selection. Ethologists therefore seek to explain how the behaviour of an animal in its natural environment may contribute to the survival of the maximum number of its relatives and offspring. This involves recognizing the stimuli that are important in nature (*see* sign stimulus) and how innate predispositions interact with *learning in the development of behaviour (*see* instinct). Studies of this sort were pioneered by Konrad Lorenz (1903–) and Niko Tinbergen (1907–).

ethylene (ethene) A colourless gaseous hydrocarbon, C_2H_4, that occurs naturally in many plants and acts as a *growth substance. Among its effects are the inhibition of longitudinal growth and the radial expansion of tissues, the promotion of fruit ripening and (in some species) flowering, and the breaking of seed dormancy in some species.

etiolation The abnormal form of growth observed when plants grow in darkness or severely reduced light. Such plants characteristically have blanched leaves and shoots, excessively long shoots, and reduced leaves and root systems.

eubacteria The true *bacteria, i.e. those typically having simple unbranched cells, rigid cell walls, and flagella for movement.

eucaryote *See* eukaryote.

euchromatin *See* chromatin.

eugenics The study of methods of improving the quality of human populations by the application of genetic principles. *Positive eugenics* would seek to do this by selective breeding programmes. *Negative eugenics* aims to eliminate harmful genes (e.g. those causing haemophilia and colour blindness) by counselling any prospective parents who are likely to be *carriers.

Euglenophyta A division containing *algae (including *Euglena*) that in certain respects resemble animals more than they do plants. For example, they lack a cell wall and some forms are colourless and thus ingest food since they cannot photosynthesize. Members of the Euglenophyta are unicellular and move by means of flagella; they usually inhabit fresh water. These algae are often classified by zoologists into the class Flagellata of the *Protozoa.

eukaryote (eucaryote) An organism consisting of cells in which the genetic material is contained within a distinct nucleus. All organisms, except bacteria and blue-green algae, are eukaryotes. *See* cell. *Compare* prokaryote.

Eustachian tube The tube that connects the *middle ear to the back of the throat (pharynx) in vertebrates. It is normally closed, but during swallowing it opens to allow air into the middle ear, which equalizes the pressure on each side of the *tympanum (eardrum). It was named after the Italian anatomist Bartolomeo Eustachio (?1520–74).

Eutheria (Placentalia) A subclass of mammals in which the embryos are retained in a uterus in the mother's body and nourished by a *placenta. The young are thus fully protected during their embryonic development and kept at a constant temperature. Placental mammals evolved during the Cretaceous period (about 100 million years ago). Modern placentals are a highly diverse group that occupy all types of habitat in all parts of the world. They include the orders *Artiodactyla, *Carnivora, *Cetacea, *Chiroptera, *Insectivora, *Perissodactyla, *Primates, *Proboscidea, and *Rodentia. *Compare* Metatheria; Prototheria.

eutrophic Describing a body of water (e.g. a lake) with an abundant supply of nutrients and a high rate of formation of organic matter by photosynthesis. *Compare* oligotrophic.

evergreen (Describing) a plant that bears leaves throughout the year, each leaf being shed independently of the others after two or three years. The leaves of evergreens are often reduced or adapted in some way to prevent excessive water loss; examples are the needles of conifers and the leathery waxy leaves of holly. *Compare* deciduous.

evocation The ability of experimental stimuli (e.g. chemicals or tissue implants) to cause unspecialized embryonic tissue to develop into specialized tissue.

evolution The gradual process by which the present diversity of plant and animal life arose from the earliest and most primitive organisms, which is believed to have been continuing for at least the past 3000 million years. Until the middle of the 18th century it was generally believed that each species was divinely created and fixed in its form throughout its existence (*see* special creation). Lamarck was the first biologist to publish a theory to explain how one species could have evolved into another (*see* Lamarckism), but it was not until the publication of Darwin's *On the Origin of Species* in 1859 that special creation was seriously challenged. Unlike Lamarck, Darwin proposed a feasible mechanism for evolution and backed it up with evidence from the fossil record and studies of comparative anatomy and embryology (*see* Darwinism; natural selection). The modern version of Darwinism, which incorporates discoveries in genetics made since Darwin's time, probably remains the most acceptable theory of species evolution. More controversial, however, and still to be firmly clarified, are the relationships and evolution of groups above the species level.

exa- Symbol E. A prefix used in the metric system to denote 10^{18} times. For example, 10^{18} metres = 1 exametre (Em).

excretion The elimination by an organism of the waste products that arise as a result of metabolic activity. These products include water, carbon dioxide, and nitrogenous compounds. In plants and simple animals waste products are excreted by simple diffusion from the body, but higher animals have specialized organs and organ systems devoted to this function. Examples of excretory organs in vertebrates are the lungs (for carbon dioxide and water), and the *kidneys (for nitrogenous compounds (urea) and water). In addition, mammals excrete small amounts of urea, salts, and water from the skin in sweat.

exocarp *See* pericarp.

exocrine gland A gland that discharges its secretion into a body cavity (such as the gut) or onto the body surface. Examples are the *sebaceous and *sweat glands, the *mammary glands, and part of the pancreas. Exocrine glands are formed in the embryo from the invagination of epithelial cells. Their secretions pass initially into a cavity (an *alveolus* or *acinus*) and then out through a duct or duct network, along which the secretion may become modified by exchange with the blood across the duct epithelium.

exodermis *See* hypodermis.

exogamy The fusion of reproductive cells from distantly related or unrelated organisms, i.e. *outbreeding. *Compare* endogamy.

exogenous Describing substances, stimuli, etc., that originate outside an organism. For example, vitamins that cannot be synthesized by an animal are said to be supplied exogenously in the diet. *Compare* endogenous.

exoskeleton A rigid external covering for the body in certain animals, such as the hard chitinous cuticle of arthropods. An exoskeleton protects and supports the body and provides points of attachment for muscles. The cuticle of arthropods must be shed at intervals to allow growth to occur (*see* ecdysis). Other examples of exoskeletons are the shells of molluscs and the bony plates of tortoises and armadillos. *Compare* endoskeleton.

exotoxin *See* toxin.

explantation The removal of cells, tissues, or organs of animals and plants for observation of their growth and development in appropriate culture media. *See also* tissue culture; organ culture.

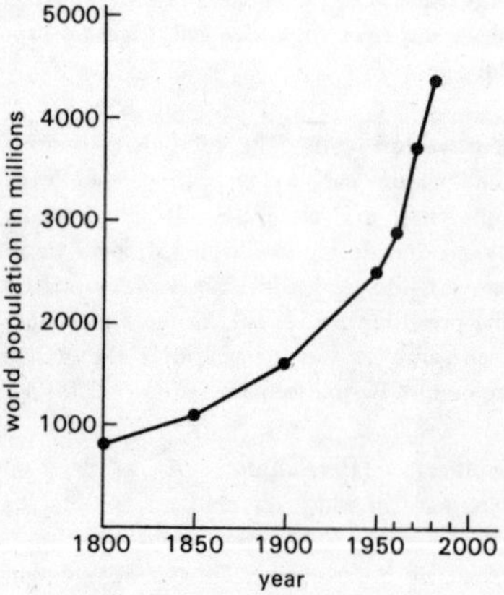

Graph showing exponential growth of the human population

exponential growth A form of population growth in which the rate of growth is related to the number of individuals present. Increase is slow when numbers are low but rises sharply as numbers increase. If population number is plotted against time on a graph a characteristic J-shaped curve results. In animal and plant populations, such factors as overcrowding, lack of nutrients, and disease limit population increase beyond a certain point and the J-shaped exponential curve tails off giving an S-shaped (sigmoid) curve.

extracellular Located or occurring outside the cell. *Cuticularization is an example of an extracellular process.

extraembryonic membranes (embryonic membranes) Tissues produced by an animal *embryo for protection and nutrition but otherwise taking no part in its development. The three membranes, which are called *fetal membranes* in man, are the *chorion, *amnion, and *allantois.

eye The organ of sight. The most primitive eyes are the *eyespots of some unicellular organisms. More advanced eyes are the *ocelli and *compound eyes of arthropods (e.g. insects). The cephalopod molluscs (e.g. the octopus and squid) and vertebrates possess the most highly developed eyes. These normally occur in pairs, are nearly spherical, and filled with fluid. Light is refracted by the *cornea through the pupil in the *iris and onto the *lens, which focuses images onto the retina. These images are received by light-sensitive cells in the retina (*see* cone; rod), which transmit impulses to the brain via the optic nerve.

eyepiece (ocular) The lens or system of lenses in an optical instrument that is nearest to the eye. It usually produces a magnified image of the previous image formed by the instrument.

eyespot (stigma) **1.** (in botany) A structure found in some free-swimming unicellular algae and in plant reproductive cells that contains orange or red pigments (carotenoids) and is sensitive to light. It enables the cell to move in relation to a light source (*see* phototaxis). **2.** (in zoology) A spot of pigment found in some lower animals, e.g. jellyfish.

eye tooth A *canine tooth in the upper jaw

F

F_1 (first filial generation) The first generation of offspring resulting from an arranged cross between selected parents in breeding experiments. *See* monohybrid.

F_2 (second filial generation) The second generation of offspring in breeding experi-

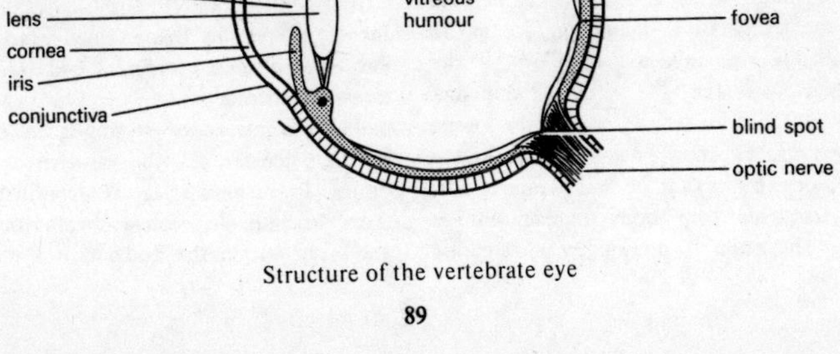

Structure of the vertebrate eye

ments, obtained by crosses between individuals of the *F_1 generation. *See* monohybrid.

facilitated diffusion The transport of molecules across the outer membrane of a living cell by a process that involves a specific carrier located within the cell membrane but does not require expenditure of energy by the cell. The carrier is believed to combine with a molecule on one side of the membrane, move through the membrane, and release the molecule on the other side. *Compare* active transport.

FAD (flavin adenine dinucleotide) A *coenzyme important in various biochemical reactions. It comprises a phosphorylated vitamin B_2 (riboflavin) molecule linked to the nucleotide adenine monophosphate (AMP). FAD is usually tightly bound to the enzyme forming a *flavoprotein*. It functions as a hydrogen acceptor in dehydrogenation reactions, being reduced to $FADH_2$. This in turn is oxidized to FAD by the *electron transport chain, thereby generating ATP (two molecules of ATP per molecule of $FADH_2$).

faeces Waste material that is eliminated from the alimentary canal through the *anus. Faeces consist of the indigestible residue of food that remains after the processes of digestion and absorption of nutrients and water have taken place, together with bacteria and dead cells shed from the gut lining.

Fahrenheit scale A temperature scale in which (by modern definition) the temperature of boiling water is taken as 212 degrees and the temperature of melting ice as 32 degrees. It was invented in 1714 by the German scientist G. D. Fahrenheit (1686–1736), who set the zero at the lowest temperature he knew how to obtain in the laboratory (by mixing ice and common salt) and took his own body temperature as 96°F. The scale is no longer in scientific use. To convert to the *Celsius scale the formula is

$$C = 5(F - 32)/9.$$

fallopian tube (oviduct) The tube that carries egg cells from the *ovary to the womb in mammals. The eggs are carried by the action of muscles and cilia. It was named after the Italian anatomist Gabriel Fallopius (1523–62).

false fruit *See* pseudocarp.

family A category used in the *classification of organisms that consists of one or several similar or closely related genera. Similar families are grouped into an order. Family names end in *-aceae* or *-ae* in botany (e.g. Cactaceae) and *-idae* in zoology (e.g. Equidae). The names are usually derived from a type genus (*Cactus* and *Equus* in the examples above) that is characteristic of the whole family (*see* type specimen). In botany, families are sometimes called *natural orders*.

fascia A sheet of fibrous connective tissue occurring beneath the skin and also enveloping glands, vessels, nerves, and forming tendon sheaths.

fascicle *See* vascular bundle.

fast green A green dye used in optical microscopy that stains cellulose, cytoplasm, collagen, and mucus green. It is frequently used to stain plant tissues, with *safranin as a counterstain. Unlike *light green*, a similar dye, it does not fade easily.

fat A mixture of lipids, chiefly *triglycerides, that is solid at normal body temperatures. Fats occur widely in plants and animals as a means of storing food energy, having twice the calorific value of carbohydrates. In mammals, fat is deposited in a layer beneath the skin (subcutaneous fat) and deep within the body as a specialized *adipose tissue (*see also* brown fat). The insulating properties of fat are also important, especially in animals lacking fur and those inhabiting cold climates (e.g. seals and whales).

Fats derived from plants and fish generally have a greater proportion of unsaturated *fatty acids than those from mammals. Their melting points thus tend to be lower, causing a softer consistency at room temperatures. Highly unsaturated fats are liquid at room temperatures and are therefore more properly called *oils.

fat body **1.** An abdominal organ in amphibians attached to the anterior of each kidney. It contains a reserve of fat that nourishes the gonads during the winter hibernation in readiness for the spring breeding season. **2.** A mass of fatty tissue spreading throughout the body cavity of insects in which fats, proteins, and glycogen are stored as a reserve for hibernation or pupation.

fatty acid An organic compound consisting of a hydrocarbon chain and a terminal carboxyl group (*see* carboxylic acids). Chain length ranges from one hydrogen atom (formic acid, HCOOH) to nearly 30 carbon atoms. Acetic, propionic, and butyric acids are important in metabolism. Long-chain fatty acids (more than 8–10 carbon atoms) most commonly occur as constituents of certain lipids, notably glycerides, phospholipids, sterols, and waxes, in which they are esterified with alcohols. These long-chain fatty acids generally have an even number of carbon atoms; unbranched chains predominate over branched chains. They may be saturated (e.g. *palmitic acid and *stearic acid) or unsaturated, with one double bond (e.g. *oleic acid) or two or more double bonds, in which case they are called *polyunsaturated fatty acids* (e.g. *linoleic acid and *linolenic acid). *See also* essential fatty acids.

The physical properties of fatty acids are determined by chain length, degree of unsaturation, and chain branching. Short-chain acids are pungent liquids, soluble in water. As chain length increases, melting points are raised and water-solubility decreases. Unsaturation and chain branching tend to lower melting points.

feathers The body covering of birds, formed as outgrowths of the epidermis and composed of the protein *keratin. Feathers provide heat insulation, they give the body its streamlined shape, and those of the wings and tail are important in flight. Basically a feather consists of a *quill*, which is embedded in the skin attached to a feather follicle and is continuous with the shaft (*rachis*) of the feather, which carries the *barbs. This basic structure is modified depending on the type of feather (*see* contour feathers; down feathers; filoplumes).

fecundity The fertility of an organism (in higher animals, generally of the female of the species). Normally all organisms, assuming they reach reproductive age, are sufficiently fecund to replace themselves several times over. Darwin noted this, together with the fact that population numbers nevertheless tended to remain fairly constant: these observations led him to formulate his theory of evolution by *natural selection.

feedback The use of part of the output of a system to control its performance. In *positive feedback*, the output is used to enhance the input; in *negative feedback*, the output is used to reduce the input. Many biological processes rely on negative feedback. As the population of a species expands, so its food supply per individual is diminished; the result is that the population then begins to fall. Many biochemical processes are controlled by feedback *inhibition.

Fehling's test A chemical test to detect reducing sugars and aldehydes in solution, devised by the German chemist H. C. von Fehling (1812–85). *Fehling's solution* consists of Fehlings A (copper(II) sulphate solution) and Fehling's B (alkaline 2,3-dihydroxybutanedioate (sodium tartrate) solution), equal amounts of which are added to the test solution. After boiling, a positive result is indicated by the formation of a brick-red precipitate of copper(I) oxide. Methanal, being a strong reducing agent, also produces copper metal; ketones do not react.

femto- Symbol f. A prefix used in the metric system to denote 10^{-15}. For example, 10^{-15} second = 1 femtosecond (fs).

femur 1. The thigh bone of terrestrial vertebrates. It articulates at one end with the pelvic girdle at the hip joint and at the other (via two *condyles) with the *tibia. **2.** The third segment of an insect's leg, attached to the *trochanter. *See also* coxa.

fenestra Either of the two delicate membranes between the *middle ear and the *inner ear. The lower membrane is the *fenestra ovalis* (*fenestra vestibuli*, or *oval window*). The stapes (the third *ear ossicle) transmits to the fenestra ovalis the vibrations it has received (via the other two ear ossicles) from the *tympanum (eardrum). Vibrations in the fenestra ovalis are transmitted to the *cochlea in the inner ear. The lower membrane, the *fenestra rotunda* (*fenestra cochleae*, or *round window*), vibrates in response to changes in pressure in the fluid (perilymph) surrounding the cochlea.

fermentation A form of *anaerobic respiration occurring in certain microorganisms, e.g. yeasts. It comprises a series of biochemical reactions by which pyruvate (the end product of *glycolysis) is converted to ethanol and carbon dioxide. Fermentation is the basis of the baking, wine, and beer industries.

ferns *See* Filicinae.

fertilization (syngamy) The fusion of the nuclei of the male and female gametes (reproductive cells) during the process of sexual reproduction to form a *zygote*. As each gamete contains only half the correct number of chromosomes, fertilization and zygote formation results in a cell with the full complement of chromosomes, half of which are derived from each of the parents. In animals the process involves fusion of the nuclei of a spermatozoan and an ovum. In most aquatic animals (e.g. fish) this takes place in the surrounding water, into which the gametes are shed. Among most terrestrial animals (e.g. insects, many mammals) fertilization occurs in the body of the female, into which the sperms are introduced. In flowering plants, after *pollination, the male gamete (pollen) produces a *pollen tube, which grows down into the female reproductive organ (carpel) to enable a pollen nucleus to fuse with the egg nucleus.

In *self-fertilization* the male and female gametes are derived from the same individual; in *cross-fertilization* they are derived from different individuals. Among plants, self-fertilization (also called *autogamy*) is common in many cultivated species, e.g. wheat and oats. In cross-fertilization (also called *allogamy*) the pollen comes either from another flower of the same plant or from a different plant (*see also* incompatibility).

fetal membranes *See* extraembryonic membranes.

fetus (foetus) The embryo of a mammal, especially a human, when development has reached a stage at which the main features of the adult form are recognizable. In humans the embryo from eight weeks to birth is called a fetus.

Feulgen's test A histochemical test in which the distribution of DNA in the chromosomes of dividing cell nuclei can be observed. It was devised by the German chemist R. Feulgen (1884–1955). A tissue section is first treated with dilute hydrochloric acid to remove the purine bases of the DNA, thus exposing the aldehyde groups of the sugar deoxyribose. The section is then immersed in *Schiff's reagent, which combines with the aldehyde groups to form a magenta-coloured compound.

fibre **1.** An elongated plant cell whose walls are extensively (usually completely) thickened with lignin. Fibres are found in the vascular tissue, usually in the xylem, where they provide structural support. The term is often used loosely to mean any kind of xylem element. The fibres of many species, e.g. flax, are of commercial importance. **2.** Any of various threadlike structures in the animal body, such as a muscle fibre, a nerve fibre, or a collagen fibre.

fibre optics *See* optical fibres.

fibrin The insoluble protein that forms fibres at the site of an injury and is the foundation of a blood clot. *See* blood clotting; clotting factors.

fibrinogen The protein dissolved in the blood plasma that, when suitably activated, is converted to insoluble *fibrin fibres. *See* clotting factors.

fibrinolysis The breakdown of the protein *fibrin by the enzyme *plasmin* (or *fibrinolysin*), which occurs when blood clots are removed from the circulation.

fibroblast A cell that secretes fibres in the intercellular substance of *connective tissue. The cells are long, flat, and star-shaped and lie close to collagen fibres.

fibula The smaller and outer of the two bones between the knee and the ankle in terrestrial vertebrates. *Compare* tibia.

field-emission microscope A type of electron microscope in which a high negative voltage is applied to a metal tip placed in an evacuated vessel some distance from a glass screen with a fluorescent coating. The tip produces electrons by *field emission*, i.e. the emission of electrons from an unheated sharp metal part as a result of a high electric field. The emitted electrons form an enlarged pattern on the fluorescent screen, related to the individual exposed planes of atoms. As the resolution of the instrument is limited by the vibrations of the metal atoms, it is helpful to cool the tip in liquid helium. Although the individual atoms forming the point are not displayed, individual adsorbed atoms of other substances can be, and their activity is observable.

field-ionization microscope (field-ion microscope) A type of electron microscope that is similar in principle to the *field-emission microscope, except that a high positive voltage is applied to the metal tip, which is surrounded by low-pressure gas (usually helium) rather than a vacuum. The image is formed in this case by *field ionization*: ionization at the surface of an unheated solid as a result of a strong electric field creating positive ions by electron transfer from surrounding atoms or molecules. The image is formed by ions striking the fluorescent screen. Individual atoms on the surface of the tip can be resolved and, in certain cases, adsorbed atoms may be detected.

filament **1.** (in zoology) A long slender hairlike structure, such as any of the *barbs of a bird's feather. **2.** (in botany) The stalk of the *stamen in a flower. It bears the anther and consists mainly of conducting tissue.

Filicinae A class of mainly terrestrial vascular plants – the ferns – belonging to the *Pteropsida. Ferns are perennial plants bearing large conspicuous leaves (fronds) usually arising from either a rhizome or a short erect stem. Bracken is a common example. Only the tree ferns have stems that reach an appreciable height. There is a characteristic uncurling of the young leaves as they expand into the adult form. Spores are borne on the underside of specialized leaves (sporophylls).

filoplumes Minute hairlike *feathers consisting of a shaft (*rachis*) bearing a few unattached barbs. They are found between the contour feathers.

filter feeding A method of feeding in which tiny food particles are ingested from the surrounding water. It is used by many aquatic invertebrates. *See also* ciliary feeding.

filtrate The clear liquid obtained by filtration.

filtration The process of separating solid particles using a filter. In vacuum filtration, the liquid is drawn through the filter by a vacuum pump.

fins The locomotory organs of aquatic vertebrates. In fish there are typically one or more *dorsal* and *ventral fins* (sometimes continuous), whose function is balance; a *caudal fin* around the tail, which is the main propulsive organ; and two paired fins: the *pectoral fins* attached to the pectoral (shoulder) girdle and the *pelvic fins* attached to the pelvic (hip) girdle, which are used in steering. These paired fins are homologous with the limbs of tetrapods. Fins are strengthened by a number of flexible fin rays, which may be cartilaginous, bony and jointed, horny, or fibrous and jointed.

fish *See* Pisces.

fission A type of asexual reproduction occurring in some unicellular organisms, e.g. diatoms, protozoans, and bacteria, in which the parent cell divides to form two (*binary fission*) or more (*multiple fission*) similar daughter cells.

fission-track dating A method of estimating the age of glass and other mineral objects by observing the tracks made in them by the fission fragments of the uranium nuclei that they contain. By irradiating the objects with neutrons to induce fission and comparing the density and number of the tracks before and after irradiation it is possible to estimate the time that has elapsed since the object solidified.

fitness (in evolution) The condition of an organism that is well adapted to its environment, as measured by its ability to reproduce itself. *See also* inclusive fitness.

fixation 1. The first stage in the preparation of a specimen for microscopical examination, in which the tissue is killed and preserved in as natural a state as possible by immersion in a chemical *fixative*. The fixative prevents the distortion of cell components by denaturing its constituent protein. Some commonly used fixatives are formaldehyde, ethanol, and Bouin's fluid (for light microscopy), and osmium tetroxide and gluteraldehyde (for electron microscopy). Fixation may also be brought about by heat. **2.** *See* nitrogen fixation.

fixed action pattern *See* instinct.

flaccid (in botany) Describing plant tissue that has become soft and less rigid than normal because the cytoplasm within its cells has shrunk and contracted away from the cell walls through loss of water (*see* plasmolysis).

flagellum A relatively long (up to 150 μm) fine whiplike structure present on the surface of certain cells, particularly motile reproductive cells (e.g. spermatozoa), bacteria, and certain protozoans. Flagella occur singly or in small groups. Beating of flagella produces movement of the cell, but some flagella (e.g. in *Hydra*) beat to cause movement of the surrounding fluid. *Compare* cilium.

flatworms *See* Platyhelminthes.

flavin adenine dinucleotide *See* FAD.

flavonoid One of a group of naturally occurring phenolic compounds many of which are plant pigments. They include the *anthocyanins, *flavonols*, and *flavones*. Patterns of flavonoid distribution have been used in taxonomic studies of plant species.

flavoprotein *See* FAD.

fleas *See* Siphonaptera.

flies *See* Diptera.

flower The structure in angiosperms (flowering plants) that bears the organs for sexual reproduction. Flowers are very variable in form, ranging from the small green insignificant wind-pollinated flowers of many grasses to spectacular brightly coloured insect-pollinated flowers. Flowers are often grouped together into *inflorescences, some of which (e.g. that of dandelion) are so compacted as to resemble a single flower. Typically flowers consist of a receptacle that bears sepals, petals, stamens, and carpels. The flower parts are adapted to bring about pollination and fertilization resulting in the formation of seeds and fruits. The sepals are usually green and leaflike and protect the flower bud. The petals of insect-pollinated flowers are adapted in many ingenious ways to attract insects and, in some instances, other animals. For example, some flowers are adapted to attract short-tongued insects by having an open shallow *corolla tube and nectar situated in an exposed position. Flowers adapted for pollination by long-tongued insects have a long corolla tube of fused petals with nectar in a concealed position. The tongue of the insect brushes against the anthers and stigma before reaching the nectar. Wind-pollinated flowers, in contrast, are inconspicuous. The anthers dangle outside the corolla and the stigmas have a feathery surface to catch the pollen grains.

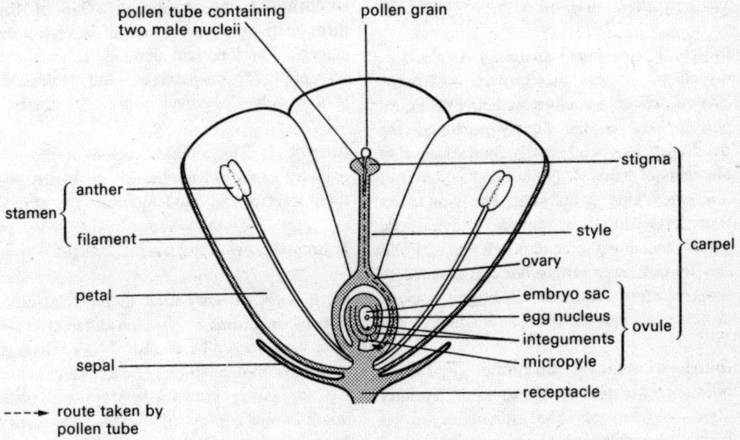

Section through a monocarpellary flower at the time of pollination

Some species are adapted for self-pollination and have small flowers, no nectar, and stamens and carpels that mature simultaneously.

flowering plants *See* Angiospermae.

flukes *See* Trematoda.

fluoridation The process of adding very small amounts of fluorine salts (e.g. sodium fluoride, NaF) to drinking water to prevent tooth decay. The fluoride becomes incorporated into the substance of the growing teeth and reduces the incidence of caries.

foetus *See* fetus.

folacin *See* folic acid.

folic acid (folacin) A vitamin of the *vitamin B complex. In its active form, tetrahydrofolic acid, it is a *coenzyme in various reactions involved in the metabolism of amino acids, purines, and pyrimidines. It is synthesized by intestinal bacteria and is widespread in food, especially green leafy vegetables. Deficiency causes poor growth and nutritional anaemia.

follicle 1. (in animal anatomy) Any enclosing cluster of cells that protects and nourishes a cell or structure within. For example, follicles in the *ovary surround the developing egg, while *hair follicles envelop the roots of hairs. **2.** (in botany) A dry fruit that, when ripe, splits along one side to release its seeds. It is formed from a single carpel containing one or more seeds. Follicles do not occur singly but are grouped to form clusters (*etaerios*). Examples include larkspur, columbine, and monk's hood.

follicle-stimulating hormone (FSH) A hormone, secreted by the anterior pituitary gland in mammals, that stimulates, in female mammals, ripening of specialized structures in the ovary (*Graafian follicles) that produce ova and, in males, the formation of sperm in the testis. It is a major constituent of fertility drugs, used to treat failure of ovulation and decreased sperm production. *See also* gonadotrophin.

food chain The transfer of energy from green plants (the primary producers) through a sequence of organisms in which each eats the one below it in the chain and is eaten by the one above. Thus plants are eaten by herbivores, which are then eaten by carnivores. These may in turn be eaten by different carnivores. The position an organism occupies in a food chain is known as its *trophic level. In practice, many animals feed at several different trophic levels, resulting in a more complex set of feeding relationships known as a *food web*. *See* bioenergetics; consumer; producer; pyramid of numbers.

foramen An aperture in an animal part or organ, especially one in a bone or cartilage. For example, the *foramen magnum* is the opening at the base of the skull through which the *spinal cord passes.

forebrain (prosencephalon) One of the three sections of the brain of a vertebrate embryo. The forebrain develops to form the *cerebrum, *hypothalamus, and *thalamus in the adult. *Compare* hindbrain; midbrain.

foregut 1. The anterior region of the alimentary canal of vertebrates, up to the anterior part of the duodenum. **2.** The anterior part of the alimentary canal of arthropods. *See also* hindgut; midgut.

form 1. A category used in the *classification of organisms into which different types of a variety may be placed. **2.** Any distinct variant within a species. Seasonal variants, e.g. the tawny brown (summer) and blue-white (winter) forms of the blue hare, may be called forms, as may the different types that constitute a *polymorphism.

fossil The remains or traces of any organism that lived in the geological past. In general only the hard parts of organisms become fossilized (e.g. bones, teeth, shells, and wood) but under certain circumstances the entire organism is preserved. For example, virtually unaltered fossils of extinct mammals, such as the woolly mammoth and woolly rhinoceros, have been found preserved in ice in the Arctic. Small organisms or parts of organisms (e.g. insects, leaves, flowers) have been preserved in *amber.

In the majority of fossils the organism has been turned to stone – a process known as *petrification*. This may take one of three forms. In *permineralization*, solutions originating underground fill the microscopic cavities in the organism. Minerals in these solutions (e.g. silica or calcite) may actually replace the original material of the organism so that even microscopic structures may be preserved; this process is known as *replacement* (or *mineralization*). A third form of petrification – *carbonization* (or *distillation*) – occurs in certain soft tissues that are composed chiefly of compounds of carbon, hydrogen, and oxygen (e.g. cellulose). After the organism has been buried, and in the absence of oxygen, carbon dioxide and water are liberated until only free carbon remains. This forms a black carbon film in the rock outlining the original organism. *Moulds* are formed when the original fossil is dissolved away leaving a mould of its outline in the solid rock. The deposition of mineral matter from underground solutions in a mould forms a *cast*. Palaeontologists often produce casts from moulds using such substances as dental wax. Moulds of thin organisms (e.g. leaves) are commonly known as *imprints*. *Trace fossils* are the fossilized remnants of the evidence of animal life, such as tracks, trails, footprints, burrows, and *coprolites* (fossilized faeces).

The ideal conditions for the formation of fossils occur in areas of rapid sedimentation, especially those parts of the seabed that lie below the zone of wave disturbance. *See also* chemical fossil; index fossil; microfossil.

fossil hominid *See* hominid.

fovea (fovea centralis) A shallow depression in the *retina of the eye, opposite the lens, that is present in some vertebrates. This area contains a large concentration of *cones with only a thin layer of overlying nerves. It is therefore specialized for the perception of colour and sharp intense images. The clarity is enhanced when light is focused on the foveae of both retinas simultaneously. *See* binocular vision.

fragmentation A method of asexual reproduction, occurring in some invertebrate animals, in which parts of the organism break off and subsequently differentiate and develop into new individuals. It occurs especially in certain coelenterates and annelids. In some, regeneration may occur before separation, producing chains of individuals budding from the parent.

fraternal twins (dizygotic twins) Two individuals that result from a single pregnancy, each having developed from a separate fertilized egg. The two egg cells contain different combinations of *alleles and so do the two sperm that fertilize them. The twins therefore have no more genetic similarity than brothers or sisters from single births. *Compare* identical twins.

free energy A measure of a system's ability to do work. The *Gibbs free energy* (or *Gibbs function*), G, is defined by $G = H - TS$, where G is the energy liberated or absorbed in a reversible process at constant pressure and constant temperature (T), H is the *enthalpy and S the *entropy of the system. Changes in Gibbs free energy, ΔG, are useful in indicating the conditions under which a chemical reaction will occur. If ΔG is positive the reaction will only occur if energy is supplied to force it away from the equilibrium position (i.e. when $\Delta G = 0$). If ΔG is negative the reaction will proceed spontaneously to equilibrium.

frogs *See* Amphibia.

frond *See* megaphyll.

fructification *See* sporophore.

fructose (fruit sugar; laevulose) A simple sugar, $C_6H_{12}O_6$, stereoisomeric with glucose. (Although natural fructose is the D-form, it is in fact laevorotatory.) Fructose occurs in green plants, fruits, and honey and tastes sweeter than sucrose (cane sugar), of which it is a constituent. Derivatives of fructose are important in the energy metabolism of living organisms. Some polysaccharide derivatives (fructans) are carbohydrate energy stores in certain plants.

fruit The structure formed from the ovary of a flower, usually after the ovules have been fertilized. It consists of the *fruit wall* (*see* pericarp) enclosing the seed(s). Other parts of the flower, such as the receptacle, may develop and contribute to the structure, resulting in a *false fruit* (*see* pseudocarp). The fruit may retain the seeds and be dispersed whole (an *indehiscent fruit*), or it may open (dehisce) to release the seeds (a *dehiscent fruit*). Fruits are divided into two main groups depending on whether the ovary wall remains dry or becomes fleshy (*succulent*). Succulent fruits are generally dispersed by animals and dry fruits by wind, water, or by some mechanical means. See illustration. *See also* composite fruit.

fruit sugar *See* fructose.

FSH *See* follicle-stimulating hormone.

fucoxanthin The major *carotenoid pigment present, with chlorophyll, in the brown algae (*see* Phaeophyta).

fumaric acid A carboxylic acid, HCOOHC:CHCOOH, that is an intermediate in the *Krebs cycle, being formed by the dehydrogenation of succinic acid.

fungi A group of simple plants lacking chlorophyll. Some authorities do not regard fungi as plants and include them with other organisms of uncertain affinities in the kingdom *Protista. Fungi are classified in the taxonomic division Mycota or Fungi. They can either exist as single cells or make up a multicellular body called a *mycelium*, which consists of filaments known as *hyphae. Most fungal cells are multinucleate and have cell walls composed chiefly of *chitin. Fungi exist primarily in damp situations on land and, because of the absence of chlorophyll, are either parasites or saprophytes on other organisms. The Mycota can be subdivided into four main classes, the *Phycomycetes, *Ascomycetes, *Basidiomycetes, and *Fungi Imperfecti. The principal criteria used in classification are the nature of the spores produced and the presence or absence of cross walls within the hyphae. *See also* slime fungi.

Fungi Imperfecti (imperfect fungi) A class of *fungi in which sexual reproduction is absent. Otherwise they are generally similar to *Ascomycetes and are consequently thought to be Ascomycetes that have lost the ability to produce ascospores. Examples of imperfect fungi are the *Penicillium* moulds.

funicle The stalk that attaches an ovule to the placenta in the ovary of a flowering plant. It contains a strand of conducting tissue leading from the placenta into the chalaza.

furanose A *sugar having a five-membered ring containing four carbon atoms and one oxygen atom.

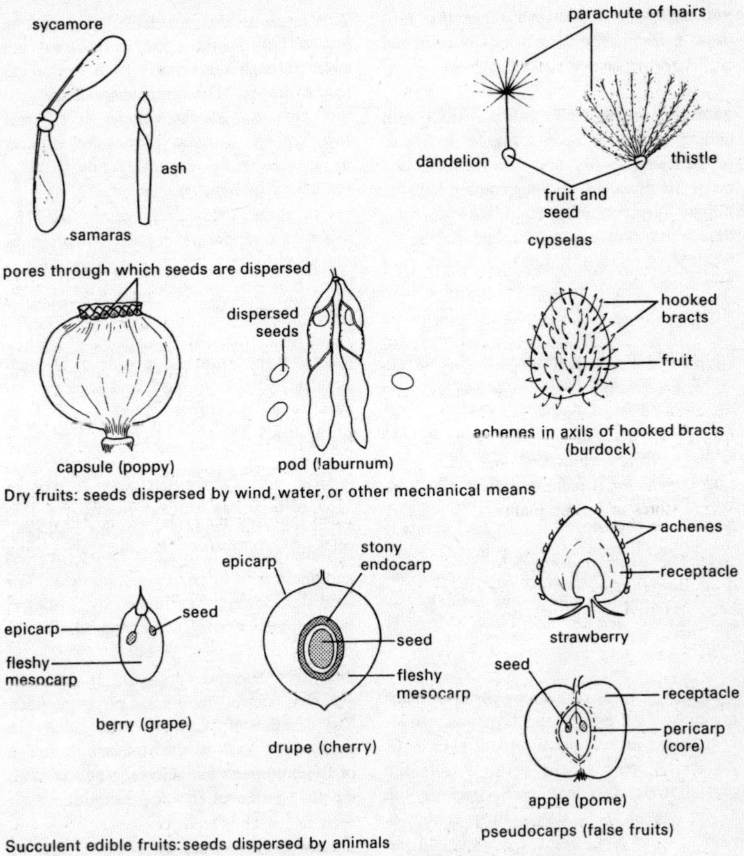

Different types of fruit and methods of seed dispersal

G

galactose A simple sugar, $C_6H_{12}O_6$, stereoisomeric with glucose, that occurs naturally as one of the products of the enzymic digestion of milk sugar (lactose) and as a constituent of gum arabic.

gall bladder A small pouch attached to the *bile duct, present in most vertebrates. *Bile, produced in the *liver, is stored in the gall bladder and released when food (especially fatty substances) enters the duodenum.

gametangium A plant sex organ that produces gametes. The term is usually restricted to the sex organs of algae and fungi.

gamete A reproductive cell that fuses with another gamete to form a zygote. Examples of gametes are ova and spermatozoa. Gametes are *haploid, i.e. they contain half the normal (diploid) number of chromosomes; thus when two fuse, the diploid number is restored (*see* fertilization). Gametes are formed by *meiosis. *See also* sexual reproduction.

gametophyte The generation in the life cycle of a plant that bears the gamete-producing sex organs. The gametophyte is *haploid. It is the dominant phase in the life cycle of mosses and other bryophytes, the *sporophyte generation depending on it either partially or completely. In pteridophytes it is the *prothallus. In seed plants it is very much reduced. For example, in angiosperms the pollen grain is the male gametophyte and the embryo sac is the female gametophyte. *See also* alternation of generations.

ganglion A mass of nervous tissue containing many *cell bodies and *synapses, usually enclosed in a connective-tissue sheath. In vertebrates most ganglia occur outside the central nervous system; exceptions are the *basal ganglia in the brain. In invertebrates ganglia occur along the nerve cords and the most anterior pair (*cerebral ganglia*) are analogous to the vertebrate brain; invertebrate ganglia constitute a part of the central nervous system.

ganoid scale *See* scales.

gas–liquid chromatography A technique for separating or analysing mixtures of gases by *chromatography. The apparatus consists of a very long tube containing the stationary phase, a nonvolatile liquid, such as a hydrocarbon oil coated on a solid support. The sample is often a volatile liquid mixture (e.g. of fatty acids), which is vaporized and swept through the column by a carrier gas (e.g. hydrogen). The components of the mixture pass through the column at different rates and are detected as they leave, either by measuring the thermal conductivity of the gas or by a flame detector.

Gas chromatography is usually used for analysis; components can be identified by the time they take to pass through the column. It is sometimes also used for separating mixtures.

gastric juice An acidic mixture of inorganic salts, hydrochloric acid, and digestive enzymes (e.g. *pepsin) produced by glands in the stomach lining.

gastric mill (proventriculus) A type of *gizzard occurring in many crustaceans. It is situated in the anterior region of the stomach and consists of a set of bones (ossicles) and muscles that grind food particles. The food particles are then filtered by bristles in the posterior section of the stomach.

gastrin A hormone, produced in the stomach, that controls the release of gastric juice. The secretion of gastrin is stimulated by the presence of food in the stomach. It is one of the hormones that integrates and controls digestive processes (*see also* secretin).

Gastropoda A class of molluscs that includes the snails, whelks, limpets, land and sea slugs, and conches. Molluscs have a well-developed head with tentacles, a large flattened foot, and a coiled twisted shell. They occupy marine, freshwater, and terrestrial habitats; in the terrestrial and some freshwater gastropods the *mantle cavity acts as a lung instead of enclosing gills.

gastrula The stage in the development of an animal embryo that succeeds the *blastula. It begins with the production of the *germ layers and the embryo becomes converted to a cup-shaped structure containing a cavity (the *archenteron).

gel A lyophilic *colloid that has coagulated to a rigid or jelly-like solid. In a gel, the disperse medium has formed a loosely-held network of linked molecules through the dispersion medium. Examples of gels are silica gel and gelatin.

gelatin(e) A colourless or pale yellow water-soluble protein obtained by boiling collagen with water and evaporating the solution. It swells when water is added and dissolves in hot water to form a solution that sets to a gel on cooling. It is used in photographic emulsions and adhesives, and in jellies and other foodstuffs.

gel filtration A type of column *chromatography in which a mixture of liquids is passed down a column containing a gel. Small molecules in the mixture can enter pores in the gel and move slowly down the column; large molecules, which cannot enter the pores, move more quickly. Thus, mixtures of molecules can be separated on the basis of their size. The technique is used particularly for separating proteins but it can also be applied to cell nuclei, viruses, etc.

gemmation A type of *vegetative propagation in which small clumps of undifferentiated cells (*gemmae*) develop on the surface of a plant. These are shed and dispersed to other areas, where they grow to produce new individuals. Gemmation is found only in certain lower plants, such as mosses and liverworts.

gene A unit of heredity composed of DNA. In classical genetics (*see* Mendelism; Mendel's laws) a gene is visualized as a discrete particle, forming part of a *chromosome, that determines a particular characteristic. It can exist in different forms called *alleles, which determine which aspect of the characteristic is shown (e.g. tallness or shortness for the characteristic of height).

A gene may be defined as the shortest length of chromosome that cannot be broken by recombination (*see* meiosis) or that can undergo *mutation, both units being recognized as a single base pair in the nucleotide sequence that makes up the DNA. The gene as a unit of function may be defined as the sequence of nucleotides concerned with the synthesis of a single polypeptide chain, corresponding to a particular sequence of the *genetic code. One or more of these *structural genes*, coding for protein, may be associated with a controlling *operator gene*; this unit is known as an *operon*. The operon itself is controlled by a third type of gene, the *regulator gene*.

gene pool All the *genes and their different alleles that are present in a population of a plant or animal species. *See also* population genetics.

generation time The interval between the beginnings of consecutive cell divisions. It may be as short as 20 minutes in bacteria. *See also* interphase.

generative nucleus One of the two male gametes in the *pollen tube of angiosperms.

gene therapy The application of *genetic engineering techniques to alter or replace defective genes. The techniques are still at the experimental stage but the eventual aim is to prevent such genetic diseases as sickle cell anaemia and thalassaemia.

genetic code The means by which genetic information in *DNA controls the manufacture of specific proteins by the cell. The code takes the form of a series of triplets (*codons*) of bases in DNA, from which is transcribed a complementary sequence of codons in messenger *RNA (*see* transcription). The sequence of these codons determines the sequence of amino acids during *protein synthesis. There are 64 possible codes from the combinations of the four bases present in DNA and messenger RNA and 20 amino acids present in body proteins: some of the amino acids are coded by more than one codon, and some codons have other functions.

genetic engineering (recombinant DNA technology) The techniques involved in altering the characters of an organism by inserting genes from another organism into its DNA. This altered DNA (known as *recombinant DNA*) is usually produced by isolating foreign genes, often by the use of *restriction enzymes, and inserting them into bacterial DNA via a *plasmid, which can enter the bacterial cell and then combine with the host's genetic material. Once inserted, the foreign gene may use the cell machinery of the bacterium to synthesize the protein that it originally coded for in the organism it was derived from. For example, the human gene for insulin production has been incorporated into bacterial DNA and such genetically engineered bacteria are used in the commercial production of this hormone.

genetics The branch of biology concerned with the study of heredity and variation. *Classical genetics* is based on the work of Gregor Mendel (*see* Mendelism). During the 20th century genetics has expanded to overlap with the fields of ecology and animal behaviour (*see* behavioural genetics; population genetics), and important advances in biochemistry and microbiology have led to clarification of the chemical nature of *genes and the ways in which they can replicate and be transmitted. *See also* genetic engineering.

genome All the genes contained in a single set of chromosomes, i.e. in a *haploid nucleus. Each parent, through its reproductive cells, contributes its genome to its offspring.

genotype The genetic composition of an organism, i.e. the combination of *alleles it possesses. *Compare* phenotype.

genus A category used in the *classification of organisms that consists of a number of similar or closely related species. The common name of an organism (especially a plant) is sometimes similar or identical to that of the genus, e.g. *Lilium* (lily), *Antirrhinum*. Similar genera are grouped into families. *See also* binomial nomenclature.

geochronology *See* varve dating.

geological time scale A time scale that covers the earth's history from its origin, estimated to be about 4600 million years ago, to the present. The chronology is divided into a hierarchy of time intervals: eras, periods, epochs, ages, and chrons (see Appendix).

geotropism The growth of plant organs in response to gravity. A main root is positively geotropic and a main stem negatively geotropic, growing downwards and upwards respectively, irrespective of the positions in which they are placed. For example, if a stem is placed in a horizontal position it will still grow upwards. Geotropism is thought to be controlled by *auxin, a plant growth substance. *See* tropism.

germ cell Any cell in the series of cells (the *germ line*) that eventually produces *gametes. In mammals the germinal epithelium of the ovaries and testes contain the germ cells.

germination 1. The initial stages in the growth of a seed to form a seedling. The embryonic shoot (plumule) and embryonic root (radicle) emerge and grow upwards and downwards respectively. Food reserves for germination come from *endosperm tissue within the seed and/or from the seed leaves (cotyledons). *See also* epigeal; hypogeal. **2.**

The first signs of growth of spores and pollen grains.

germ layers The layers of cells in an animal embryo at the *gastrula stage, from which are derived the various organs of the animal's body. There are two or three germ layers: an outer layer (*see* ectoderm), an inner layer (*see* endoderm), and in most animal groups a middle layer (*see* mesoderm).

germ plasm *See* Weismannism.

gestation The period in animals bearing live young (especially mammals) from the fertilization of the egg to birth of the young (parturition). In man gestation is known as *pregnancy* and takes about nine months (40 weeks).

giant fibre A nerve fibre with a very large diameter, found in many types of invertebrate (e.g. earthworms and squids). Its function is to allow extremely rapid transmission of nervous impulses and hence rapid escape movements in emergencies.

gibberellin Any of a group of plant *growth substances chemically related to gibberellic acid. Gibberellins promote shoot elongation in certain plants, often overcoming genetic dwarfism. They are also involved in the release of buds from dormancy and in promoting seed germination.

giga- Symbol G. A prefix used in the metric system to denote one thousand million times. For example, 10^9 joules = 1 gigajoule (GJ).

gill 1. (in zoology) A respiratory organ used by aquatic animals to obtain oxygen from the surrounding water. A gill consists essentially of a membrane or outgrowth from the body, with a large surface area and a plentiful blood supply, through which diffusion of oxygen and carbon dioxide between the water and blood occurs. Fishes have *internal gills*, formed as outgrowths from the pharynx wall and contained within *gill slits. Water entering the mouth is pumped out through these slits and over the gills. The gills of most aquatic invertebrates and amphibian larvae are *external gills*, which project from the body so that water passes over them as the animal moves. **2.** (in botany) One of the ridges of tissue that radiate from the centre of the underside of the cap of mushrooms. The spores are produced on these gills.

gill bar A cartilaginous support for the tissue between the gill slits in lower chordates, such as *Amphioxus*.

gill slit An opening leading from the pharynx to the exterior in aquatic vertebrates and *Amphioxus*. In *Amphioxus* they function in *filter feeding. In fish they contain the *gills and are usually in the form of a series of long slits. They are absent in adult tetrapod vertebrates (except for some amphibians) but their presence in some form in the embryos of all vertebrates is a characteristic of the phylum *Chordata.

gingiva (gum) The part of the epithelial tissue lining the mouth that covers the jaw bones. It is continuous with the sockets surrounding the roots of the teeth.

gizzard A muscular compartment of the alimentary canal of many animals that is specialized for breaking up food. In birds the gizzard lies between the *proventriculus and the duodenum and contains small stones and grit, which assist in breaking up the food when the gizzard contracts. *See also* gastric mill.

gland A group of cells or a single cell in animals or plants that is specialized to secrete a specific substance. In animals there are two types of glands, both of which synthesize their secretions. *Endocrine glands discharge their products directly into the blood vessels; *exocrine glands secrete through a duct or network of ducts into a body cavity or onto the body surface. Secretory cells are characterized by having droplets (*vesicles*) containing their products. *See also* secretion.

In plants glands are specialized to secrete certain substances produced by the plant. The secretions may be retained within a single cell, secreted into a special cavity or duct, or secreted to the outside. Examples are the water glands (*hydathodes) of certain leaves, nectaries (*see* nectar), and the digestive glands of certain insectivorous plants.

glenoid cavity The socket-shaped cavity in the *scapula (shoulder blade) that holds the head of the *humerus in a ball-and-socket joint.

globulin Any of a group of globular proteins that are generally insoluble in water and present in blood, eggs, milk, and as a reserve protein in seeds. Blood serum globulins comprise four types: α_1-, α_2-, and β-globulins, which serve as carrier proteins; and γ-globulins, which include the *immunoglobulins responsible for immune responses.

glomerulus A tangled mass of blood capillaries enclosed by the cup-shaped end (*Bowman's capsule*) of a kidney tubule (*see* nephron). Waste products and water pass from these capillaries into the Bowman's capsule and down the nephron.

glottis The opening from the pharynx to the trachea (windpipe). In mammals it also serves as the space for the *vocal cords. *See also* epiglottis; larynx.

glucagon A hormone, secreted by the *islets of Langerhans in the pancreas, that increases the concentration of glucose in the blood by stimulating the metabolic breakdown of glycogen. It thus antagonizes the effects of *insulin (*see* antagonism).

glucocorticoid *See* corticosteroid.

gluconeogenesis *See* glycolysis.

gluconic acid An optically active hydroxycarboxylic acid, $CH_2(OH)(CHOH)_4COOH$. It is the carboxylic acid corresponding to the aldose sugar glucose, and can be made by the action of certain moulds.

glucose (dextrose; grape sugar) A white crystalline sugar, $C_6H_{12}O_6$, occurring widely in nature. Like other *monosaccharides, glucose is optically active: most naturally occurring glucose is dextrorotatory. Glucose and its derivatives are crucially important in the energy metabolism of living organisms. It is a major energy source, being transported around the body in blood, lymph, and cerebrospinal fluid to the cells, where energy is released in the process of *glycolysis. Glucose is present in the sap of plants, in fruits, and in honey and is also a constituent of many polysaccharides, most notably starch and cellulose. These yield glucose when broken down, for example by enzymes during digestion.

glutamic acid *See* amino acid.

glutamine *See* amino acid.

glutathione A *peptide comprising the amino acids glutamic acid, cysteine, and glycine. It occurs widely in plants, animals, and microorganisms, serving chiefly as an antioxidant. Reduced glutathione reacts with potentially harmful oxidizing agents and is itself oxidized. This is important in ensuring the proper functioning of proteins, haemoglobin, membrane lipids, etc. Glutathione is also involved in amino acid transport across cell membranes.

gluten A mixture of two proteins, gliadin and glutenin, occurring in the endosperm of wheat grain. Their amino acid composition varies but glutamic acid (33%) and proline (12%) predominate. The composition of wheat glutens determines the 'strength' of the flour and whether or not it is suitable for biscuit or bread making. Sensitivity of the lining of the intestine to gluten occurs in *coeliac disease*, a condition that must be treated by a gluten-free diet.

glyceride (acylglycerol) A fatty-acid ester of glycerol. Esterification can occur at one, two, or all three hydroxyl groups of the glycerol molecule producing mono-, di-, and triglycerides respectively. *Triglycerides are the major constituent of fats and oils found in living organisms. Alternatively, one of the hydroxyl groups may be esterified with a phosphate group forming a phosphoglyceride (*see* phospholipid) or to a sugar forming a *glycolipid*.

glycerine *See* glycerol.

glycerol (glycerine; propane-1,2,3,-triol) A trihydric alcohol, $HOCH_2CH(OH)CH_2OH$. Glycerol is a colourless sweet-tasting viscous liquid, miscible with water but insoluble in ether. It is widely distributed in all living organisms as a constituent of the *glycerides, which yield glycerol when hydrolysed.

glycine *See* amino acid.

glycogen (animal starch) A *polysaccharide consisting of a highly branched polymer of glucose occurring in animal tissues, especially in liver and muscle cells. It is the major store of carbohydrate energy in animal cells and is present as granular clusters of minute particles.

glycolysis (Embden–Meyerhof pathway) The series of biochemical reactions in which glucose is broken down to pyruvate with the release of usable energy in the form of *ATP. One molecule of glucose undergoes two phosphorylation reactions and is then split to form two triose-phosphate molecules. Each of these is converted to pyruvate. The net energy yield is two ATP molecules per glucose molecule. In *aerobic respiration pyruvate then enters the *Krebs cycle. Alternatively, when oxygen is in short supply or absent, the pyruvate is converted to various products by *anaerobic respiration. Other simple sugars, e.g. fructose and galactose, and glycerol (from fats) enter the

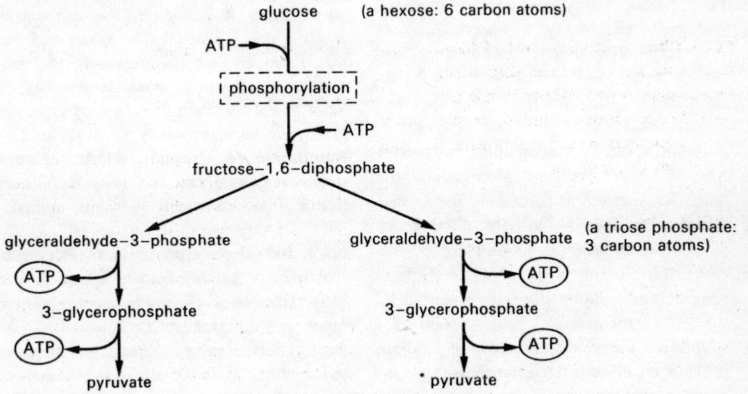

The principal stages of glycolysis

glycolysis pathway at intermediate stages. Many of the glycolytic reactions are reversible and are used to reform glucose from pyruvate in the process of *gluconeogenesis*.

Gnathostomata A subphylum consisting of all vertebrates that possess jaws. It includes the fishes, amphibians, reptiles, birds, and mammals. *Compare* Agnatha.

gnotobiotic Designating germ-free conditions, especially those in which experimental animals are inoculated with known strains of microorganisms.

Golgi apparatus An assembly of vesicles and folded membranes within the cytoplasm of plant and animal *cells that stores and transports secretory products (such as enzymes and hormones) and plays a role in formation of a cell wall (when this is present). It is named after its discoverer, the Italian cytologist Camillo Golgi (1843–1926). *See also* endoplasmic reticulum.

gonad Any of the usually paired organs in animals that produce reproductive cells (gametes). The most important gonads are the male *testis, which produces spermatozoa, and the female *ovary, which produces ova (egg cells). The gonads also produce hormones that control secondary sexual characteristics.

gonadotrophin (gonadotrophic hormone) Any of several hormones, secreted by the mammalian anterior *pituitary gland, that stimulate reproductive activity of the testes or ovaries (the gonads). Gonadotrophins include *follicle-stimulating hormone, *luteinizing hormone, and a hormone produced by the placenta (*chorionic gonadotrophin*) that suppresses the menstrual cycle and whose presence in urine in large amounts is an indication of pregnancy.

Graafian follicle The fluid-filled cavity that surrounds and protects the developing egg cell in the ovary of a mammal. After the release of the ovum it develops into a *corpus luteum. It is named after the Dutch anatomist Reinier de Graaf (1641–73).

graft An isolated portion of living tissue that is joined to another tissue, either in the same or a different organism, the consequent growth resulting in fusion of the tissues. (The word is also used for the process of joining the tissues.)

Grafting of plant tissues is a horticultural practice used to propagate plants, especially certain bushes and fruit trees, artificially. A shoot or bud of the desired variety (the *scion*) is grafted onto a rootstock of either a common or a wild related species (the *stock*). The scion retains its desirable characteristics (e.g. flower form or fruit yield) and supplies the stock with food made by photosynthesis. The stock supplies the scion with water and mineral salts and affects only the size and vigour of the scion.

Animal and human grafts are used to replace faulty or damaged parts of the body. An *autograft* is taken from one part of the body and transferred to another part of the same individual, e.g. a skin graft used for severe burns. A *homograft* (*allograft*) is taken from one individual (the *donor*) and implanted in another (the *recipient*), the process being known as *transplantation*, e.g. a heart or kidney transplant. In such cases the graft may be regarded by the body as foreign (a state of *incompatibility*): an *immune response follows and the graft is rejected (*see also* HL-A system).

graft hybrid A type of plant *chimaera that may be produced when a part of one plant (the scion) is grafted onto another plant of a different genetic constitution (the stock). Shoots growing from the point of union of the graft contain tissues from both the stock and the scion.

gram Symbol g. One thousandth of a kilogram. The gram is the fundamental unit of mass in *c.g.s. units and was formerly used in such units as the *gram-atom*, *gram-molecule*, and *gram-equivalent*, which have now been replaced by the *mole.

Gram's stain A staining method used to differentiate bacteria. The bacterial sample is smeared on a microscope slide, stained with a violet dye, treated with acetone-alcohol (a decolourizer), and finally counterstained with a red dye. *Gram-positive* bacteria retain the first dye, appearing blue-black under the microscope. In *Gram-negative* bacteria, the acetone-alcohol washes out the violet dye and the counterstain is taken up, the cells appearing red. It is named after the Danish bacteriologist H. C. J. Gram (1853–1938), who first described the technique (since modified) in 1884.

granum (*pl.* **grana**) A stack of platelike bodies (thylakoids), many of which are found in plant *chloroplasts. Grana bear the light-receptive pigment chlorophyll and contain the enzymes responsible for the light reactions of photosynthesis.

grape sugar *See* glucose.

graptolites A group of extinct marine colonial animals that were common in the Palaeozoic era. Graptolites are generally regarded as being related to the *Coelenterata. They had chitinous outer skeletons in the form of simple or branched stems, the individual polyps occupying minute cups (*thecae*) along these stems. Fossils of these skeletons are found in Palaeozoic rocks of all continents; they are particularly abundant in Ordovician and Silurian rock strata, for which they are used as *index fossils. At the end of the Silurian many graptolites became extinct but a few groups continued into the early Carboniferous.

gray Symbol Gy. The derived SI unit of absorbed *dose of ionizing radiation (*see* radiation units). It is named after the British radiobiologist L. H. Gray (1905–65).

green algae *See* Chlorophyta.

greenhouse effect 1. The effect within a greenhouse in which solar radiation mainly in the visible range of the spectrum passes through the glass roof and walls and is absorbed by the floor, earth, and contents, which re-emit the energy as infrared radiation. Because the infrared radiation cannot escape through the glass, the temperature inside the greenhouse rises. **2.** A similar effect in which the earth's atmosphere behaves like the greenhouse and the surface of the earth absorbs most of the solar radiation, re-emitting it as infrared radiation. This is absorbed by carbon dioxide, water, and ozone in the atmosphere as well as by clouds and reradiated back to earth. At night this absorption prevents the temperature falling rapidly after a hot day, especially in regions with a high atmospheric water content.

grey matter Part of the tissue that makes up the central nervous system of vertebrates. It is brown-grey in colour, consisting largely of nerve *cell bodies, *synapses, and *dendrites. The grey matter is the site of coordination between nerves of the central nervous system. *Compare* white matter.

grooming The actions of an animal of rearranging fur or feathers and cleaning the body surface by biting, scratching, licking, etc., which is important for removing parasites and spreading essential oils over the body surface. In many mammals, especially primates, grooming between individuals (*allogrooming*) has an important role in maintaining social cohesion.

ground tissues All the plant tissues formed by the *apical meristems except the epidermis and vascular tissue. The principal ground tissues are the *cortex, *pith, and primary *medullary rays, and they consist chiefly of *parenchyma.

Collenchyma is a form of ground tissue less frequently observed. It consists of cells with limited amounts of additional cellulose thickening in the walls and is most commonly found in the stem cortex.

growth An increase in the dry weight or volume of an organism through cell division and cell enlargement. Growth may continue throughout the life of the organism, as occurs in woody plants, or it may cease at maturity, as in man and other mammals. *See also* allometric growth; exponential growth.

growth hormone (GH; somatotrophin) A hormone, secreted by the mammalian pituitary gland, that stimulates protein synthesis and growth of the long bones in the legs and arms. Production of growth hormone is greatest during early life. In man, overproduction results in gigantism and underproduction in dwarfism.

growth ring (annual ring) Any of the rings that can be seen in a cross-section of a woody stem (e.g. a tree trunk). It represents the *xylem formed in one year as a result of fluctuating activity of the vascular *cambium. In temperate climates pale soft *spring wood*, characterized by large xylem vessels, is formed in spring and early summer. Growth slows down in late summer and a darker dense *autumn wood* with smaller xylem vessels is formed. The age of a tree can be determined by counting the rings. Under certain circumstances two or more growth rings may form in one year, giving rise to false annual rings.

growth substance (phytohormone; plant hormone) Any of a number of organic chemicals that are synthesized by plants and regulate growth and development. They are usually made in a particular region, such as the shoot tip, and transported to other regions, where they take effect. *See* abscisic acid; auxin; cytokinin; ethene (ethylene); gibberellin.

guanidine A crystalline basic compound related to urea $HN{:}C(NH_2)_2$.

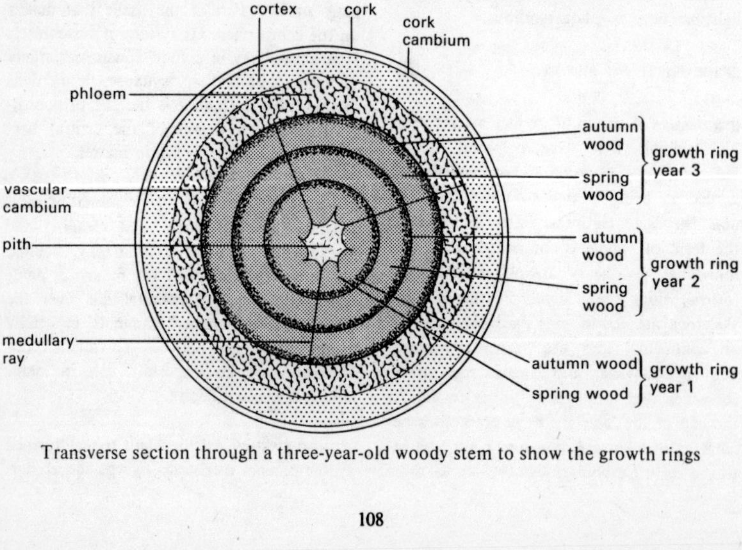

Transverse section through a three-year-old woody stem to show the growth rings

guanine A *purine derivative. It is one of the major component bases of *nucleotides and the nucleic acids *DNA and *RNA.

guard cell *See* stoma.

gum 1. Any of a variety of substances obtained from plants. Typically they are insoluble in organic solvents but form gelatinous or sticky solutions with water. Gum resins are mixtures of gums and natural resins. Gums are produced by the young xylem vessels of some plants (mainly trees) in response to wounding or pruning. The exudate hardens when it reaches the plant surface and thus provides a temporary protective seal while the cells below divide to form a permanent repair. Excessive gum formation is a symptom of some plant diseases. *See also* mucilage. **2.** *See* gingiva.

gut *See* alimentary canal.

guttation *See* hydathode.

Gymnospermae A subdivision of the *Spermatophyta or a class of the *Pteropsida containing the conifers (*see* Coniferales) and related species, such as cycads (*see* Cycadales) and ginkgos, as well as the extinct seed ferns (*see* Pteridospermales). Gymnosperms have an extensive fossil record going back to the late Devonian. The gametes are carried in male and female *cones, fertilization usually being achieved by wind-borne pollen. The ovules and the seeds into which they develop are borne unprotected (rather than enclosed in a carpel, as are those of the *Angiospermae). (The term gymnosperm means naked seed.) Internal tissue and cell structure of these species is not as advanced as in the angiosperms.

gynaecium (gynoecium) The female sex organs (*carpels) of a flower. *Compare* androecium.

gynoecium *See* gynaecium.

H

habitat The place in which an organism lives, which is characterized by its physical features or by the dominant plant types. Freshwater habitats, for example, include streams, ponds, rivers, and lakes.

habituation 1. A simple type of learning consisting of a gradual waning in the response of an animal to a continuous or repeated stimulus that is not associated with *reinforcement. **2.** The condition of being psychologically, but not physically, dependent on a drug, with a desire to continue its use but not to increase the dosage.

haem (heme) An iron-containing molecule that binds with proteins as a *cofactor or *prosthetic group to form the *haemoproteins*. These are *haemoglobin, *myoglobin, and the *cytochromes. Essentially, haem comprises a *porphyrin with its four nitrogen atoms holding the iron(II) atom as a chelate. This iron can reversibly bind oxygen (as in haemoglobin and myoglobin) or (as in the cytochromes) conduct electrons by conversion between the iron(II) and iron (III) series.

haematoxylin A compound used in its oxidized form (*haematein*) as a blue dye in optical microscopy, particularly for staining smears and sections of animal tissue. It stains nuclei blue and is frequently used with *eosin as a counterstain for cytoplasm. Haematoxylin requires a mordant, such as iron alum, which links the dye to the tissue. Different types of haematoxylin can be made up depending on the mordant used, the method of oxidation, and the pH. Examples are *Delafield's haematoxylin* and *Ehrlich's haematoxylin*.

haemocyanin (hemocyanin) Any of a group of copper-containing respiratory proteins found in solution in the blood of certain arthropods and molluscs. Haemocyanins contain two copper atoms that reversibly bind oxygen, changing between the colourless deoxygenated form (CuI) and the blue oxygenated form (CuII). In some species, haemocyanin molecules form giant polymers with molecular weights of several million.

haemoglobin One of a group of globular proteins occurring widely in animals as oxygen carriers in blood. Vertebrate haemoglobin comprises two pairs of polypeptide chains (forming the *globin* protein) with each chain folded to provide a binding site for a *haem group. Each of the four haem groups binds one oxygen molecule to form *oxyhaemoglobin*. Dissociation occurs in oxygen-depleted tissues: oxygen is released and haemoglobin is reformed. The haem groups also bind other inorganic molecules, including carbon monoxide. This binds more strongly than oxygen and competes with it (hence its toxicity). In vertebrates, haemoglobin is contained in the red blood cells (*erythrocytes).

haemolysis The breakdown of red blood cells. It may be due to the action of disease-causing microorganisms, poisons, antibodies in mismatched blood transfusions, or certain allergic reactions. It produces anaemia.

hair 1. A multicellular threadlike structure, consisting of many dead keratinized cells, that is produced by the epidermis in mammalian *skin. The section of a hair below the skin surface (the *root*) is contained within a *hair follicle, the base of which produces the hair cells. Hair assists in maintaining body temperature by reducing heat loss from the skin. Bristles and whiskers are specialized types of hair. **2.** Any of various threadlike structures on plants, such as a *trichome.

hair follicle A narrow tubular depression in mammalian skin containing the root of a *hair. It is lined with epidermal cells and extends down through the epidermis and dermis to its base in the subcutaneous tissue. The ducts of *sebaceous glands empty into hair follicles.

hallux The innermost digit on the hind limb of a tetrapod vertebrate. In man it is the big toe and contains two phalanges. The hallux is absent in some mammals and in many birds it is directed backwards as an adaptation to perching. *Compare* pollex.

halophyte A plant that can tolerate a high concentration of salt in the soil. Such conditions occur in salt marshes and mudflats. Examples of halophytes are thrift (*Armeria*), sea lavender (*Limonium*), and rice grass (*Spartina*).

haploid Describing a nucleus, cell, or organism with a single set of unpaired chromosomes. The haploid number is designated as *n*. Reproductive cells, formed as a result of *meiosis, are haploid. Fusion of two such cells (*see* fertilization) restores the normal (*diploid) number.

haptotropism *See* thigmotropism.

hardwood *See* wood.

haustorium A specialized structure of certain parasitic plants that penetrates the cells of the host plant to absorb nutrients. In parasitic fungi haustoria are formed from enlarged hyphae and in parasitic flowering plants, such as the dodder (*Cuscuta*), they are outgrowths of the stem.

Haversian canals Narrow tubes within compact *bone containing blood vessels and nerves. They generally run parallel to the bone surface. Each canal surrounded by a series of rings of bone (*lamellae*) is known as a *Haversian system.* Haversian systems are joined to each other by bone material.

health physics The branch of medical physics concerned with the protection of medical, scientific, and industrial workers from the hazards of ionizing radiation and other dangers associated with atomic physics. Establishing the maximum permissible *dose of radiation, the disposal of radioactive waste, and the shielding of dangerous equipment are the principal activities in this field.

heart A hollow muscular organ that, by means of regular contractions, pumps blood through the circulatory system. Mammals have a four-chambered heart consisting of two atria and two ventricles; the right and left sides are completely separate from each other so there is no mixing of oxygenated and deoxygenated blood. Oxygenated blood from the pulmonary veins enters the heart through the left atrium, passes to the left ventricle, and leaves the heart through the *aorta. Deoxygenated blood from the *venae cavae enters the right atrium and is pumped through the right ventricle to the pulmonary artery, which conveys it to the lungs for oxygenation. The mitral and bicuspid valves ensure that there is no backflow of blood. The contractions of the heart are initiated and controlled by the *pacemaker; in an average adult human the heart contracts about 70 times per minute.

The hearts of other vertebrates are similar except in the number of atria and ventricles (there may be one or two) and in the degree of separation of oxygenated and deoxygenated blood. Invertebrates, however, show

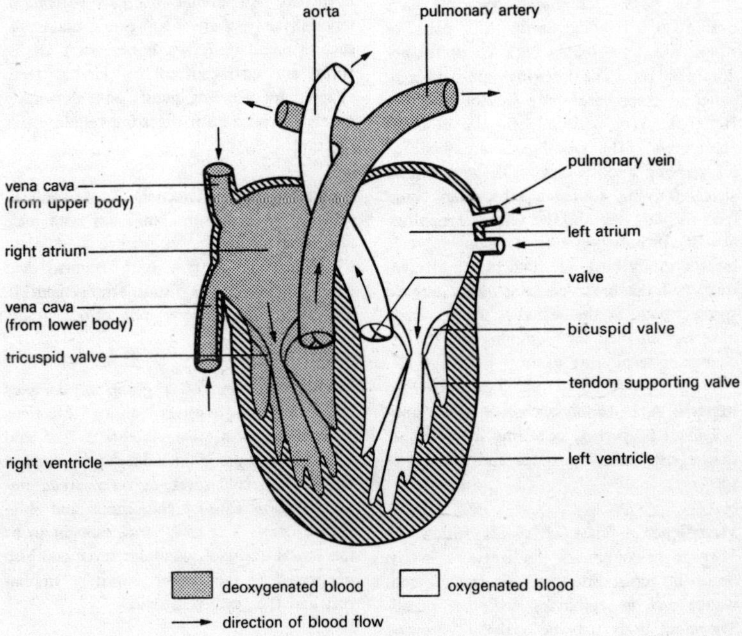

Structure of the mammalian heart

great variation in the form and functioning of the heart.

heartwood (duramen) The wood at the centre of a tree trunk or branch. It consists of dead *xylem cells heavily thickened with lignin and provides structural support. Many heartwood cells contain oils, gums, and resins, which darken the wood. *Compare* sapwood.

hecto- Symbol h. A prefix used in the metric system to denote 100 times. For example, 100 coulombs = 1 hectocoulomb (hC).

heliotropism *See* phototropism.

heme *See* haem.

Hemiptera An order of insects comprising the true bugs. Hemipterans typically have oval flattened bodies with two pairs of wings, which are folded back across the abdomen at rest. The forewings are hardened, either at their bases only (in the suborder Heteroptera) or uniformly (in the suborder Homoptera). The mouthparts are modified for piercing and sucking, with long slender stylets forming a double tube. Many bugs feed on plant sap and are serious agricultural pests, including aphids, leaf-hoppers, scale insects, mealy bugs, etc. Others are carnivorous, and the order contains many aquatic species, such as the water boatmen, which have legs adapted for swimming and the exchange of respiratory gases.

heparin A mucopolysaccharide with *anticoagulant properties, occurring in vertebrate tissues, especially the lungs and blood vessels.

Hepaticae A class of plants within the *Bryophyta comprising the liverworts, which occur in moist situations (including fresh water) and as epiphytes on other plants. The plant body may be a thallus, growing closely pressed to the ground (thallose liverworts, e.g. *Pellia*), or it may bear many leaflike lobes (leafy liverworts). It gives rise to leafless stalks bearing capsules. Spores formed in the capsules are released and grow to produce new plants.

hepatic portal system The vein (*hepatic portal vein*) or veins that transport blood containing the absorbed products of digestion from the intestine directly to the liver.

herbaceous Describing a plant that contains little permanent woody tissue. The aerial parts of the plant die back after the growing season. In *annuals the whole plant dies; in *biennials and herbaceous *perennials the plant has organs (e.g. bulbs or corms) that are modified to survive beneath the soil in unfavourable conditions.

herbivore An animal that eats vegetation, especially any of the plant-eating mammals, such as ungulates (cows, horses, etc.). Herbivores are characterized by having teeth adapted for grinding plants and alimentary canals specialized for digesting cellulose (*see* caecum).

hermaphrodite (bisexual) **1.** An animal, such as the earthworm, that has both male and female reproductive organs. **2.** A plant whose flowers contain both stamens and carpels. This is the usual arrangement in most plants. *Compare* monoecious; dioecious.

herpesvirus One of a group of complex DNA-containing viruses causing infections in man and most other vertebrates that tend to recur. The group includes *herpes simplex*, the agent of cold sores; *herpes varicella/zoster*, the virus causing chickenpox and shingles; *Epstein–Barr (EB) virus*, thought to be the causal agent of glandular fever and also implicated in the cancer Burkitt's lymphoma; and the *cytomegalovirus.

hesperidium *See* berry.

heterochromatin *See* chromatin.

heterodont Describing animals that possess teeth of more than one type (i.e. *incisors, *canine teeth, *premolars, and *molars), each with a particular function. Most mammals are heterodont. *Compare* homodont.

heterogametic sex The sex that is determined by possession of two dissimilar *sex chromosomes (i.e. XY). In humans and many other mammals this is the male sex. The heterogametic sex produces reproductive cells (gametes) of two kinds, half containing an X chromosome and half a Y chromosome. *Compare* homogametic sex.

heterosis *See* hybrid vigour.

heterotrophism A type of nutrition in which energy is derived from the intake and digestion of organic substances, normally plant or animal tissues. The breakdown products of digestion are used to synthesize the organic materials required by the organism. All animals obtain their food this way; i.e. they are *heterotrophs*. *Compare* autotrophism.

heterozygous Describing an organism that has two different *alleles controlling a particular feature. The aspect of the feature displayed by the organism will be that determined by the *dominant allele. Heterozygous organisms do not breed true. *Compare* homozygous.

hexose A *monosaccharide that has six carbon atoms in its molecules.

hibernation A sleeplike state in which some animals pass the winter months as a way of surviving food scarcity and cold weather. Various physiological changes occur, such as lowering of the body temperature and slowing of the pulse rate and other vital processes, and the animal lives on its reserve of body fat. Animals that hibernate include bats, hedgehogs, and many fish, amphibians, and reptiles. *See also* dormancy. *Compare* aestivation.

hilum A scar on the seed coat of a plant marking the point at which the seed was attached to the fruit wall by the *funicle. It is a feature that distinguishes seeds from fruits.

hindbrain (rhombencephalon) One of the three sections of the brain of a vertebrate embryo. It develops to form the *cerebellum, *pons, and *medulla oblongata, which control and coordinate fundamental physiological processes (including respiration and circulation of blood). *Compare* forebrain; midbrain.

hindgut **1.** The posterior part of the alimentary canal of vertebrates, comprising the posterior section of the colon. **2.** The posterior section of the alimentary canal of arthropods. *See also* foregut; midgut.

hip girdle *See* pelvic girdle.

hippocampus A part of the vertebrate brain consisting of two ridges, one over each of the two lateral *ventricles. It is highly developed in advanced mammals (primates and whales) and its function appears to be related to the expression of responses that generate emotion (such as fear and anger).

Hirudinea A class of freshwater and terrestrial annelid worms that comprises the leeches. They have suckers at both anterior and posterior ends but no bristles. Some are blood-sucking parasites of vertebrates and invertebrates but the majority are predators.

histamine A substance that is released during allergic reactions, e.g. hay fever. Formed from the amino acid histidine, histamine can occur in various tissues but is concentrated in connective tissue. It causes dilation and increased permeability of small blood vessels, which results in such symptoms as localized swelling, itching, sneezing, and runny eyes and nose. The effects of histamine can be countered by the administration of *antihistamine drugs.

histidine *See* amino acid.

histochemistry The study of the distribution of the chemical constituents of tissues by means of their chemical reactions. It utilizes such techniques as *staining, light and electron microscopy, *autoradiography, and *chromatography.

histocompatibility antigen *See* antigen; HL-A system.

histology The microscopic study of the tissues of living organisms. The study of cells, a specialized branch of histology, is known as *cytology.

histone Any of a group of water-soluble proteins found in association with the *DNA of plant and animal chromosomes. They contain a large proportion of the basic (positively charged) amino acids lysine, arginine, and histidine. They are believed to be involved in the condensation and coiling of chromosomes during cell division and have also been implicated in nonspecific suppression of gene activity. Histones do not occur in vertebrate sperm cells (*see* protamine) or in bacteria and blue-green algae.

HL-A system (histocompatibility lymphocyte-A system) A group of *antigens in man, present on the surface of cell membranes, that is important in determining the acceptance or rejection by the body of a tissue or organ transplant (*see* graft). These antigens are one group of the so-called *histocompatibility* (or *transplantation*) *antigens*. Successful transplantation requires a minimum number of HL-A differences between the donor's and recipient's tissues.

Holocene (Recent) The most recent geological epoch of the *Quaternary period, comprising roughly the past 10 000 years since the end of the *Pleistocene up to the present. It follows the final glacial of the Pleistocene and thus is sometimes known as the *Postglacial* epoch. Some geologists consider the Holocene to be an interglacial phase of the Pleistocene that will be followed by another glacial.

holocrine secretion *See* secretion.

holoenzyme A complex comprising an enzyme molecule and its *cofactor. Only in this state is an enzyme catalytically active. *Compare* apoenzyme.

holophytic Describing organisms that feed like plants, i.e. that are photoautotrophic. *See* autotrophism.

holotype *See* type specimen.

holozoic Describing organisms that feed like animals, i.e. that are heterotrophic. *See* heterotrophism.

homeostasis The regulation by an organism of the chemical composition of its blood and body fluids and other aspects of its *internal environment so that physiological processes can proceed at optimum rates. Examples of homeostatic regulation are the maintenance of the *acid–base balance and body temperature (*see* homoiothermy; poikilothermy).

hominid Any member of the primate family Hominidae, which includes man and his fossil ancestors (*fossil hominids*) in the genus **Homo*.

Homo The genus of primates that includes modern man (*H. sapiens*, the only living representative) and various extinct species. The oldest *Homo* fossils are those of *H. habilis*, found in the Olduvai Gorge in

Tanzania in 1960 and estimated to be about 1.75 million years old. The find was significant as it had been thought that man emerged in Asia rather than Africa. *H. habilis* appears to have been 1–1.5 m tall and had more manlike features and a larger brain than **Australopithecus*. However, the direct ancestor of modern man is thought to be *H. erectus*, which includes Java man and Peking man and appeared about 1.5 million years ago. Fossils of *H. erectus*, which are sometimes called *Pithecanthropus* (ape man), are similar to present-day man except that there was a prominent ridge above the eyes and no forehead or chin. They had crude stone tools and used fire. *H. sapiens* appeared about 100 000 years ago as *Neanderthal man and subsequently (about 35 000 years ago) as *Cromagnon man.

homodont Describing animals whose teeth are all of the same type. Most vertebrates except mammals are homodont. *Compare* heterodont.

homogametic sex The sex that is determined by possession of two similar *sex chromosomes (i.e. XX). In humans and many other mammals this is the female sex. All the reproductive cells (gametes) produced by the homogametic sex have the same kind of sex chromosome (i.e. an X chromosome). *Compare* heterogametic sex.

homogamy The condition in a flower in which the male and female reproductive organs mature at the same time, thereby allowing self-fertilization. *Compare* dichogamy.

homoiothermy (warm-bloodedness) The maintenance by an animal of its internal body temperature at a relatively constant value by using metabolic processes to counteract fluctuations in the temperature of the environment. Homoiothermy occurs in birds and mammals. The heat produced by their tissue metabolism and the heat lost to the environment are balanced by various means to keep body temperature constant: 36–38°C in mammals and 38–40°C in birds. The *hypothalamus in the brain monitors blood temperature and controls thermoregulation by both nervous and hormonal means. This produces both short-term responses, such as shivering or sweating, and long-term adjustments to metabolism according to seasonal changes in climate (acclimatization). Warm-blooded animals generally possess insulating feathers or fur. Their relatively high internal temperature permits fast action of muscles and nerves and enables them to lead highly active lives even in cold climates. However, in certain animals, homoiothermy is abandoned during periods of *hibernation. *Compare* poikilothermy.

homologous Describing features of organisms that have the same evolutionary origin but have developed different functions. For example the wings of a bat, the flippers of a dolphin, and the arms of a man are homologous organs, having evolved from the paired pectoral fins of a fish ancestor. *Compare* analogous.

homologous chromosomes Chromosomes having the same structural features. In *diploid nuclei, pairs of homologous chromosomes can be identified at the start of meiosis. One member of each pair comes from the female parent and the other from the male. Homologous chromosomes have the same pattern of genes along the chromosome but the nature of the genes may differ (*see* allele).

homoplastic Denoting similarity resulting from *convergent evolution. *Compare* patristic.

homozygous Describing an organism having two identical *alleles controlling a particular feature (these may be either dominant or recessive). Homozygous organisms breed true when crossed with genetically identical organisms. *Compare* heterozygous.

hormone 1. A substance that is manufactured and secreted in very small quantities into the bloodstream by an *endocrine gland or a specialized nerve cell (*see* neurohormone) and regulates the growth or functioning of a specific tissue or organ in a distant part of the body. For example, the hormone *insulin controls the rate and manner in which glucose is used by the body. Other hormones include the *sex hormones, *corticosteroids, *adrenaline, *thyroxine, and *growth hormone. **2.** A plant *growth substance.

horsetails *See* Sphenopsida.

host An organism whose body provides nourishment and shelter for another. *See* parasitism.

humerus The long bone of the upper arm of tetrapod vertebrates. It articulates with the *scapula (shoulder blade) at the *glenoid cavity and with the *ulna and *radius (via a *condyle) at the elbow.

humus The dark-coloured amorphous colloidal material that constitutes the organic component of soil. It is formed by the decomposition of plant and animal remains and excrement and has a complex and variable chemical composition. Being a colloid, it can hold water and therefore improves the water-retaining properties of soil; it also enhances soil fertility and workability. Acidic humus (*mor*) is found in regions of coniferous forest, where the decay is brought about mainly by fungi. Alkaline humus (*mull*) is typically found in grassland and deciduous forest: it supports an abundance of microorganisms and small animals (e.g. earthworms).

hybrid The offspring of a mating in which the parents differ in at least one characteristic. The term is usually used of offspring of widely different parents, e.g. different varieties or species. Hybrids between different animal species are usually sterile, as is the mule (a cross between a horse and a donkey). *See also* hybrid vigour.

hybrid vigour (heterosis) The increased vigour displayed by the offspring from a cross between genetically different parents. Hybrids from crosses between different crop varieties (F_1 hybrids) are often stronger and produce better yields than the original varieties. Mules, the offspring of mares crossed with donkeys, have greater strength and resistance to disease and a longer lifespan than either parent.

hydathode A pore found in the *epidermis of the leaves of certain plants. Like *stomata, hydathodes are surrounded by two crescent-shaped cells but these, unlike guard cells, do not regulate the size of the aperture. Hydathodes are used by the plant to secrete water under conditions in which *transpiration is inhibited; for example, when the atmosphere is very humid. This process of water loss is called *guttation*.

hydrocortisone *See* cortisol.

hydrogen bond A type of electrostatic interaction between molecules occurring in molecules that have hydrogen atoms bound to electronegative atoms (F, N, O). It is a strong dipole–dipole attraction caused by the electron-withdrawing properties of the electronegative atom. Thus, in the water molecule the oxygen atom attracts the electrons in the O–H bonds. The hydrogen atom has no inner shells of electrons to shield the nucleus, and there is an electrostatic interaction between the hydrogen proton and a lone pair of electrons on an oxygen atom in a neighbouring molecule. Each oxygen atom has two lone pairs and can make hydrogen bonds to two different hydrogen atoms. The strengths of hydrogen bonds are about one tenth of the strengths of normal covalent bonds. Hydrogen bonding does, however, have significant effects on physical properties. Thus it accounts for the unusual properties of water and for its relatively high boiling point. It is also of great importance in living organisms. Hydrogen bonding occurs between bases in the chains of DNA (*see* base pairing). It also occurs between the C=O and N–H groups in proteins, and is responsible for maintaining the secondary structure.

hydrophily A rare form of pollination in which pollen is carried to a flower by water. It occurs by one of two methods. In Canadian pondweed (*Elodea canadensis*) the male flowers break off and float downstream until they contact the female flowers. In *Zostera*, a marine species, the filamentous pollen grains are themselves carried in the water. *Compare* anemophily; entomophily.

hydrophyte Any plant that lives either in very wet soil or completely or partially submerged in water. Structural modifications of hydrophytes include the reduction of mechanical and supporting tissues and vascular tissue, the absence or reduction of a root system, and specialized leaves that may be either floating or finely divided, with little or no cuticle. Examples of hydrophytes are waterlilies and certain pondweeds. *Compare* mesophyte; xerophyte.

hydroponics A commercial technique for growing certain crop plants in culture solutions rather than in soil. The roots are immersed in an aerated solution containing the correct proportions of essential mineral salts. The technique is based on various water culture methods used in the laboratory to assess the effects of the absence of certain mineral elements on plant growth.

hydrotropism The growth of a plant part in response to water. Roots, for example, grow towards water in the soil. *See* tropism.

5-hydroxytryptamine *See* serotonin.

hygroscopic Describing a substance that can take up water from the atmosphere. *See also* deliquescence.

Hymenoptera An order of insects that includes the ants, bees, wasps, ichneumon flies, and sawflies. Hymenopterans generally have a narrow waist between thorax and abdomen. The smaller hindwings are interlocked with the larger forewings by a row of tiny hooks on the leading edges of the hindwings. Some species are wingless. The mouthparts are typically adapted for biting, although some advanced forms (e.g. bees) possess a tubelike proboscis for sucking liquid food, such as nectar. The long slender *ovipositor can serve for sawing, piercing, or stinging. Metamorphosis occurs via a pupal stage to the adult form. *Parthenogenesis is common in the group.

Ants and some bees and wasps live in colonies, often comprising numerous individuals divided into *castes and organized into a coordinated and complex society. The colony of the honeybee (*Apis mellifera*), for example, consists of workers (sterile females), drones (fertile males), and usually a single fertile female – the queen. The sole concern of the queen is egg laying. She determines the gender of the egg by either withholding or releasing stored sperm. Unfertilized eggs become males; fertilized eggs become females. The workers fulfil a variety of tasks, including nursing the developing larvae, building the wax cells (combs) of the hive, guarding the colony, and foraging for nectar and pollen. The single function of the larger drones is to mate with the young queen on her nuptial flight.

hyoid arch The second of seven bony V-shaped arches that support the gills of primitive vertebrates. In advanced vertebrates, part of the hyoid arch has evolved to form the stapes (one of the *ear ossicles). The rest of it forms the *hyoid bone*, which supports the tongue.

hyper- A prefix denoting over, above, high; e.g. hypertensive.

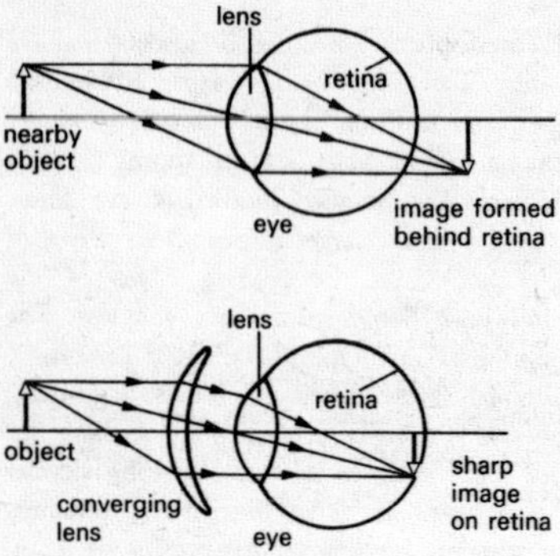

Hypermetropia

hypermetropia (hyperopia) Long-sightedness. A vision defect in which the lens of the eye is unable to accommodate sufficiently to throw the image of near objects onto the retina. It is caused usually by shortness of the eyeball rather than any fault in the lens system. The subject requires spectacles with converging lenses to bring the image from behind the retina back on to its surface.

hyperplasia Increase in the size of a tissue or organ due to an increase in the number of its component cells. *Compare* hypertrophy.

hypertension *See* blood pressure.

hypertonic solution A solution that has a higher osmotic pressure than some other solution.

hypertrophy An increase in the size of a tissue or organ due to an increase in the size of its component cells. Hypertrophy often occurs in response to an increased workload in an organ, which may result from malfunction or disease. *Compare* hyperplasia.

hypha A delicate filament in fungi many of which may form either a loose network (*mycelium) or a tightly packed interwoven mass of *pseudoparenchyma, as in the fruiting body of mushrooms. Hyphae may be branched or unbranched and may or may not possess cross walls. The cell wall consists of either fungal cellulose or a nitrogenous compound called *chitin. The cell wall is lined with cytoplasm, which often contains oil globules and glycogen, and there is a central vacuole. The hyphae produce enzymes that in parasitic fungi digest the host tissue, and in saprophytic fungi digest dead organic matter.

hypo- A prefix denoting under, below, low; e.g. hypotonic.

hypocotyl The region of the stem beneath the stalks of the seed leaves (*cotyledons) and directly above the young root of an embryo plant. It grows rapidly in seedlings showing *epigeal germination and lifts the cotyledons above the soil surface. In this region (the *transition zone*) the arrangement of vascular bundles in the root changes to that of the stem. *Compare* epicotyl.

hypodermis (exodermis) The outermost layer of cells in the plant *cortex, lying immediately below the epidermis. These cells are sometimes modified to give additional structural support or to store food materials or water. After the loss of the *piliferous layer of the root the hypodermis takes over the protective functions of the epidermis.

hypogeal 1. Describing seed germination in which the seed leaves (cotyledons) remain below ground. Examples of hypogeal germination are seen in oak and runner bean. *Compare* epigeal. **2.** Describing fruiting bodies that develop underground, such as truffles and peanuts.

hypophysis *See* pituitary gland.

hypothalamus Part of the vertebrate brain that is derived from the *forebrain and located on the ventral surface below the *thalamus and the *cerebrum. The hypothalamus regulates a wide variety of physiological processes, including maintenance of body temperature, water balance, sleeping, and feeding, via both the *autonomic nervous system (which it controls) and the *neuroendocrine system. Its endocrine functions are largely mediated by the *pituitary gland. The pituitary responds to hormone-like substances released from the hypothalamus, which thereby indirectly controls hormone production in other glands.

hypotonic solution A solution that has a lower osmotic pressure than some other solution.

I

IAA (indoleacetic acid) *See* auxin.

ice age A period in the earth's history during which ice advanced towards the equator and a general lowering of temperatures occurred. The last major ice age, that of the Pleistocene period (sometimes known as the *Ice Age*), ended about 10 000 years ago. At least four major ice advances (glacials) occurred during the Pleistocene period; these were separated by interglacials during which the ice retreated and temperatures rose. At present it is not known if the earth is between ice ages or is in an interglacial. It has been established that ice ages also occurred during the Precambrian (about 500 million years ago) and during the Permo-Carboniferous (about 250 million years ago).

ICSH *See* luteinizing hormone.

identical twins (monozygotic twins) Two individuals that develop from a single fertilized egg cell by its division into two genetically identical parts. Each part eventually gives rise to a separate individual and these twins are therefore identical in every respect. *Compare* fraternal twins.

idiogram *See* karyogram.

ileum The portion of the mammalian *small intestine that follows the *jejunum and precedes the *large intestine. It is a site of digestion and absorption. The internal lining of the ileum bears numerous small outgrowths (*see* villus), which increase its absorptive surface area.

ilium The largest of the three bones that make up each half of the *pelvic girdle. The ilium bears a flattened wing of bone that is attached by ligaments to the sacrum (*see* sacral vertebrae). *See also* ischium; pubis.

imago The adult sexually mature stage in the life cycle of an insect after metamorphosis.

imbibition The uptake of water by substances that do not dissolve in water, so that the process results in swelling of the substance. Imbibition is a property of many biological substances, including cellulose (and other constituents of plant cell walls), starch, and some proteins. It occurs in dry seeds before they germinate and – together with osmosis – is responsible for the uptake of water by growing plant cells.

immersion objective An optical microscope objective in which the front surface of the lens is immersed in a liquid on the cover glass of the microscope specimen slide. Cedar-wood oil (for an *oil-immersion lens*) or sugar solution is frequently used. It has the same refractive index as the glass of the slide, so that the object is effectively immersed in it. The presence of the liquid increases the effective aperture of the objective, thus increasing the resolution.

immune response The reaction of the body to foreign or potentially dangerous substances (*antigens), particularly disease-producing microorganisms. The response involves the production by specialized white blood cells (*lymphocytes) of proteins known as *antibodies, which react with the antigens to render them harmless. The antibody–antigen reaction is highly specific. *See also* anaphylaxis; immunity.

immunity The state of relative insusceptibility of an animal to infection by disease-producing organisms or to the harmful effects of their poisons (toxins). Immunity depends on the presence in the blood of *antibodies and white blood cells (*lymphocytes), which produce an *immune response. *Inherited (natural) immunity* is that with which an individual is born. *Acquired immunity* is of two types: *active immunity* arises when the body produces antibodies against an invading foreign substance (*antigen), either through infection or *immunization; *passive immunity* is induced by injection of serum taken from an individual already immune to a particular antigen. Active immunity tends to be long-lasting; passive immunity is short-lived. *See also* autoimmunity.

immunization The production of *immunity in an individual by artificial means. *Active immunization* (*vaccination*) involves the introduction, either orally or by injection (*inoculation*), of specially treated bacteria, viruses, or their toxins to stimulate the production of *antibodies (*see* vaccine). *Passive immunization* is induced by the injection of preformed antibodies.

immunoglobulin One of a group of proteins (*globulins) in the body that act as *antibodies. They are produced by specialized white blood cells (*lymphocytes) and are present in blood serum and other body fluids.

imperfect fungi *See* Fungi Imperfecti.

Imperial units The British system of units based on the pound and the yard. The former f.p.s. system was used in engineering and was loosely based on Imperial units; for all scientific purposes *SI units are now used. Imperial units are also being replaced for general purposes by metric units.

implantation (nidation) (in embryology) The embedding of a fertilized mammalian egg into the wall of the uterus (womb) where it will continue its development. After fertilization in the fallopian tube the egg passes into the womb in the form of a ball of cells (*blastocyst*). Its outer cells destroy cells of the uterine wall, forming a cavity into which the blastocyst sinks.

imprinting A specialized form of *learning in which young animals, during a critical period in their early development, learn to recognize and approach some large moving object nearby. In nature this is usually the mother, though simple models or individuals of a different species (including man) may suffice. Imprinting was first described by Konrad Lorenz (1903–), working with young ducks and geese.

impulse (nerve impulse) The signal that travels along the length of a *nerve fibre and is the means by which information is transmitted through the nervous system. It is marked by the flow of ions across the membrane of the *axon caused by changes in the permeability of the membrane, producing a reduction in potential difference that can be detected as the *action potential. The strength of the impulse produced in any nerve fibre is constant (*see* all-or-none response).

inbreeding Mating between closely related individuals, the extreme condition being self-fertilization, which occurs in many plants and some primitive animals. A population of inbreeding individuals generally shows less variation than an *outbreeding population. Continued inbreeding among a normally outbreeding population leads to *inbreeding depression* (the opposite of *hybrid vigour) and an increased incidence of harmful characteristics. For example, in humans, certain mental and other defects tend to occur more often in families with a history of cousin marriages.

incisor A sharp flattened chisel-shaped tooth in mammals that is adapted for biting food and – in rodents – for gnawing. In humans there are normally two pairs of incisors (central and lateral) in each jaw.

inclusive fitness The quality that organisms attempt (unconsciously) to maximize as the result of natural selection acting on genes that are influential in controlling their behaviour and physiology. It includes the individual's own *reproductive success* (usually taken as the number of its offspring that survive to adulthood) and also the effects of the individual's actions on the reproductive success of its relatives, because relatives have a higher probability of sharing some identical genes with the individual than do other members of the population. When interactions between relatives are likely to occur (which happens during the lives of many animals and plants) *kin selection will operate.

incompatibility **1.** The condition that exists when foreign grafts or blood transfusions evoke a marked *immune response and are rejected. **2.** The phenomenon in which pollen from one flower fails to fertilize other flowers on the same plant (*self-incompatibility*) or on other genetically similar plants. This genetically determined mechanism prevents self-fertilization (breeding between likes) and promotes cross-fertilization (breeding between individuals with different genetic compositions). *See also* allogamy; fertilization; pollination.

incomplete dominance The condition that arises when neither *allele controlling a characteristic is dominant and the aspect displayed by the organism results from the partial influence of both alleles. For example, a snapdragon plant with alleles for red and for white flowers produces pink flowers. *Compare* co-dominance.

incus (anvil) The middle of the three *ear ossicles of the mammalian middle ear.

indefinite inflorescence *See* racemose inflorescence.

independent assortment The separation of the alleles of one gene into the reproductive cells (gametes) independently of the way in which the alleles of other genes have segregated. By this process all possible combinations of alleles should occur equally frequently in the gametes. In practice this does not happen because alleles situated on the same chromosome tend to be inherited together. However, if the allele pairs Aa and Bb are on different chromosomes, the combinations AB, Ab, aB, and ab will normally be equally likely to occur in the gametes. *See* meiosis; Mendel's laws.

index fossil (zone fossil) An animal *fossil of a group that existed continuously during a particular span of geological time and can therefore be used to date the rock in which it is found. Index fossils are found chiefly in sedimentary rocks. They are an essential tool in stratigraphy for comparing the geological ages of sedimentary rock formations. Examples are *ammonites and *graptolites.

indicator species A plant or animal species that is very sensitive to a particular environmental factor, so that its presence (or absence) in an area can provide information about the levels of that factor. For example, some lichens are very sensitive to the concentration of sulphur dioxide (a major pollutant) in the atmosphere. Examination of the lichens present in an area can provide a good indication of the prevailing levels of sulphur dioxide.

indigenous Describing a species that occurs naturally in a certain area, as distinct from one introduced by man; native.

indoleacetic acid (IAA) *See* auxin.

induction 1. (in embryology) The ability of natural stimuli to cause unspecialized embryonic tissue to develop into specialized tissue. **2.** (in obstetrics) The initiation of childbirth by artificial means; for example, by injection of the hormone *oxytocin.

indusium The kidney-shaped covering of the *sorus of certain ferns that protects the developing sporangia. It withers when the sorus ripens to expose the sporangia.

industrial melanism The increase of melanic (dark) forms of an animal in areas darkened by industrial pollution. The example most often quoted is that of the peppered moth (*Biston betularia*), melanic forms of which markedly increased in the industrial north of England during the 19th century. Experiments have shown that the dark forms increase in polluted regions because they are less easily seen by birds against a dark background; conversely the paler forms survive better in unpolluted areas.

infection The invasion of any living organism by disease-causing microorganisms (*see* pathogen), which proceed to establish themselves, multiply, and produce various symptoms in their host. Pathogens may invade via a wound or (in animals) through the mucous membranes lining the alimentary, respiratory, and reproductive tracts, and may be transmitted by an infected individual, a *carrier, or an insect *vector. Symptoms in animals appear after an initial symptomless *incubation period* and typically consist of localized *inflammation, often with pain and fever. Infections are combatted by the body's natural defences (*see* immune response). Treatment with drugs (*see* antibiotic; antiseptic) is effective against most bacterial, fungal, and protozoan infections but not against viral infections. *See also* immunization.

inferior Describing a structure that is positioned below or lower than another structure in the body. For example, in flowering plants the ovary is described as inferior when it is located below the other organs of the flower (*see* epigyny). *Compare* superior.

inflammation The defence reaction of tissue to injury, infection, or irritation by chemicals or physical agents. Cells in the affected tissue release a chemical (*histamine) that causes localized dilatation of blood vessels so that fluid leaks out and blood flow is increased. This leads to swelling, redness, heat, and often pain. White blood cells, particularly *phagocytes, enter the tissue and an *immune response is stimulated. A gradual healing process usually follows.

inflorescence A particular arrangement of flowers on a single main stalk of a plant. There are many different types of inflorescence, which are classified into two main groups depending on whether the tip of the flower axis goes on producing new flower buds during growth (*see* racemose inflorescence) or loses this ability (*see* cymose inflorescence).

inhibition 1. (in biochemistry) A reduction in the rate of an enzyme-catalysed reaction by substances called *inhibitors*. If the inhibitor molecules resemble the substrate molecules they may bind to the active site of the enzyme, so preventing normal enzymatic activity. Alternatively they may form a complex with the substrate–enzyme intermediate or irreversibly destroy the enzyme configuration and active-site properties. The toxic effects of many substances are produced in this way. Inhibition by reaction products (*feedback inhibition*) is important in the control of enzyme activity. **2.** (in physiology) The prevention or reduction of the activity of effectors (such as muscles) by means of certain nerve impulses. Inhibitory activity often provides a balance to stimulation of a process; for example, the impulse to stimulate contraction of a voluntary muscle may be accompanied by an inhibitory impulse to prevent contraction of its antagonist.

initial One of a group of cells (or, in lower plants, a single cell) that divides to produce the cells of a plant tissue or organ. The cells of the apical meristem, cambium, and cork cambium are initials.

innate behaviour An inherited pattern of behaviour that appears in a similar form in all normally reared individuals of the same sex and species. *See* instinct.

inner ear The structure in vertebrates, surrounded by the temporal bone of the skull, that contains the organs of balance and hearing. It consists of soft hollow sensory structures (the *membranous labyrinth*), containing fluid (*endolymph*), surrounded by fluid (*perilymph*), and encased in a bony cavity (the *bony labyrinth*). It consists of two chambers, the *sacculus* and *utriculus*, that bear the *cochlea and *semicircular canals respectively.

innervation The supply of nerve fibres to and from an organ.

innominate artery A short artery that branches from the aorta to divide into the *subclavian artery (the main artery to the arm) and the right *carotid artery (which supplies blood to the head).

innominate bone One of the two bones that form each half of the *pelvic girdle in adult vertebrates. This bone is formed by the fusion of the *ilium, *ischium, and *pubis.

Insecta (Hexapoda) A class of arthropods comprising some 700 000 known species, but many more are thought to exist. They are distributed worldwide in nearly all terrestrial habitats. Ranging in length from 0.5 to over 300 mm, an insect's body consists of a head, a thorax of three segments and usually bearing three pairs of legs and one or two pairs of wings, and an abdomen of eleven segments. The head possesses a pair of sensory *antennae and a pair of large *compound eyes, between which are three simple eyes (*ocelli). The *mouthparts are variously adapted for either chewing or sucking, enabling insects to feed on a wide range of plant and animal material. Insects owe much of their success to having a highly waterproof *cuticle (to resist desiccation) and wings – outgrowths of the body wall that confer the greater mobility of flight. Breathing occurs through a network of tubes (*see* trachea).

Most insect species have separate sexes and undergo sexual reproduction. In some, this may alternate with asexual *parthenogenesis and in a few, males are unknown and reproduction is entirely asexual. The newly hatched young grow by undergoing a series of moults. In the more primitive groups, e.g. *Dermaptera, *Dictyoptera, and *Hemiptera, the young (called a *nymph) resembles the adult. More advanced groups, e.g. *Coleoptera, *Diptera, *Lepidoptera, and *Hymenoptera, undergo *metamorphosis, in which the young (called a *larva) is transformed into a quiescent *pupa from which the fully formed adult emerges. Insects are of vital importance in many ecosystems and many are of economic significance to man – as animal or plant pests or disease vectors or beneficially as crop pollinators or producers of silk, honey, etc.

Insectivora An order of small, mainly nocturnal, mammals that includes the hedgehogs, moles, and shrews. They have long snouts covered with stiff tactile hairs and their teeth are specialized for seizing and crushing insects and other small prey. The insectivores have changed very little since they evolved in the Cretaceous period, 130 million years ago.

insectivore An animal that eats insects, especially a mammal of the order Insectivora (hedgehogs, shrews, etc.).

insectivorous plant *See* carnivorous plant.

instar A stage in the larval development of an insect between two moults (ecdyses). There are usually a number of larval instars before the pupal stage and metamorphosis.

instinct An innate tendency to behave in a particular way, which does not depend critically on particular learning experiences for its development and therefore is seen in a similar form in all normally reared individuals of the same sex and species. Much instinctive behaviour takes the form of *fixed action patterns*. These are movements that – once started – are performed in a stereotyped way unaffected by external stimuli. For example, a frog's prey-catching tongue flick is performed in the same way whether or not anything is caught. Some complex instinctive behaviour, however, requires some learning by the animal before it is perfected. Birdsong, for example, consists of an innate component that is modified and made more complex by the influence of other birds, the habitat, etc.

insulin A protein hormone, secreted by the *islets of Langerhans in the pancreas, that promotes the uptake of glucose by body cells and thereby controls its concentration in the blood. Insulin was the first protein whose amino-acid sequence was fully determined (in 1955). Underproduction of insulin results in the accumulation of large amounts of glucose in the blood and its subsequent excretion in the urine. This condition, known as *diabetes mellitus*, can be treated successfully by insulin injections.

integration (in physiology) The coordination within the brain of separate but related nervous processes. For example, sensory information from the inner ear and the eye are both necessary for the sense of balance. These stimuli must be integrated by the brain not only with each other but also with various motor nerves, which coordinate the muscles that control posture.

integument **1.** The outermost body layer of an animal, characteristically comprising a layer of living cells – the *epidermis – together with a superficial protective coat, which may be a secreted hardened *cuticle, as in arthropods, or dead keratinized cells, as in vertebrates (*see* skin). **2.** The outer protective covering of a plant *ovule. It is perforated by a small pore, the *micropyle. Usually two integuments are present in angiosperms and one in gymnosperms. After fertilization the integuments form the *testa of the seed.

intercellular (in biology) Located or occurring between cells. *Compare* intracellular.

intercostal muscles The muscles located between the *ribs, surrounding the lungs. They play an essential role in breathing (*see* respiratory movement).

interferon Any of a number of proteins that increase the resistance of a cell to attack by viruses by unmasking genes that synthesize antiviral proteins. In humans, three groups of interferons have been discovered: alpha interferons from white blood cells; beta interferons from connective tissue fibroblasts; and gamma interferons from lymphocytes. Interferons are also produced by lymphocyte 'killer cells', which attack altered tissue cells, such as cancer cells. This converts other normal lymphocytes to killer cells and effects other changes in the immune system. Interferon is therefore of great potential use in treating both acute viral infections and cancer.

internal environment The conditions that prevail within the body of an organism, particularly with respect to the composition of the intercellular fluid. Selective absorption of materials across cell membranes plays a large part in controlling the internal environment of both animals and plants. Animals in addition can regulate their body fluids by the action of hormones and the nervous system. *See* homeostasis.

internode 1. (in botany) The part of a plant stem between two *nodes. **2.** (in neurology) The myelinated region of a nerve fibre between two nodes of Ranvier. *See* myelin sheath.

interphase The period following the completion of *cell division, when the nucleus is not dividing. During this period changes in both the nucleus and the cytoplasm result in the complete development of the daughter cells.

intersex An organism displaying characteristics that are intermediate between those of the typical male and typical female of its species. For example, a human intersex may have testes that fail to develop, so that although he is technically a man he has the external appearance of a woman. Intersexes may be produced in various ways; for example, by malfunctioning of the sex hormones. *See also* hermaphrodite.

interstitial-cell-stimulating hormone *See* luteinizing hormone.

intestine The portion of the *alimentary canal posterior to the stomach. Its major functions are the final digestion of food matter from the stomach, the absorption of soluble food matter, the absorption of water, and the production of *faeces. *See* large intestine; small intestine.

intracellular (in biology) Located or occurring within cells. *Compare* intercellular.

inulin A polysaccharide, made up from fructose molecules, that is stored as a food reserve in the roots or tubers of many plants, such as the dahlia.

in vitro Describing biological processes that are made to occur outside the living body, in laboratory apparatus (literally 'in glass', i.e. in a test tube). *Compare in vivo.*

in vivo Describing biological processes as they are observed to occur in their natural environment, i.e. within living organisms. *Compare in vitro.*

involucre A protective structure in some flowering plants and bryophytes. In flowering plants it consists of a ring of *bracts arising beneath the flower cluster of those species with a *capitulum (i.e. members of the dandelion family) or an *umbel (i.e. members of the carrot family). In bryophytes the involucre is a projection of tissue from the thallus that arches over the developing *archegonium.

involuntary muscle (smooth muscle) Muscle whose activity is not under the control of the will; it is supplied by the *autonomic nervous system. Involuntary muscle comprises long spindle-shaped cells without striations. These cells occur singly, in groups, or as sheets in the skin, around hair follicles, and in the digestive tract, respiratory tract, urinogenital tract, and the circulatory system. The cells contract slowly in spontaneous rhythms or when stretched;

they may show sustained contraction (tonus) for long periods without fatigue. *Compare* voluntary muscle.

involution 1. A decrease in the size of an organ or the body. It may be associated with functional decline, as occurs in the ageing process, or follow enlargement, as when the uterus returns to its normal size after pregnancy. **2.** The turning or rolling inwards of cells that occurs during the development of some vertebrate embryos.

ionizing radiation Radiation of sufficiently high energy to cause ionization in the medium through which it passes. It may consist of a stream of high-energy particles (e.g. electrons, protons, alpha-particles) or short-wavelength electromagnetic radiation (ultraviolet, X-rays, gamma-rays). This type of radiation can cause extensive damage to the molecular structure of a substance either as a result of the direct transfer of energy to its atoms or molecules or as a result of the secondary electrons released by ionization. In biological tissue the effect of ionizing radiation can be very serious, usually as a consequence of the ejection of an electron from a water molecule and the oxidizing or reducing effects of the resulting highly reactive species:

$$H_2O \rightarrow e^- + H_2O^* + H_2O^+ \rightarrow \cdot OH + H_3O^+ + \cdot H,$$

where the dot before a radical indicates an unpaired electron and * denotes an excited species.

iridium anomaly The occurrence of unusually high concentrations of the relatively scarce metal iridium at the boundaries of certain geological strata. Two such layers have been discovered, one at the end of the Cretaceous, 65 million years ago, and the second at the end of the Eocene, 34 million years ago. One theory to account for these suggests that on each occasion a huge iridium-containing meteorite may have collided with the earth, producing a cloud of dust that settled out to form an iridium-rich layer. The environmental consequences of such an impact, notably in causing a general warming of the earth by the *greenhouse effect, may have led to the extinction of the dinosaurs at the end of the Cretaceous and the extinction of many radiolarians at the end of the Eocene.

iris The pigmented ring of muscular tissue, lying between the cornea and the lens, in the eyes of vertebrates and some cephalopod molluscs. It has a central hole (the *pupil*) through which light enters the eye and it contains both circular and radial muscles. Reflex contraction of the former occurs in bright light to reduce the diameter of the pupil; contraction of the radial muscles in dim light increases the pupil diameter and therefore the amount of light entering the eye. Colour is determined by the amount of the pigment melanin in the iris. Blue eyes result from relatively little melanin; grey and brown eyes from increasingly larger amounts.

irradiation Exposure to any form of radiation; often exposure to *ionizing radiation is implied.

irritability One of the fundamental properties of all organisms: the capacity to detect, interpret, and respond to changes in the environment (e.g. the stimuli of light, touch, chemicals, etc.). Multicellular animals have specialized *sense organs and *effector organs for this purpose; in unicellular animals and plants, which lack a nervous system, the reception of, and response to, a stimulus occur in the same cell.

ischium The most posterior of the three bones that make up each half of the *pelvic girdle. *See also* ilium; pubis.

islets of Langerhans Small groups of cells in the pancreas that function as an endocrine gland to secrete the hormones *insulin and *glucagon. They are named after their discoverer, the German anatomist and microscopist Paul Langerhans (1847–88).

isogamy Sexual reproduction involving the production and fusion of gametes that are similar in size and structure. It occurs only in some lower animals and plants, e.g. certain protozoans and algae. *Compare* anisogamy.

isolating mechanism Any of the factors that prevent interbreeding (and therefore exchange of genes) between all the members of a species. Probably the most important isolating mechanism is geographical isolation, when different populations of the same species occupy different habitats and therefore never come into contact with each other. Ultimately this may lead to the populations becoming separate species by means of *adaptive radiation.

isoleucine *See* amino acid.

isomerism *See* stereoisomerism.

isoprene A colourless liquid diene, CH_2:C(CH_3)CH:CH_2. The systematic name is *2-methylbuta-1,3-diene*. It is the structural unit in *terpenes and natural rubber, and is used in making synthetic rubbers.

J

jejunum The portion of the mammalian *small intestine that follows the *duodenum and precedes the *ileum. The surface area of the lining of the jejunum is greatly increased by numerous small outgrowths (*see* villus). This facilitates the absorption of digested material, which is the prime function of the jejunum.

jellyfish *See* Coelenterata.

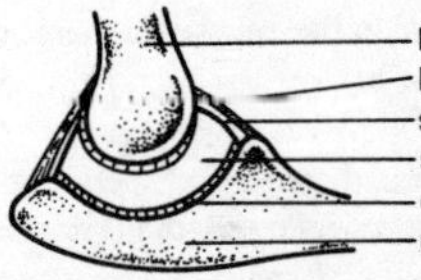

a hinge joint (the elbow)

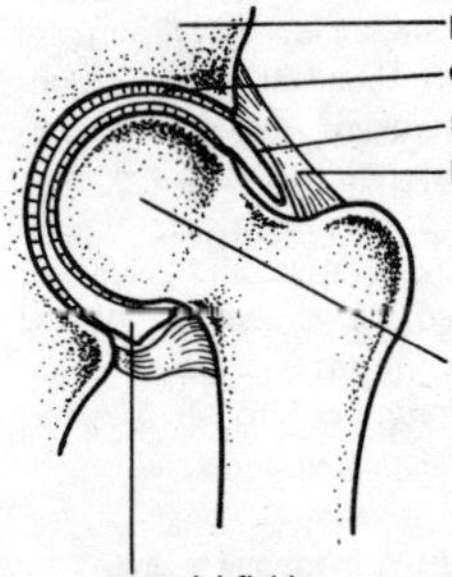

a ball-and-socket joint (the hip)

Types of freely movable joints

joint The point of contact between two (or more) bones, together with the tissues that surround it. Joints fall into three classes that differ in the degree of freedom of movement they allow: (1) *immovable joints*, e.g. the *sutures between the bones that form the cranium; (2) *slightly movable joints*, e.g. the *symphyses between the vertebrae of the spinal column; and (3) *freely movable* or *synovial joints*, e.g. those that occur between the limb bones. Synovial joints include the *ball-and-socket joints* (between the limbs and the hip and shoulder girdles), which allow movement in all directions; and the *hinge joints* (e.g. at the knee and elbow), which allow movement in one plane only. A synovial joint is bound by ligaments and lined with *synovial membrane.

jugular vein A paired vein in the neck of mammals that returns blood from the head to the heart. It joins the subclavian vein at the base of the neck.

Jurassic The second geological period of the Mesozoic era. It followed the Triassic, which ended about 190 million years ago, and extended until the beginning of the Cretaceous period, about 139 million years ago. It was named in 1829 by A. Brongniart after the Jura Mountains on the borders of France and Switzerland. Jurassic rocks include clays and limestones in which fossil flora and fauna are abundant. Plants included ferns, cycads, ginkgos, rushes, and conifers. Important invertebrates included *ammonites (on which the Jurassic is zoned), corals, brachiopods, bivalves, and echinoids. Reptiles dominated the vertebrates and the first flying reptiles – the pterosaurs – appeared. The first primitive bird, *Archaeopteryx*, also made its appearance.

juvenile hormone A hormone secreted by insects from a pair of endocrine glands (*corpora allata*) close to the brain. It inhibits metamorphosis and maintains the presence of larval features.

K

Kainozoic *See* Cenozoic.

kallidin *See* kinin.

karyogram (idiogram) A diagram representing the characteristic features of the *chromosomes of a species.

karyokinesis The division of a cell nucleus. *See* meiosis; mitosis.

karyotype The number and structure of the *chromosomes in the nucleus of a cell. The karyotype is identical in all the *diploid cells of an organism.

keel (carina) The projection of bone from the sternum (breastbone) of a bird or bat, to which the powerful flight muscles are attached. The sterna of flightless birds (e.g. ostrich and emu) lack keels.

kelp Any large brown seaweed (*see* Phaeophyta) or its ash, used as a source of iodine.

kelvin Symbol K. The *SI unit of thermodynamic temperature equal to the fraction 1/273.16 of the thermodynamic temperature of the triple point of water. The magnitude of the kelvin is equal to that of the degree Celsius (centigrade), but a temperature expressed in degrees Celsius is numerically equal to the temperature in kelvins less 273.15 (i.e. $°C = K - 273.15$). The unit is named after Lord Kelvin (1824–1907).

keratin Any of a group of fibrous *proteins occurring in hair, feathers, hooves, and horns. Keratins have coiled polypeptide chains that combine to form supercoils of several polypeptides linked by disulphide bonds between adjacent cysteine amino acids. Aggregates of these supercoils form microfibrils, which are embedded in a protein matrix. This produces a strong but elastic structure.

keratinization (cornification) The process in which the cytoplasm of the outermost cells of the mammalian *epidermis is replaced by *keratin. Keratinization occurs in the *stratum corneum, feathers, hair, claws, nails, hooves, and horns.

ketohexose *See* monosaccharide.

ketone body Any of three compounds, acetoacetic acid (3-oxobutanoic acid, CH_3COCH_2COOH), β-hydroxybutyric acid (3-hydroxybutanoic acid, $CH_3CH(OH)CH_2COOH$), and acetone (propanone, CH_3COCH_3), produced by the liver as a result of the metabolism of body fat deposits.

Ketone bodies are normally used as energy sources by peripheral tissues. However, if carbohydrate supply is limited (e.g. during starvation or in diabetics), the blood level of ketone bodies rises and they may be present in urine, giving it a characteristic 'pear drops' odour. This condition is called *ketosis*.

ketopentose *See* monosaccharide.

ketose *See* monosaccharide.

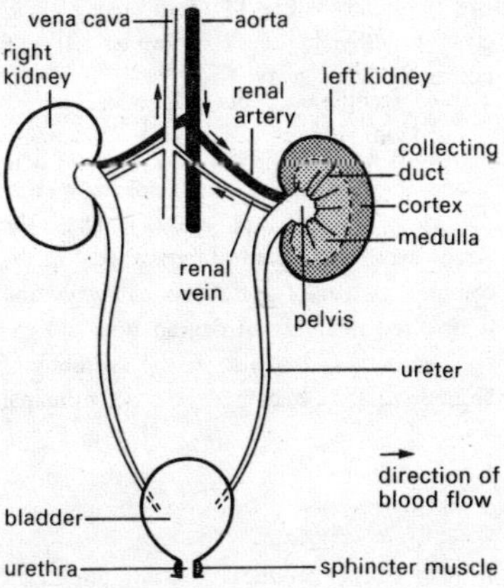

The kidneys of a mammal

kidney The main organ of *excretion of vertebrates, through which nitrogenous waste material (usually in the form of *urine) is eliminated from the body. In mammals there is a pair of kidneys situated in the abdomen. They are made up of tubular units called *nephrons, through which nitrogenous waste is filtered from the blood, with the formation of urine. The nephrons drain into a basin-like cavity in the kidney (the *renal pelvis*), which leads to the *ureter and *bladder.

kilo- Symbol k. A prefix used in the metric system to denote 1000 times. For example, 1000 volts = 1 kilovolt (kV).

kilogram Symbol kg. The *SI unit of mass defined as a mass equal to that of the international platinum–iridium prototype kept by the International Bureau of Weights and Measures at Sèvres, near Paris.

kinesis The movement of a cell or organism in response to a stimulus in which the rate of movement depends on the intensity (rather than the direction) of the stimulus. For example, a woodlouse moves slowly in a damp atmosphere and quickly in a dry one.

kinetochore Two parallel brushlike filaments situated in the *centromere of a chromosome. Kinetochores attach chromatids to the *spindle fibres during nuclear division.

kingdom The highest category into which organisms are classified. Traditionally two kingdoms have been recognized, the Plantae (*see* plant) and the Animalia (*see* animal). More recently a number of other kingdoms have been suggested. For example, the fungi are considered by some to be sufficiently distinct from green plants to merit being placed in a separate kingdom, Fungi. The differences between *prokaryotes and *eukaryotes have led some to recognize the two kingdoms Prokaryota and Eukaryota. *See also* Protista.

kinin 1. One of a group of peptides, occurring in blood, that are involved in inflammation. Kinins are formed by the enzymatic splitting of blood plasma globulins (*kininogens*) at the site of inflammation. Kinins so far identified include *bradykinin* and *kallidin*. They cause local increases in the permeability of small blood vessels. **2.** *See* cytokinin.

kinomere *See* centromere.

kin selection Natural selection of genes that tend to cause the individuals bearing them to be altruistic to close relatives. These relatives therefore have a higher probability of bearing identical copies of those same genes than do other members of the population. Thus kin selection for a gene that tends to cause an animal to share food with a close relative will result in the gene being spread through the population because it (unconsciously) benefits itself. The more closely two animals are related, the higher the probability that they share some identical genes and therefore the more closely their interests coincide. Parental care is a special case of kin selection. *See* inclusive fitness.

Krebs cycle (citric acid cycle; tricarboxylic acid cycle; TCA cycle) A cyclical series of biochemical reactions that is fundamental to the metabolism of aerobic organisms, i.e. animals, plants, and many microorganisms. The enzymes of the Krebs cycle are located in the *mitochondria and are in close association with the components of the *electron transport chain. The two-carbon acetyl coenzyme A (acetyl CoA) reacts with the four-carbon oxaloacetate to form the six-carbon citric acid. In a series of seven reactions, this is reconverted to oxaloacetate and produces two molecules of carbon dioxide. Most importantly, the cycle generates one molecule of guanosine triphosphate (GTP – equivalent to 1 ATP) and reduces three molecules of the coenzyme *NAD to NADH and one molecule of the coenzyme *FAD to $FADH_2$. NADH and $FADH_2$ are then oxidized by the electron transport chain to generate three and two molecules of ATP respectively. This gives a net yield of 12 molecules of ATP per molecule of acetyl CoA.

Acetyl CoA can be derived from carbohydrates (via *glycolysis), fats, or certain amino acids. (Other amino acids may enter the cycle at different stages.) Thus the Krebs cycle is the central 'crossroads' in the complex system of metabolic pathways and is involved not only in degradation and energy production but also in the synthesis of biomolecules. It is named after its principal

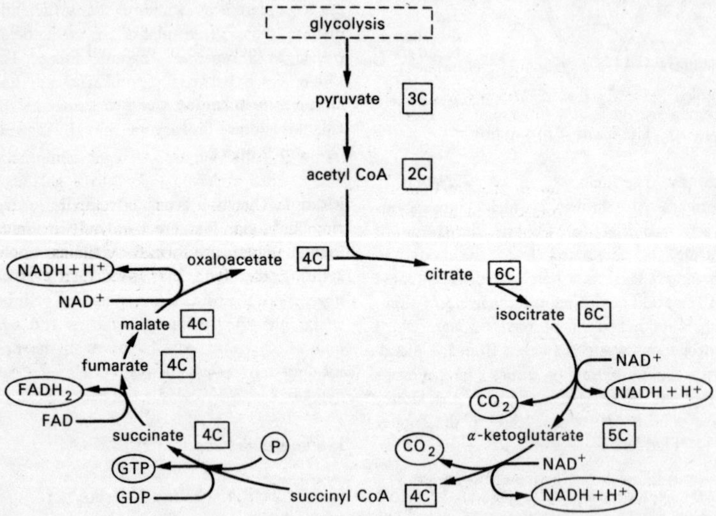

Krebs cycle

discoverer, Sir Hans Adolf Krebs (1900–81).

L

labelling The process of replacing a stable atom in a compound with a radioisotope of the same element to enable its path through a biological system to be traced by the radiation it emits. In some cases a different stable isotope is used and the path is detected by means of a mass spectrometer. A compound containing either a radioactive or stable isotope is called a *labelled compound*. If a hydrogen atom in each molecule of the compound has been replaced by a tritium atom, the compound is called a *tritiated compound*. A radioactive labelled compound will behave chemically and physically in the same way as an otherwise identical stable compound, and its presence can easily be detected. This process of *radioactive tracing* is widely used in biology and medicine.

labium The lower lip in the *mouthparts of an insect, which is used in feeding and is formed by the fusion of a pair of appendages (the second *maxillae).

labrum The upper lip in the *mouthparts of an insect. It is formed from a plate of cuticle hinged to the head above the mouth and is used in feeding.

labyrinth The system of cavities and tubes that comprises the *inner ear of vertebrates. It consists of a system of membranous structures (*membranous labyrinth*) housed in a similar shaped bony cavity (*bony labyrinth*).

lachrimal gland (lachrymal gland) The tear gland, present in the eyelids of some vertebrates. The fluid (tears) produced by this gland cleanses and lubricates the exposed surface of the eye; it drains into the nose through the lachrimal duct.

lactation The discharge of milk from the *mammary glands. This generally only occurs after birth of the young and is stimulated by the sucking action of the infants. The production of milk is under the control of hormones, notably *prolactin.

lacteal A minute blind-ended lymph vessel that occurs in each *villus of the small intestine. Digested fats are absorbed into the lacteals (*see* chyle) and transported to the bloodstream through the *thoracic duct.

lactic acid (2-hydroxypropanoic acid) An alpha hydroxy carboxylic acid, $CH_3CH(OH)COOH$, with a sour taste. Lactic acid is produced from pyruvic acid in active muscle tissue when oxygen is limited (*see* oxygen debt) and subsequently removed for conversion to glucose by the liver. During strenuous exercise it may build up in the muscles, causing cramplike pains. It is also produced by fermentation in certain bacteria and is characteristic of sour milk.

lactogenic hormone *See* prolactin.

lactose (milk sugar) A sugar comprising one glucose molecule linked to a galactose molecule. Lactose is manufactured by the mammary gland and occurs only in milk. For example, cows' milk contains about 4.7% lactose. It is less sweet than sucrose (cane sugar).

lacuna A gap or cavity in the tissues of an organism; for example, the hollow centre of certain plant stems or any of the small cavities in bone in which the bone-forming cells are found.

laevorotatory Designating a chemical compound that rotates the plane of plane-polarized light to the left (anticlockwise for someone facing the oncoming radiation). *See* optical activity.

laevulose *See* fructose.

Lamarckism One of the earliest superficially plausible theories of *evolution, proposed by the French biologist Jean-Baptiste de Lamarck (1744–1829) in 1809. He suggested that changes in an individual are acquired during its lifetime, chiefly by increased use or disuse of organs in response to "a need that continues to make itself felt", and that these changes are inherited by its offspring. Thus the long neck and limbs of a giraffe are explained as having evolved by the animal stretching its neck to browse on the foliage of trees. This so-called inheritance of acquired characteristics has never unquestionably been demonstrated to occur and the theory was largely displaced by *Darwinism. Lamarckism is also incompatible with the *Central Dogma of molecular biology. *See also* Lysenkoism.

lamella 1. (in botany) **a.** Any of the paired folds of membranes seen between the *grana in a plant chloroplast. **b.** Any of the spore-bearing gills on the underside of the cap of many mushrooms and toadstools. *See also* middle lamella. **2.** (in zoology) Any of various thin layers of membranes, especially any of the thin layers of tissue of which compact bone is formed.

Lamellibranchia A class of aquatic molluscs (the bivalves) that include the oysters, mussels, and clams. They are characterized by a laterally flattened body and a shell consisting of two hinged shells (i.e. a bivalved shell). The enlarged gills are covered with cilia and have the additional function of filtering microscopic food particles from the water flowing over them. Bivalves live on the sea bed or lake bottom and are sedentary, so the head and foot are reduced.

lamina 1. The thin and usually flattened blade of a leaf, in which photosynthesis and transpiration occurs. The bulk of the lamina is made up of *mesophyll cells interspersed by a network of veins (*vascular bundles). The mesophyll is enclosed by a protective epidermis that produces a waxy cuticle. **2.** The leaflike part of the thallus of certain algae, notably kelps. *See also* stipe.

large intestine The portion of the alimentary canal of vertebrates between the *small intestine and the *anus. It consists of the *caecum, *colon, and *rectum and its principal function is the absorption of water and formation of faeces.

larva The juvenile stage in the life cycle of most invertebrates, amphibians, and fish, which hatches from the egg, is unlike the adult in form, and is usually incapable of sexual reproduction (*see* paedogenesis). It develops into the adult by undergoing *metamorphosis. Larvae can feed themselves and are otherwise self-supporting. Examples are the tadpoles of frogs, the caterpillars of butterflies, and the ciliated planktonic larvae of many marine animals. *Compare* nymph.

larynx The anterior portion of the *trachea (windpipe) of tetrapod vertebrates, which in amphibians, reptiles, and mammals contains the *vocal cords. Movement of the cartilage in the walls of the larynx (by means of the laryngeal muscles) alters the tension of the vocal cords. This changes the pitch of the sound emitted by the vocal cords when they vibrate. The final voiced sound is modified by resonance within the oral and nasal cavities.

latex A milky fluid of mixed composition found in some herbaceous plants and trees. Its function is not clear but it may assist in protecting wounds (*compare* gum) and it may be involved in the nutrition of the plant. The latex of some species, notably rubber trees, is collected for commercial purposes.

lauric acid (dodecanoic acid) A white crystalline *fatty acid, $CH_3(CH_2)_{10}COOH$. Glycerides of the acid are present in natural fats and oils (e.g. coconut and palm-kernel oil).

LD_{50} Lethal dose 50, or median lethal dose: the amount of a pharmacological or toxic substance (such as ionizing radiation) that causes death in 50% of a group of experimental animals. For each LD_{50} the species and weight of the animal and the route of administration of the substance is specified. LD_{50}s are used both in toxicology and in the *bioassay of therapeutic compounds.

L-dopa *See* dopa.

leaf A flattened structure that develops from a superficial group of tissues, the leaf buttress, on the side of the stem apex. Each leaf has a lateral bud in its axil. Leaves are arranged in a definite pattern (*see* phyllotaxis) and usually show limited growth. Each consists of a broad flat *lamina (leaf blade)

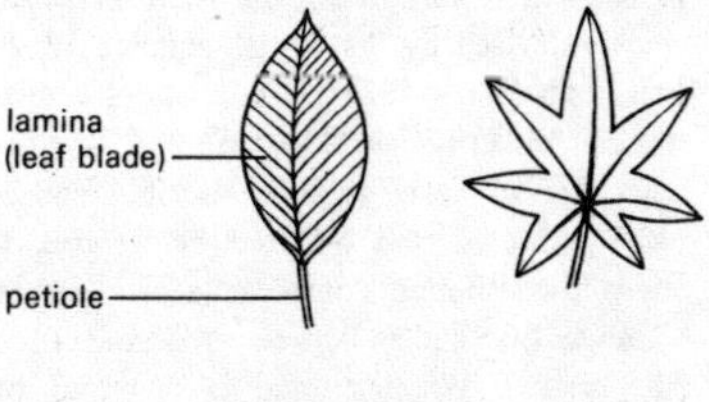

Simple leaves

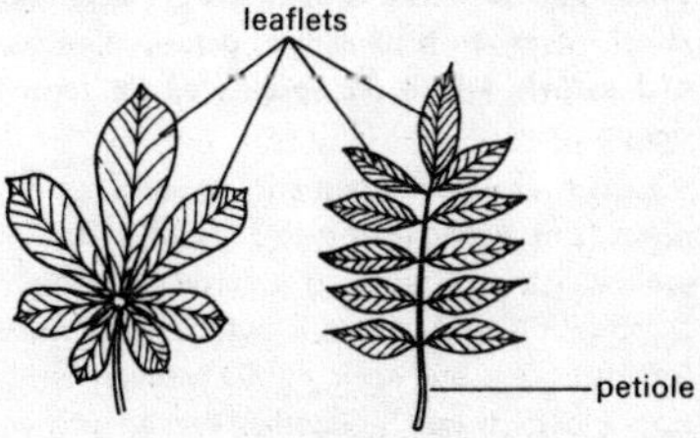

Compound leaves

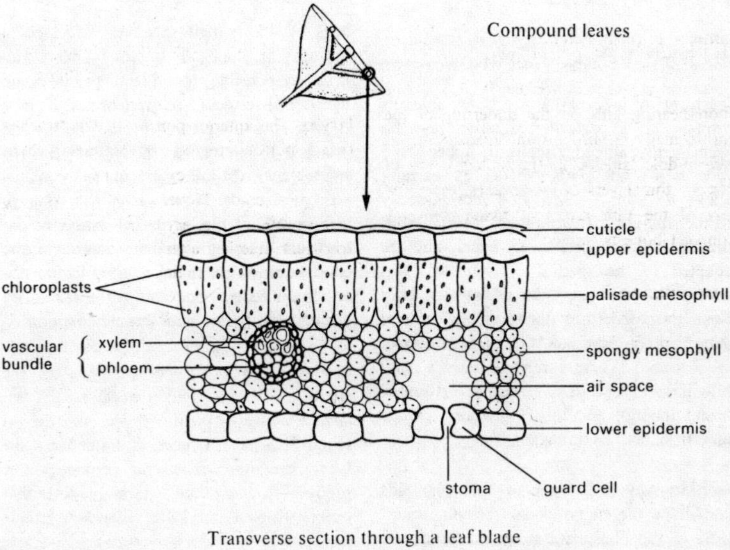

Transverse section through a leaf blade

and a leaf base, which attaches the leaf to the stem; a leaf stalk (*petiole*) may also be present. The leaves of bryophytes are simple appendages, which are not homologous with the leaves of vascular plants as they develop on the gametophyte generation.

Leaves show considerable variation in size, shape, arrangement of veins, type of attachment to the stem, and texture. They may be *simple* or divided into *leaflets*, i.e. *compound*. Types of leaf include: *cotyledons (seed leaves); *scale leaves*, which lack chlorophyll and develop on rhizomes or protect the inner leaves of a bud; *foliage leaves*, which are the main organs for photosynthesis and transpiration; and *bracts and *floral leaves*, such as sepals, petals, stamens, and carpels, which are specialized for reproduction.

Leaves may be modified for special purposes. For example the leaf bases of bulbs are swollen with food to survive the winter. In some plants leaves are reduced to spines for protection and their photosynthetic function is carried out by another organ, such as a *cladode.

leaf buttress *See* primordium.

learning A process by which an animal's response to a particular situation may be permanently altered, usually in a beneficial way, as a result of its experience. Learning allows an animal to respond more flexibly to the situations it encounters: learning abilities in different species vary widely and are adapted to the species' environment. Numerous different categories of learning have been proposed, but there is no general agreement on how many different processes are involved. These categories include *habituation, associative learning (through *conditioning), and insight learning (which may include *imprinting).

lecithin Any of a group of *phospholipids containing the amino alcohol *choline esterified to the phosphate group. Lecithins are the most abundant animal phospholipids and also occur in higher plants, but rarely in microorganisms.

leeches *See* Hirudinea.

legume (pod) A dry fruit formed from a single carpel and containing one or more seeds, which are shed when mature. It is the characteristic fruit of the Leguminosae (pea family). It splits, often explosively, along both sides and the two halves of the fruit move apart to expose the seeds. A special form of the legume is the *lomentum.

Leishman's stain A neutral stain for blood smears devised by the British surgeon W. B. Leishman (1865–1926). It consists of a mixture of *eosin (an acidic stain), and *methylene blue (a basic stain) in alcohol and is usually diluted and buffered before use. It stains the different components of blood in a range of shades between red and blue. The similar *Wright's stain* is favoured by American workers.

lens A transparent biconvex structure in the eyes or analogous organs of many animals, responsible for directing light onto light-sensitive cells. In vertebrates it is a flexible structure centred behind the iris and attached to the *ciliary body. In terrestrial species its main function is to focus images onto the retina. To focus on near objects, the circular muscles in the ciliary body contract and the lens becomes more convex; contraction of the radial muscles in the ciliary body flattens the lens for focusing on distant objects (*see also* accommodation).

lenticel Any of the raised pores in the stems of woody plants that allow gas exchange between the atmosphere and the internal tissues. The pore is formed by the *cork cambium, which, at certain points, produces a loose bulky form of cork that pushes through the outer tissues to create the lenticel.

Lepidoptera An order of insects comprising the butterflies and moths, found mainly in tropical regions. Adults possess two pairs of membranous wings, often brightly coloured and usually coupled together. The wings, body, and legs are covered with minute scales. Adult mouthparts are generally modified to form a long proboscis for sucking nectar, fruit juices, etc. Butterflies are typically small-bodied, active during daylight, and rest with their wings folded vertically; moths have larger bodies, are nocturnal, and rest with their wings in various positions. The larvae (caterpillars) have a prominent head and a segmented wormlike body, most segments bearing a pair of legs. They chew leaves and stems, sometimes causing considerable damage to crop plants. The larvae undergo metamorphosis via a *pupa (chrysalis) to the adult form. In some groups, the pupa is enclosed in a cocoon of silk derived from silk glands (modified salivary glands); others use leaves, etc. to build a cocoon.

leptotene The beginning of the first prophase of *meiosis, when the chromatids can be seen and *pairing begins.

lethal dose 50 *See* LD_{50}.

leucine *See* amino acid.

leucocyte (white blood cell) A colourless cell with a nucleus, found in blood and lymph. Leucocytes are formed in lymph nodes and red bone marrow and are capable of amoeboid movement. They can produce *antibodies and move through the walls of vessels to migrate to the sites of injuries, where they surround and isolate dead tissue, foreign bodies, and bacteria. There are two major types: those without granules in the cytoplasm (*lymphocytes and *monocytes) and those with granular cytoplasm (*granulocytes*).

leucoplast Any of various organelles in plant cells that contain no pigment and are therefore colourless. Leucoplasts are usually found in tissues not normally exposed to light and frequently contain reserves of starch, protein, or oil. *See* plastid. *Compare* chromoplast.

***l*-form** *See* optical activity.

LH *See* luteinizing hormone.

lice *See* Mallophaga (bird lice); Siphunculata (sucking lice).

Lichenes (lichens) A group consisting of organisms that are symbiotic associations (*see* symbiosis) between a fungus (usually one of the *Ascomycetes) and an alga (from the *Chlorophyta or *Cyanophyta). The fungus usually makes up most of the plant body and the cells of the alga are distributed within it. The alga photosynthesizes and passes most of its food to the fungus and the fungus protects the algal cells. The lichen reproduces by means of fungal spores, which must find a suitable alga on germination. Lichens are slow growing but can live in regions that are too cold or exposed for other plants. They may form a flattened crust or be erect and branching. Many grow as *epiphytes, especially on tree trunks. Some species are very sensitive to air pollution and have been used as *indicator species.

life cycle The complete sequence of events undergone by organisms of a particular species from the fusion of gametes in one generation to the same stage in the following generation. In most animals gametes are formed by *meiosis of germ cells in the reproductive organs of the parents. The zygote, formed by the fusion of two gametes, eventually develops into an organism essentially similar to the parents. In plants, however, the products of meiosis are spores, which develop into plants (the *gametophyte generation) often very different in form from the spore-forming (*sporophyte) generation. The sporophyte generation is restored when gametes, formed by the gametophyte generation, fuse. *See* alternation of generations.

ligament A resilient but flexible band of tissue (chiefly *collagen) that holds two or more bones together at a movable *joint. Ligaments restrain the movement of bones at a joint and are therefore important in preventing dislocation.

ligase Any of a class of enzymes that catalyse the formation of covalent bonds using the energy released by the cleavage of ATP. Ligases are important in the synthesis and repair of many biological molecules, including DNA.

light green *See* fast green.

light reactions *See* photosynthesis.

lignin A complex organic polymer that is deposited within the cellulose of plant cell walls during secondary thickening. Lignification makes the walls woody and therefore rigid.

lignite *See* coal.

ligule 1. A membranous scalelike outgrowth from the leaves of certain flowering plants. Many grasses have a ligule at the base of the leaf blade. **2.** A small membranous structure that develops on the upper surface of a young leaf base in certain pteridophytes, for example *Selaginella*. It withers as the plant matures. **3.** A strap-shaped extension from the corolla tube in certain florets of a *capitulum. They are termed *ligulate* (or *ray*) *florets*.

liming The application of lime (calcium hydroxide) to soils to increase levels of calcium and decrease acidity.

linear energy transfer (LET) The energy transferred per unit path length by a moving high-energy charged particle (such as an electron or a proton) to the atoms and molecules along its path. It is of particular importance when the particles pass through living tissue as the LET modifies the effect of a specific dose of radiation. LET is proportional to the square of the charge on the particle and increases as the velocity of the particle decreases.

linkage The association of a number of characteristics that are inherited together because the genes responsible for them are situated on the same chromosome. Linkage can be broken by *crossing over during mitosis, when sections of chromosomes are exchanged and new combinations of genes are produced. *See also* sex linkage.

linkage map *See* chromosome map.

Linnaean system *See* binomial nomenclature.

linoleic acid A liquid polyunsaturated *fatty acid with two double bonds, $CH_3(CH_2)_4CH{:}CHCH_2CH{:}CH(CH_2)_7COOH$. Linoleic acid is abundant in many plant fats and oils, e.g. linseed oil, groundnut oil, and soya-bean oil. It is an *essential fatty acid.

linolenic acid A liquid polyunsaturated *fatty acid with three double bonds in its structure: $CH_3CH_2CH{:}CHCH_2CH{:}CHCH_2CH{:}CH(CH_2)_7COOH$. Linolenic acid occurs in certain plant oils, e.g. linseed and soya-bean oil, and in algae. It is one of the *essential fatty acids.

lipase An enzyme secreted by the pancreas and the glands of the small intestine of vertebrates that catalyses the breakdown of fats into fatty acids and glycerol.

lipid Any of a diverse group of organic compounds, occurring in living organisms, that are insoluble in water but soluble in organic solvents, such as chloroform, benzene, etc. Lipids are broadly classified into two categories: *complex lipids*, which are esters of long-chain fatty acids and include the *glycerides (which constitute the *fats and *oils of animals and plants), glycolipids, *phospholipids, and *waxes; and *simple lipids*, which do not contain fatty acids and include the *steroids and *terpenes.

Lipids have a variety of functions in living organisms. Fats and oils are a convenient and concentrated means of storing food energy in plants and animals. Phospholipids and *sterols, such as cholesterol, are major components of cell membranes. Waxes provide vital waterproofing for body surfaces. Terpenes include vitamins A, E, and K, and phytol (a component of chlorophyll) and occur in essential oils, such as menthol and camphor. Steroids include the adrenal hormones, sex hormones, and bile acids.

Lipids can combine with proteins to form *lipoproteins*, e.g. in cell membranes. In bacterial cell walls, lipids may associate with polysaccharides to form *lipopolysaccharides*.

lipoic acid A vitamin of the *vitamin B complex. It is one of the *coenzymes involved in the decarboxylation of pyruvate by the enzyme pyruvate dehydrogenase. This reaction has to take place before carbohydrates can enter the *Krebs cycle during aerobic respiration. Good sources of lipoic acid include liver and yeast.

lipolysis The breakdown of storage lipids in living organisms. Most long-term energy reserves are in the form of triglycerides in fats and oils. When these are needed, e.g. during starvation, lipase enzymes convert the triglycerides into glycerol and the component fatty acids. These are then transported to tissues and oxidized to provide energy

lipoprotein *See* lipid.

liposome A microscopic spherical membrane-enclosed vesicle or sac (20–30 nm in diameter) made artificially in the laboratory by the addition of an aqueous solution to a phospholipid gel. The membrane resembles a cell membrane and the whole vesicle is similar to a cell organelle. Liposomes can be incorporated into living cells and are used to transport relatively toxic drugs into diseased cells, where they can exert their maximum effects. For example, liposomes containing the drug methotrexate, used in the treatment of cancer, can be injected into the patient's blood. The cancerous organ is heated to a temperature higher than body temperature, so that when the liposome passes through its blood vessels, the membrane melts and the drug is released. The study of the behaviour of liposome membranes is used in research into membrane function, particularly to observe the behaviour of membranes during anaesthesia with respect to permeability changes.

litre Symbol l. A unit of volume in the metric system regarded as a special name for the cubic decimetre. It was formerly defined as the volume of 1 kilogram of pure water at 4°C at standard pressure, which is equivalent to 1.000 028 dm^3.

littoral Designating or occurring in the marginal shallow-water zone of a sea or lake, especially (in the sea) between high and low tide lines. In this zone enough light penetrates to the bottom to support rooted aquatic plants. *Compare* profundal; sublittoral.

liver A large lobed organ in the abdomen of vertebrates that plays an essential role in many metabolic processes by regulating the composition and concentration of nutrients and toxic materials in the blood. It receives the products of digestion dissolved in the blood via the *hepatic portal vein and its most important functions are to convert excess glucose to the storage product *glycogen, which serves as a food reserve; to break down excess amino acids to ammonia, which is converted to *urea and excreted via the kidneys; and to store and break down fats (*see* lipolysis). Other functions of the liver are (1) the production of *bile; (2) the breakdown (detoxification) of poisonous substances in the blood; (3) the removal of damaged red blood cells; (4) the synthesis of vitamin A and the blood-clotting substances prothrombin and fibrinogen; and (5) the storage of iron.

liverworts *See* Hepaticae.

living fossil Any organism whose closest relatives are extinct and that was once itself thought to be extinct. An example is the coelacanth, a primitive fish that was common in the Devonian era, the first recent living specimen of which was discovered in 1938.

lizards *See* Squamata.

loculus A small cavity in a plant or animal body. In plants the loculus of the ovary is the cavity containing the ovules and the loculi of the anther contain the developing pollen grains.

locus The position of a gene on a chromosome. The two alleles of a gene occupy the same locus on *homologous chromosomes.

logarithmic scale A scale of measurement in which an increase or decrease of one unit represents a tenfold increase or decrease in the quantity measured. Decibels and pH measurements are common examples of logarithmic scales of measurement.

lomentum A type of dry dehiscent fruit formed from a single carpel but divided into one-seeded compartments by constrictions between the seeds. *Legumes (e.g. those of *Acacia*) and *siliquas (e.g. those of wild radish) can be divided in this way.

long-day plant A plant in which flowering can be induced or enhanced by long days, usually of more than 12 hours of daylight. Examples are spinach and spring barley. *See* photoperiodism. *Compare* day-neutral plant; short-day plant.

long-sightedness *See* hypermetropia.

lumbar vertebrae The *vertebrae in the region of the lower back. They occur below the *thoracic vertebrae and above the *sacral vertebrae. In mammals they bear processes for the attachment of back muscles.

lumen 1. The space enclosed by a vessel, duct, or other tubular or saclike organ. The central cavity of blood vessels and of the digestive tract are examples. **2.** Symbol lm. The SI unit of luminous flux equal to the flux emitted by a uniform point source of 1 candela in a solid angle of 1 steradian.

lung The *respiratory organ of air-breathing vertebrates. A pair of lungs is situated in the thorax, within the ribcage. Each consists essentially of a thin moist membrane that is folded to increase its surface area. Exchange of oxygen and carbon dioxide takes place between blood capillaries on one side of the membrane and air on the other. The lung is supplied with air through a *bronchus. In mammals and reptiles the membrane of the lung takes the form of numerous sacs (*see* alveolus) that are connected to the bronchus via *bronchioles. The lungs themselves contain no muscular tissue and are ventilated by *respiratory movements, the mechanisms of which vary with the species.

lungfish *See* Dipnoi.

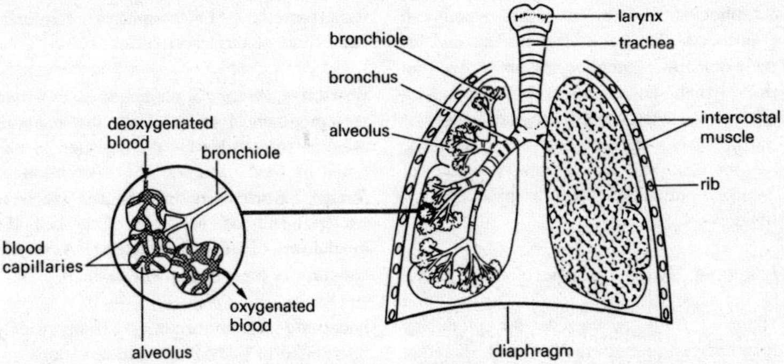

The lungs and air passages of a mammal (right lung cut open to show internal structure)

luteinizing hormone (LH; interstitial-cell-stimulating hormone; ICSH) A hormone, secreted by the anterior pituitary gland in mammals, that stimulates in males the production of sex hormones (*androgens) by secretory cells in the testes (known as *interstitial cells*) and in females ovulation, *progesterone synthesis, and *corpus luteum formation.

luteotrophic hormone *See* prolactin.

Lycopsida A subdivision of the *Tracheophyta or a class of the *Pteridophyta containing the clubmosses (*Lycopodium*) and related genera (including *Selaginella*) as well as numerous extinct forms, which reached their peak in the Carboniferous period with giant coal-forming tree species. Lycopsids have roots and their stems are covered with numerous small leaves.

lymph The colourless liquid found within the *lymphatic system, into which it drains from the spaces between the cells. Lymph (called *tissue fluid* in the intercellular spaces) resembles *blood plasma, consisting mostly of water with dissolved salts and proteins. Fats are found in suspension and their presence varies wth food intake. The lymph eventually enters the bloodstream near the heart.

lymphatic system The network of vessels that conveys *lymph from the tissue fluids to the bloodstream. Tiny lymph capillaries drain into larger tubular vessels that converge to form the right lymphatic duct and the *thoracic duct, which connect with the venous blood supply to the heart. Associated with the lymphatic vessels at intervals along the system are the *lymph nodes. The lymph capillary walls are very permeable, so lymph bathing the body's tissues can drain away molecules that are too large to pass through blood capillary walls. Lymph is pumped by cycles of contraction and relaxation of the lymphatic vessels and also by the action of adjoining muscles.

lymph node A mass of *lymphoid tissue, many of which occur at intervals along the *lymphatic system. Lymph in the lymphatic vessels flows through the lymph nodes, which filter out bacteria and other foreign particles, so preventing them from entering the bloodstream and causing infection. The lymph nodes also produce *lymphocytes. In humans, major lymph nodes occur in the neck, under the arms, and in the groin.

lymphocyte A type of white blood cell (*leucocyte) that has a large nucleus and little cytoplasm. Lymphocytes are formed in the *lymph nodes and provide about a quarter of all leucocytes. They are important in the body's defence and are responsible for immune reactions as the presence of *antigens stimulates them to produce *antibodies.

lymphoid tissue The type of tissue found in the *lymph nodes, *tonsils, *spleen, and *thymus. It is responsible for producing lymphocytes and antibodies and therefore contributes to the body's defence against infection.

Lysenkoism The official Soviet science policy governing the work of geneticists in the USSR from about 1940 to 1960. It was named after its chief promoter, the agriculturalist Trofim Lysenko (1898–1976). Lysenkoism dismissed all the advances that had been made in classical genetics, denying the existence of genes, and held that the variability of organisms was produced solely by environmental changes. There was also a return to a belief in the inheritance of acquired characteristics (*see* Lamarckism). This state of affairs continued, despite overwhelming conflicting evidence from Western scientists, because it provided support for communist theory.

lysigeny The localized disruption of plant cells to form a cavity (surrounded by remnants of the broken cells) in which secretions accumulate. Examples are the oil cavities in the leaves of citrus trees. *Compare* schizogeny.

lysine *See* amino acid.

lysis The destruction of a living cell. This may be effected by *lysosomes or *phagocytes, either as part of the normal metabolic process (as when cells are damaged or worn out) or as a reaction against invading cells (e.g. bacteria). *Bacteriophages eventually cause lysis of their host cells.

lysosome A membrane-bound sac within the cytoplasm of animal *cells that contains enzymes responsible for the digestion of material in food vacuoles, the dissolution of foreign particles entering the cell (such as bacteria) and, on the death of the cell, the breakdown of all cell structures. A similar structure is present in plant cells.

lysozyme An antibacterial enzyme widely distributed in body fluids and secretions, including tears and saliva. It disrupts the polysaccharide components of bacterial cell walls, leaving them susceptible to destruction.

M

macromolecule A very large molecule. Natural and synthetic polymers have macromolecules, as do such substances as haemoglobin. *See also* colloids.

macronutrient A nutrient required by an organism in relatively large amounts. Macronutrients are typically used as sources of energy or structural components and include water, carbon dioxide, carbohydrates, proteins, lipids, and certain minerals (*see also* essential element). *Compare* micronutrient.

macrophage (histiocyte; clasmatocyte) A large phagocytic cell (*see* phagocyte) normally found in tissues that produce blood cells. Macrophages ingest bacteria and cell debris. They develop from *monocytes and assume different shapes depending on their position. *Reticuloendothelial cells* are similar but are found only lining blood vessels and do not move around the body.

macrophyll *See* megaphyll.

macula 1. One of a number of patches of sensory cells in the *utriculus and *sacculus of the inner ear that are involved in maintaining balance and may also assist in assessing the degree of acceleration and deceleration when the body is moving. **2.** An area of the *retina of the vertebrate eye with increased visual acuity. Maculae occur in some animals that lack *foveae and often surround foveae in those animals that possess them.

malic acid (2-hydroxybutanedioic acid) A crystalline solid, HOOCCH(OH)CH_2COOH. L-malic acid occurs in living organisms as an intermediate metabolite in the *Krebs cycle and also (in certain plants) in photosynthesis. It is found especially in the juice of unripe fruits, e.g. green apples.

malleus (hammer) The first of the three *ear ossicles of the mammalian *middle ear.

Mallophaga An order of wingless insects comprising the bird lice. Bird lice are minute with dorsoventrally flattened ovoid bodies, reduced eyes, and biting mouthparts. They are ectoparasites of birds, feeding on particles of dead skin, feather fragments, and sometimes blood. The eggs hatch to form nymphs resembling the adults.

Malpighian body (Malpighian corpuscle) The part of a *nephron in the kidney that consists of its cup-shaped end together with the *glomerulus that it encloses. It is named after its discoverer, the Italian anatomist M. Malpighi (1628–94).

Malpighian layer (stratum germinativum) The innermost layer of the *epidermis of mammalian *skin, separated from the underlying dermis by a fibrous *basement membrane*. It is only in this layer of the epidermis that active cell division (*mitosis) occurs. As the cells produced by these divisions age and mature, they migrate upwards through the layers of the epidermis to replace the cells being continuously worn away at the surface.

maltose (malt sugar) A sugar consisting of two linked glucose molecules that results from the action of the enzyme amylase on starch. Maltose occurs in barley seeds following germination and drying, which is the basis of the malting process used in the manufacture of beer and malt whisky.

malt sugar *See* maltose.

Mammalia A class of vertebrates containing some 4250 species. Mammals are warm-blooded animals (*see* homoiothermy), typically having sweat glands whose secretion cools the skin and an insulating body covering of hair. All mammals have *mammary glands, which secrete milk to nourish the young. Mammalian teeth are differentiated into incisors, canines, premolars, and molars and the middle ear contains three sound-conducting *ear ossicles. The four-chambered heart enables complete separation of oxygenated and deoxygenated blood and a muscular *diaphragm takes part in breathing movements, both of which ensure that the tissues are well supplied with oxygen. This, together with well-developed sense organs and brain, have enabled mammals to pursue an active life and to colonize a wide variety of habitats.

Mammals evolved from carnivorous reptiles in the Triassic period about 225 million years ago. There are three subclasses: the primitive egg-laying *Prototheria (monotremes), the pouched *Metatheria (marsupials), and the *Eutheria (placental mammals), which includes most of the living orders.

mammary glands The milk-producing organs (possibly modified sweat glands) of female mammals, which provide food for the young (*see* milk; colostrum). Their number (2 to 20) and position (on the chest or abdomen) vary according to the species. In most mammals the gland openings project as a nipple or teat. Nipples have a number of milk-duct openings; teats have one duct leading from a storage cavity.

mandible **1.** One of a pair of horny *mouthparts in insects, crustaceans, centipedes, and millipedes. The mandibles lie in front of the weaker *maxillae and their lateral movements assist in biting and crushing the food. **2.** The lower jaw of vertebrates. **3.** Either of the two parts of a bird's beak.

mannitol A polyhydric alcohol, $CH_2OH(CHOH)_4CH_2OH$, derived from mannose or fructose. It is the main soluble sugar in fungi and lichens and an important carbohydrate reserve in brown algae. Mannitol is used as a sweetener in certain foodstuffs.

mannose A *monosaccharide, $C_6H_{12}O_6$, stereoisomeric with glucose, that occurs naturally only in polymerized forms called *mannans*. These are found in plants, fungi, and bacteria, serving as food energy stores.

manometer A device for measuring pressure differences, usually by the difference in height of two liquid columns. The simplest type is the U-tube manometer, which consists of a glass tube bent into the shape of a U. If a pressure to be measured is fed to one side of the U-tube and the other is open to the atmosphere, the difference in level of the liquid in the two limbs gives a measure of the unknown pressure.

mantle The fold of skin covering the dorsal surface of the body of molluscs, which extends into lateral flaps that protect the gills in the *mantle cavity* (the space between the body and mantle). The outer surface of the mantle secretes the shell (in species that have shells).

marsupials *See* Metatheria.

mast cells Large amoeboid cells that are found in vertebrate *connective tissue. They may be connected with the deposition of fat. In certain situations they produce *histamine, responsible for allergy symptoms. They also produce *heparin, an anticoagulant, which may prevent fibrin from forming clots in the intercellular spaces.

mastoid process An outgrowth from the temporal bone of the skull containing air cavities that communicate with the cavity of the middle ear. In man it is a route through which infection may spread from the middle ear.

matrix (in histology) The component of tissues (e.g. bone and cartilage) in which the cells of the tissue are embedded.

maxilla **1.** One of a pair of *mouthparts in insects, crustaceans, centipedes, and millipedes. They lie behind the *mandibles and their lateral movements assist in feeding. Crustaceans have two pairs of maxillae but in insects the second pair are fused together forming the *labium. **2.** One of a pair of large tooth-bearing bones in the upper jaw of vertebrates. In mammals they carry all the upper teeth except the incisors.

maximum permissible dose *See* dose.

meatus A small canal or passage in the body. An example is the *external auditory meatus* of the *outer ear in mammals, which connects the exterior opening to the eardrum.

mechanoreceptor A *receptor that responds to such mechanical stimuli as touch, sound, and pressure. The skin is rich in mechanoreceptors.

median eye (pineal eye) An eyelike structure, with a lens and retina, found on the top of the head of some lizards, *Sphenodon*, and the Cyclostomata (lampreys) as well as in many fossil vertebrates. Its function is unknown.

median lethal dose *See* LD_{50}.

mediastinum **1.** A membrane in the mid-line of the *thorax of mammals that separates the lungs. **2.** The space between the two lungs, which is occupied by the heart and oesophagus.

medulla **1.** (in zoology) The central tissue of various organs, including the adrenal glands (*adrenal medulla*) and kidneys (*renal medulla*). **2.** (in botany) *See* pith.

medulla oblongata Part of the vertebrate *brainstem, derived from the *hindbrain, that is continuous with the spinal cord. Its function is to regulate the reflex responses controlling respiration, heart beat, blood pressure, and other involuntary processes. It gives rise to many of the *cranial nerves.

medullary ray (ray) Any of the vertical plates of *parenchyma cells running radially through the cylinder of vascular tissue in the stems and roots of plants. Each may be one to many cells in width. *Primary medullary rays* occur in young plants and in those not showing secondary thickening; they pass from the cortex through to the pith. *Secondary medullary rays* are produced by the vascular *cambium and terminate in xylem and phloem tissues. Medullary rays store and transport food materials.

medullated nerve fibre A nerve fibre that is characterized by a *myelin sheath, which insulates the axon.

medusa The free-swimming stage in the life cycle of the *Coelenterata. Medusae are umbrella-shaped, with tentacles round the edge and the mouth in the centre underneath. They swim by pulsations of the body and reproduce sexually. In the Hydrozoa (e.g. *Hydra*) they alternate in the life cycle with *polyps, from which they are produced by budding. In the Scyphozoa, which includes all the common jellyfish, the medusa is the dominant form and the polyp is reduced or absent.

mega- Symbol M. A prefix used in the metric system to denote one million times. For example, 10^6 volts = 1 megavolt (MV).

megaphyll A type of foliage leaf in ferns and seed plants that has branched or parallel vascular bundles running through the lamina. The megaphylls of ferns are large pinnate leaves called *fronds*. A megaphyll was formerly called a *macrophyll*. *Compare* microphyll.

megaspore *See* sporophyll.

megasporophyll *See* sporophyll.

meiosis (reduction division) A type of cell division that gives rise to four reproductive cells (gametes) each with half the chromosome number of the parent cell. Two consecutive divisions occur (see illustration). In the first, *homologous chromosomes become paired and may exchange genetic material (*see* crossing over) before moving away from each other into separate daughter nuclei. This is the actual reduction division because each of the two nuclei so formed contains only half of the original chromosomes. The daughter nuclei then divide by mitosis and four *haploid cells are produced. *See also* prophase; metaphase; anaphase; telophase.

melanin Any of a group of polymers, derived from the amino acid tyrosine, that cause pigmentation of eyes, skin, and hair in vertebrates. Melanins are produced by specialized epidermal cells called *melanocytes*. Certain invertebrates, fungi, and microorganisms also produce melanin pigments. The 'ink' of the octopus and squid is a notable example. Hereditary *albinism is caused by the absence of the enzyme tyrosinase, which is necessary for melanin production.

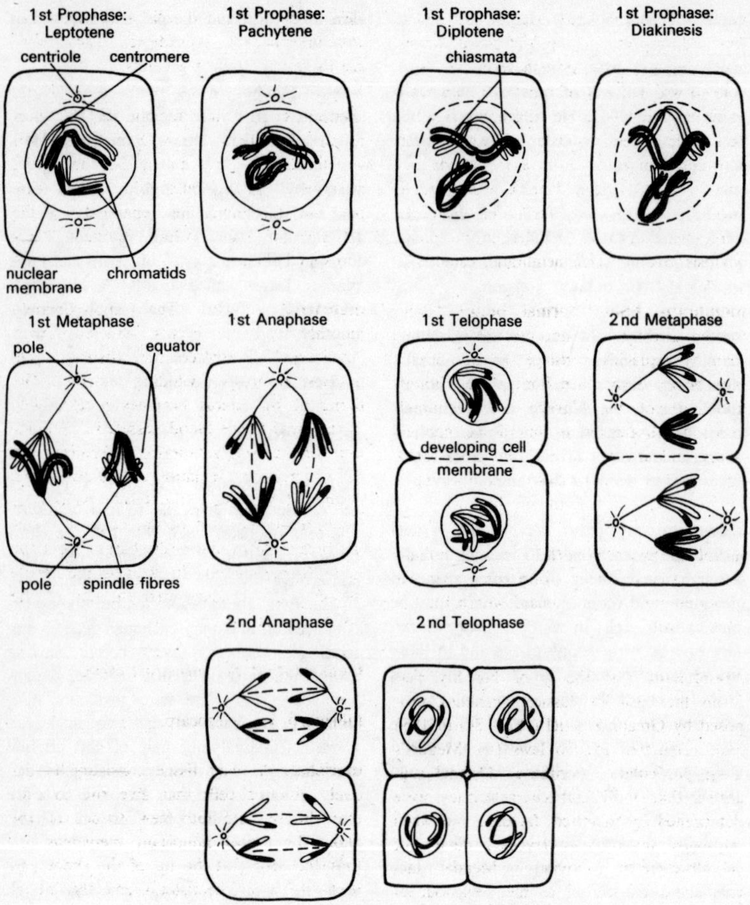

The stages of meiosis in a cell containing two pairs of homologous chromosomes

melanism Black coloration of the body caused by overproduction of the pigment melanin, often as a reaction to the environment. There are several species of melanic moths in industrially polluted areas (*see* industrial melanism) and the panther is a melanic form of leopard.

membrane A sheetlike tissue that covers, connects, or lines biological cells and their organelles, organs, and other structures. Membranes consist of polar lipids (predominantly the phospholipids lecithin and cephalin) and protein: the lipids form a double layer (bilayer) with protein molecules suspended within it. Membranes are permeable to water and fat-soluble substances but impermeable to polar substances (e.g. sugars). Movement of ions across a membrane occurs by carrier-mediated processes, including *active transport and *facilitated diffusion. Large molecules may be taken up by *phagocytosis. *See also* cell membrane; osmosis.

membrane bone (dermal bone) *Bone formed directly in connective tissue, rather than by replacing cartilage (*compare* cartilage bone). Some face bones, skull bones, and part of the clavicle are membrane bones. Small areas of membrane become jelly-like and attract calcium salts. Bone-forming cells break down these areas forming a bone lattice, which eventually fills in.

membranous labyrinth The soft tubular sensory structures that form the *inner ear of vertebrates and are housed within the bony labyrinth.

Mendelism The theory of heredity that forms the basis of classical *genetics, proposed by Gregor Mendel (1822–84) in 1866 and formulated in two laws (*see* Mendel's laws; particulate inheritance). Mendel suggested that individual characteristics were determined by inherited 'factors', and when improved microscopes revealed details of cell structure the behaviour of Mendel's factors could be related to the behaviour of chromosomes during *meiosis.

Mendel's laws Two laws summarizing Gregor Mendel's theory of inheritance (*see also* Mendelism). The *Law of Segregation* states that each hereditary characteristic is controlled by two 'factors' (now called *alleles), which segregate (separate) and pass into separate germ (reproductive) cells. The *Law of Independent Assortment* states that pairs of 'factors' segregate independently of each other when germ cells are formed (*see also* independent assortment). These laws are the foundation of genetics.

meninges The three membranes that surround the brain and spinal cord of vertebrates: the *pia mater, the *arachnoid membrane, and the outer *dura mater. The pia and arachnoid are separated by the *subarachnoid space*, which contains *cerebrospinal fluid.

menstrual cycle The approximately monthly cycle of events associated with *ovulation that replaces the *oestrous cycle in most primates (including humans). The lining of the uterus becomes progressively thicker with more blood vessels in preparation for the *implantation of a fertilized egg cell (blastocyst). Ovulation occurs during the middle of the cycle (the fertile period). If fertilization does not occur the uterine lining breaks down and is discharged from the body (*menstruation*). In women the fertile period is 11–15 days after the end of the last menstruation.

menstruation *See* menstrual cycle.

mericarp *See* schizocarp.

meristem A plant tissue consisting of actively dividing cells that give rise to cells that differentiate into new tissues of the plant. The most important meristems are those occurring at the tip of the shoot and root (*see* apical meristem) and the lateral meristems in the older parts of the plant (*see* cambium; cork cambium).

merocrine secretion *See* secretion.

mesencephalon *See* midbrain.

mesentery A thin sheet of tissue, bounded on each side by *peritoneum, that supports the gut and other organs in the body cavities of animals. Vertebrates have a well-developed dorsal mesentery that anchors the stomach and intestine and contains blood vessels and nerves supplying the gut. The reproductive organs and their ducts are also supported by mesenteries.

mesocarp *See* pericarp.

mesoderm The layer of cells in the *gastrula that lies between the *ectoderm and *endoderm. It develops into the muscles, circulatory system, and sex organs and in vertebrates also into the excretory system and skeleton. *See also* germ layers.

mesoglea The gelatinous noncellular layer between the endoderm and ectoderm in the body wall of coelenterates. It may be thin, as in *Hydra*, or tough and fibrous, as in the larger jellyfish and sea anemones. It often contains cells that have migrated from the two body layers but these do not form tissues and organs and the mesoglea is not homologous with the mesoderm of *triploblastic animals.

mesophyll The internal tissue of a leaf blade (lamina), consisting of *parenchyma cells. There are two distinct forms. *Palisade mesophyll* lies just beneath the upper epidermis and consists of cells elongated at right angles to the leaf surface. They contain a large number of *chloroplasts and their principal function is photosynthesis. *Spongy mesophyll* occupies most of the remainder of the lamina. It consists of spherical loosely arranged cells containing fewer chloroplasts than the palisade mesophyll. Between these cells are air spaces leading to the *stomata.

mesophyte Any plant adapted to grow in soil that is well supplied with water and mineral salts. Such plants wilt easily when exposed to drought conditions as they are not adapted to conserve water. The majority of flowering plants are mesophytes. *Compare* hydrophyte; xerophyte.

mesothelium A single layer of thin plate-like cells covering the surface of the inside of the abdominal cavity and thorax and surrounding the heart, forming part of the *peritoneum and *pleura (*see* serous membrane). It is derived from the *mesoderm. *Compare* endothelium; epithelium.

Mesozoic The geological era that extended from the end of the *Palaeozoic era, about 225 million years ago, to the beginning of the *Cenozoic era, about 65 million years ago. It comprises the *Triassic, *Jurassic, and *Cretaceous periods. The Mesozoic era is often known as the *Age of Reptiles* as these animals, which included the dinosaurs, pterosaurs, and ichthyosaurs, became the dominant lifeform; most became extinct before the end of the era.

messenger RNA *See* RNA.

metabolism The sum of the chemical reactions that occur within living organisms. The various compounds that take part in or are formed by these reactions are called *metabolites*. In animals many metabolites are obtained by the digestion of food, whereas in plants only the basic starting materials (carbon dioxide, water, and minerals) are externally derived. The synthesis (*anabolism) and breakdown (*catabolism) of most compounds occurs by a number of reaction steps, the reaction sequence being termed a *metabolic pathway*. Some pathways (e.g. *glycolysis) are linear; others (e.g. the *Krebs cycle) are cyclic. The changes at each step in a pathway are usually small and are promoted by efficient biological catalysts – the enzymes. In this way the amounts of energy required or released at any given stage are minimal, which helps in maintaining a constant *internal environment. Various *feedback mechanisms exist to govern metabolic rates.

metacarpal One of the bones in the *metacarpus.

metacarpus The hand (or corresponding part of the forelimb) in terrestrial vertebrates, consisting of a number of rod-shaped bones (*metacarpals*) that articulate with the bones of the wrist (*see* carpus) and those of the fingers (*see* phalanges). The number of metacarpals varies between species: in the basic *pentadactyl limb there are five, but this number is reduced in many species.

metameric segmentation (metamerism; segmentation) The division of an animal's body (except at the head region – *see* cephalization) into a number of compartments (*segments* or *metameres*) each containing the same organs. Metameric segmentation is most strongly marked in annelid worms (e.g. earthworms), in which the muscles, blood vessels, nerves, etc. are repeated in each segment. In these animals the segmentation is obvious both externally and internally. It also occurs internally in arthropods and in the embryonic development of all vertebrates, in which it is confined to parts of the muscular, skeletal, and nervous systems and does not show externally.

metamorphosis The rapid transformation from the larval to the adult form that occurs in the life cycle of many invertebrates and amphibians. Examples are the changes from a tadpole to an adult frog and from a pupa to an adult insect. Metamorphosis often involves considerable destruction of larval tissues by lysosomes, and in both insects and amphibians it is controlled by hormones.

metaphase The second stage of cell division, during which the membrane around the nucleus breaks down, the *spindle forms, and centromeres attach the chromosomes to the equator of the spindle. In the first metaphase of *meiosis pairs of chromosomes (bivalents) are attached, while in *mitosis and the second metaphase of meiosis, individual chromosomes are attached.

metaplasia The transformation of a tissue into a different type. This is an abnormal process; for example, metaplasia of the epithelium of the bronchi may be an early sign of cancer.

metatarsal One of the bones in the *metatarsus.

metatarsus The foot (or corresponding part of the hindlimb) in terrestrial vertebrates, consisting of a number of rod-shaped bones (*metatarsals*) that articulate with the bones of the ankle (*see* tarsus) and those of the toes (*see* phalanges). The number of metatarsals varies between species: in the basic *pentadactyl limb there are five, but this number is reduced in some species.

Metatheria A subclass of mammals containing the marsupials. The female bears an abdominal pouch (*marsupium*) into which the newly born young, which are in a very immature state, move to complete their development. They obtain nourishment from the mother's mammary teats. Modern marsupials are restricted to Australasia (where they include the kangaroos, koala bears, phalangers, and bandicoots) and America (the opossums). Marsupials evolved during the late Cretaceous period, 80 million years ago. In Australia, where the marsupials have been isolated for millions of years, they show the greatest diversity of form, having undergone *adaptive radiation to many of the niches occupied by placental mammals elsewhere. *Compare* Eutheria; Prototheria.

Metazoa The subkingdom of animals in which the body is composed of many cells grouped into tissues and coordinated by a nervous system. It comprises all vertebrates and invertebrates except the *Protozoa and *Porifera (sponges).

methionine *See* amino acid.

methylene blue A blue dye used in optical microscopy to stain nuclei of animal tissues. It is also suitable as a vital stain and a bacterial stain.

metre Symbol m. The SI unit of length that is equal to 39.37 inches. It is formally defined as the length of the path travelled by light in vacuum during a time interval of 1/299 792 458 of a second. This definition, adopted by the General Conference on Weights and Measures in October, 1983, replaced the 1967 definition based on the krypton lamp, i.e. 1 650 763.73 wavelengths in a vacuum of the radiation corresponding to the transition between the levels $2p^{10}$ and $5d^5$ of the nuclide krypton-86. This definition (in 1958) replaced the older definition of a metre based on a platinum-iridium bar of standard length.

micelle An aggregate of molecules in a *colloid. For example, phospholipids in aqueous solution form micelles – small clusters of molecules in which the nonpolar hydrocarbon groups are in the centre and the hydrophilic polar groups are on the outside.

micro- Symbol μ. A prefix used in the metric system to denote one millionth. For example, 10^{-6} metre = 1 micrometre (μm).

microbiology The scientific study of microorganisms (e.g. bacteria, viruses, and fungi). Originally this was directed towards their effects (e.g. in causing disease and decay), but during the 20th century the emphasis has shifted to their physiology and biochemistry. Microbes are now recognized as important vehicles for the study of biochemical processes common to all living organisms, and their rapid growth enables their laboratory culture in large numbers for studies in genetics.

microdissection (micromanipulation) A technique used for the dissection of living cells under the high power of an optical microscope. It utilizes minute mechanically manipulated instruments, such as needles, scalpels, and pipettes. For example, the instruments may be used to remove a single nucleus from one cell and to implant it in another.

microfossil A *fossil that is so small that it can only be studied under a microscope. Microfossils include bacteria, diatoms, and Protozoa and parts of organisms, such as plant pollen and skeletal fragments. Microfossils are important in the correlation of rocks where only small samples are available. The study of microfossils, particularly pollen, is known as *palynology.

micromanipulation *See* microdissection.

micronutrient A nutrient required by organisms in relatively small quantities. Micronutrients are typically found in cofactors and coenzymes. They include *vitamins and trace elements (*see* essential element). *Compare* macronutrient.

microorganism (microbe) Any organism that can be observed only with the aid of a microscope. Microorganisms include bacteria, viruses, rickettsiae, protozoans, and some algae and fungi. *See* microbiology.

microphyll A type of foliage leaf seen in certain pteridophytes (e.g. *Selaginella*, *Lycopodium*) that has a single unbranched midrib. Such leaves are generally no more than a few millimetres long. *Compare* megaphyll.

micropyle A small opening in the surface of a plant ovule through which the pollen tube passes prior to fertilization. It results from the incomplete covering of the nucellus by the integuments. It remains as an opening in the testa of most seeds through which water is absorbed.

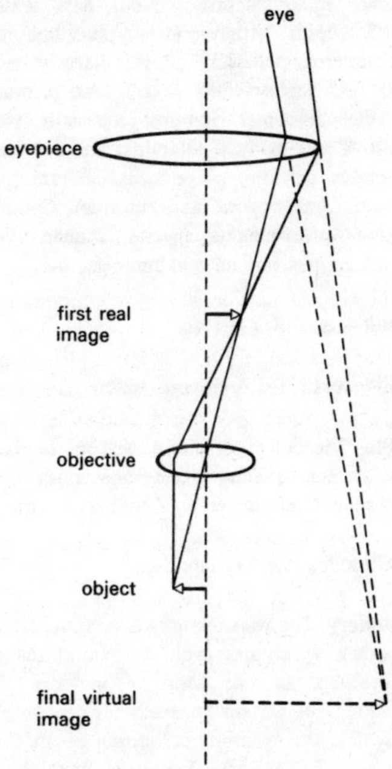

Compound microscope

microscope A device for forming a magnified image of a small object. The *simple microscope* consists of a biconvex magnifying glass or an equivalent system of lenses, either hand-held or in a simple frame. The *compound microscope* uses two lenses or systems of lenses, the second magnifying the real image formed by the first. The lenses are usually mounted at the opposite ends of a tube that has mechanical controls to move it in relation to the object. An optical condenser and mirror, often with a separate light source, provide illumination of the object. The widely used *binocular microscope* consists of two separate instruments fastened together so that one eye looks through one while the other eye looks through the other. This gives stereoscopic vision. *See also* electron microscope; phase-contrast microscope; ultraviolet microscope.

microspore *See* sporophyll.

microsporophyll *See* sporophyll.

microtome A machine used to cut thin sections (3–5 μm thick) of plant or animal tissue for microscopical observation. There are various designs of microtome, each basically consisting of a steel knife, a block for supporting the specimen, and a device for moving the specimen towards the knife. The specimen is usually supported by being embedded in wax; if a *freezing microtome* is used, the specimen is frozen. An *ultramicrotome* is used to cut much thinner sections (20–100 nm thick) for electron microscopy. The biological material is embedded in plastic or resin, sectioned with a glass or diamond knife, and the cut sections are allowed to float on the surface of water in an adjacent water bath.

microtubule A minute protein filament in living cells that occurs singly or in pairs, triplets, or bundles. Microtubules help certain cells (e.g. nerve cells) to maintain their shape; they also occur in *cilia, *flagella, and the *centrioles and form the *spindle during nuclear division.

microvillus One of a number of minute finger-like projections on the free surfaces of epithelial cells. Microvilli are covered with cell membrane and their cytoplasm is continuous with the main cell cytoplasm. Their purpose is probably to increase the absorptive or secretory surface area of the cell, and they are abundant on the villi of the intestine.

midbrain (mesencephalon) One of the three sections of the brain of a vertebrate embryo. Unlike the *forebrain and the *hindbrain, the midbrain does not undergo further subdivision to form additional zones.

In mammals it becomes part of the *brainstem, but in amphibians, reptiles, and birds the roof of the midbrain becomes enlarged as the *tectum*, a dominant centre for integration, and may include a pair of *optic lobes*.

middle ear (tympanic cavity) The air-filled cavity within the skull of vertebrates that lies between the *outer ear and the *inner ear. It is linked to the pharynx (and therefore to outside air) via the *Eustachian tube and in mammals contains the three *ear ossicles, which transmit auditory vibrations from the outer ear (via the *tympanum) to the inner ear (via the *fenestra ovalis).

middle lamella A thin layer of material, consisting mainly of pectins, that binds together the walls of adjacent plant cells.

midgut 1. The middle section of the alimentary canal of vertebrates, which is concerned with digestion and absorption. It comprises most of the small intestine. **2.** The middle section of the alimentary canal of arthropods. *See also* foregut; hindgut.

migration The seasonal movement of complete populations of animals to a more favourable environment. It is usually a response to lower temperatures resulting in a reduced food supply, and is often triggered by a change in day length (*see* photoperiodism). Migration is common in mammals (e.g. porpoises), fish (e.g. eels and salmon), and some insects but is most marked in birds. The Arctic tern, for example, migrates annually from its breeding ground in the Arctic circle to the Antarctic – a distance of some 17 600 km. Migrating animals appear to possess considerable powers of orientation, but the precise mechanisms by which they navigate are not known.

milk The fluid secreted by the *mammary glands of mammals. It provides a balanced and highly nutritious food for offspring. Cows' milk comprises about 87% water, 3.6% lipids (triglycerides, phospholipids, cholesterol, etc.), 3.3% protein (largely casein), 4.7% lactose (milk sugar), and, in much smaller amounts, vitamins (especially vitamin A and many B vitamins) and minerals (notably calcium, phosphorus, sodium, potassium, magnesium, and chlorine). Composition varies among species; human milk contains less protein and more lactose.

milk sugar *See* lactose.

milk teeth *See* deciduous teeth.

milli- Symbol m. A prefix used in the metric system to denote one thousandth. For example, 0.001 metre = 1 millimetre (mm).

millipedes *See* Myriapoda.

mimicry The resemblance of one animal to another, which has evolved as a means of protection. In one form of mimicry the markings of certain harmless insects closely resemble the *warning coloration of another insect (the *model*). Predators that have learnt to avoid the model will also avoid good mimics of it. This phenomenon is often found among butterflies. A second form of mimicry involves the mutual resemblance of a group of animals, all harmful, such as the wasp, bee, and hornet, so that a predator, having experienced one, will subsequently avoid them all.

mineralocorticoid *See* corticosteroid.

mitochondrion A structure within the cytoplasm of plant and animal *cells that carry out aerobic respiration: it is the site of the *Krebs cycle and *electron transport chain, and therefore the cell's energy production. Mitochondria vary greatly in shape, size, and number but are typically oval or sausage-shaped and bounded by two membranes, the inner one being folded into finger-like projections (*cristae*). They are most numerous in cells with a high level of metabolic activity.

mitosis The division of a cell to form two daughter cells each having a nucleus containing the same number and kind of chromosomes as the mother cell. The changes during divisions are clearly visible with a light microscope. Each chromosome divides lengthwise into two *chromatids, which separate and form the chromosomes of the resulting daughter nuclei. The process is divided into four stages, *prophase, *metaphase, *anaphase, and *telophase, which merge into each other (see illustration). Mitotic divisions ensure that all the cells of an individual are genetically identical to each other and to the original fertilized egg.

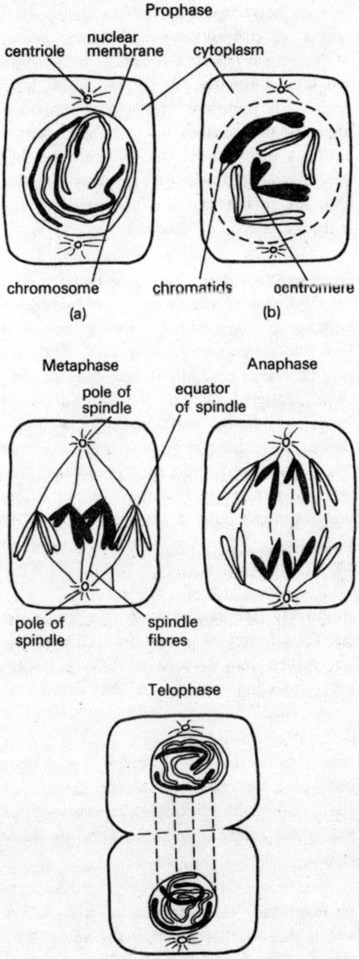

The stages of mitosis in a cell containing two pairs of homologous chromosomes

mitral valve (biscuspid valve) A valve, consisting of two flaps, situated between the left atrium and the left ventricle of the heart of birds and mammals. When the left ventricle contracts, forcing blood into the aorta, the mitral valve closes the aperture to the left atrium, thereby preventing any backflow of blood. The valve reopens to allow blood to flow from the atrium into the ventricle. *Compare* tricuspid valve.

mmHg A unit of pressure equal to that exerted under standard gravity by a height of one millimetre of mercury, or 133.322 pascals.

modern synthesis *See* neo-Darwinism.

molality *See* concentration.

molar 1. A broad ridged tooth in the adult dentition of mammals, found at the back of the jaws behind the premolars. There are two or more molars on each side of both jaws; their surfaces are raised into ridges or *cusps for grinding food during chewing. In man the third (and most posterior) molar does not appear until young adulthood: these molars are known as *wisdom teeth.* **2.** Denoting that an extensive physical property is being expressed per *amount of substance, usually per mole. For example, the molar heat capacity of a compound is the heat capacity of that compound per unit amount of substance, i.e. it is usually expressed in $J\ K^{-1}\ mol^{-1}$.

molarity *See* concentration.

mole Symbol mol. The SI unit of *amount of substance. It is equal to the amount of substance that contains as many elementary units as there are atoms in 0.012 kg of carbon–12. The elementary units may be atoms, molecules, ions, radicals, electrons, etc., and must be specified. 1 mole of a compound has a mass equal to its *relative molecular mass expressed in grams.

molecular biology The study of the structure and function of large molecules associated with living organisms, in particular proteins and the nucleic acids *DNA and *RNA.

Molisch's test *See* alpha-naphthol test.

Mollusca A phylum of soft-bodied invertebrates characterized by an unsegmented body differentiated into a *head*, a ventral muscular *foot* used in locomotion, and a dorsal *visceral hump* covered by a fold of skin – the *mantle – which secretes a protective shell in many species. Respiration is by means of gills or a lunglike organ and the feeding organ is a *radula. Molluscs occur in marine, freshwater, and terrestrial habitats and there are six classes, including the *Gastropoda (snails, slugs, etc.), *Lamellibranchia (bivalves), and *Cephalopoda (squids and octopuses).

Monera In some systems of classification, the kingdom that includes the bacteria, blue-green algae, and other microorganisms with prokaryotic *cells. *See also* Protista.

mongolism *See* Down's syndrome.

monochasium *See* cymose inflorescence.

monoclonal antibody A specific *antibody produced by one of numerous identical cells derived from a single parent cell. (The population of these cells comprises a *clone and each cell is said to be *monoclonal*.) The parent cell is obtained by the fusion of a normal antibody-producing cell (a lymphocyte) with a cell derived from a malignant tumour of *lymphoid tissue of a mouse. This hybrid cell then multiplies rapidly and yields large amounts of antibody. Monoclonal antibodies are used to identify a particular antigen within a mixture and can therefore be used for identifying blood groups; they also enable the production of highly specific, and therefore effective, *vaccines.

Monocotyledonae One of the two classes or subclasses of plants within the *Angiospermae, distinguished by having one seed leaf (*cotyledon) within the seed. The monocotyledons generally have parallel leaf veins, scattered vascular bundles within the stems, and flower parts in threes or multiples of three. Monocotyledon species include some crop plants (e.g. cereals, onions, fodder grasses), ornamentals (e.g. tulips, orchids, lilies), and a very limited number of trees (e.g. the palms). *Compare* Dicotyledonae.

monocyte The largest form of white blood cell (*leucocyte) in vertebrates. It has a kidney-shaped nucleus and is actively phagocytic, ingesting bacteria and cell debris (*see* phagocyte).

monoecious Describing plant species that have separate male and female flowers on the same plant. Examples of monoecious plants are maize and birch. *Compare* dioecious.

monohybrid The offspring of a cross between parents that differ in the alleles they possess for one particular gene, one parent having two dominant alleles and the other two recessives. All the offspring have one

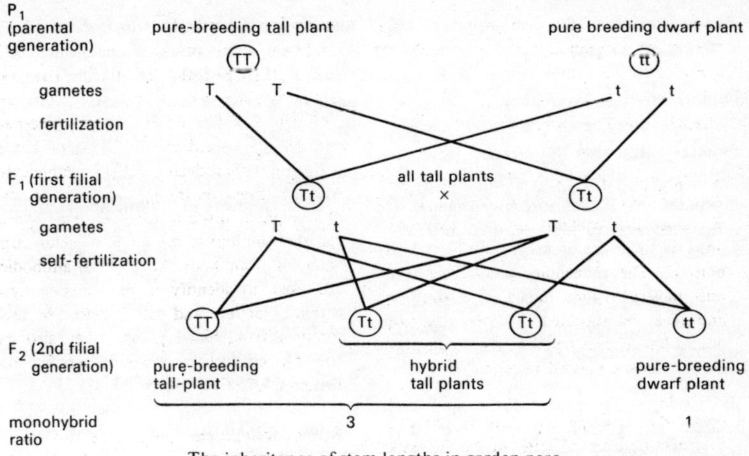

The inheritance of stem lengths in garden peas

dominant and one recessive allele for that gene (i.e. they are hybrid at that one locus). Crossing between these offspring yields a characteristic 3:1 (monohybrid) ratio in the following generation of dominant:recessive phenotypes (see illustration).

monophyletic Denoting any group of organisms that are assumed to have originated from the same ancestor, i.e. any family, class, etc., of a natural classification. Sometimes the term has a more limited meaning and designates only those groups that include *all* the descendants of a common ancestor. In this restricted sense the birds are considered monophyletic because they are the sole descendants of a group of arboreal Triassic reptiles but the modern reptiles are not, because their common amphibian ancestor also gave rise to the birds and mammals. Such groups as the reptiles are described as *paraphyletic*. *Compare* polyphyletic.

monophyodont Describing a type of dentition that consists of a single set of teeth that last for the entire lifespan of an animal. *Compare* diphyodont; polyphyodont.

monopodium The primary axis of growth in such plants as pine trees. It consists of a single main stem that continues to grow from the tip and gives rise to lateral branches. *Compare* sympodium.

monosaccharide (simple sugar) A carbohydrate that cannot be split into smaller units by the action of dilute acids. Monosaccharides are classified according to the number of carbon atoms they possess: *trioses* have three carbon atoms; *tetroses*, four; *pentoses*, five; *hexoses*, six; etc. Each of these is further divided into *aldoses* and *ketoses*, depending on whether the molecule contains an aldehyde group (–CHO) or a ketone group (–CO–). For example glucose, having six carbon atoms and an aldehyde group, is an *aldohexose* whereas fructose is a *ketohexose*. These aldehyde and ketone groups confer reducing properties on monosaccharides: they can be oxidized to yield sugar acids. They also react with phosphoric acid to produce phosphate esters (e.g. in *ATP), which are important in cell metabolism. Monosaccharides can exist as either straight-chain or ring-shaped molecules. They also exhibit *optical activity, giving rise to both dextrorotatory and laevorotatory forms.

monotremes *See* Prototheria.

monozygotic twins *See* identical twins.

morphine An alkaloid present in opium. It is an analgesic and narcotic, used medically for the relief of severe pain.

morphogenesis The development, through growth and differentiation, of form and structure in an organism.

morphology The study of the form and structure of organisms, especially their external form. *Compare* anatomy.

mosses *See* Musci.

moths *See* Lepidoptera.

motivation The internal conditions responsible for temporary reversible changes in the responsiveness of an animal to external stimulation. Thus an animal that has been deprived of food will accept less palatable food than one that has not been deprived: the difference is attributed to a change in feeding motivation. Changes in responsiveness due to maturation, *learning, or injury are not usually readily reversible and are therefore not considered to be due to changes in motivation. Early attempts to describe motivation in terms of a number of separate 'drives' (e.g. food drive, sex drive) have not found general favour, partly because 'drives' interact with one another; for example, water deprivation often affects an animal's willingness to feed.

motor neurone A *neurone that transmits nerve impulses from the central nervous system to an effector organ (such as a muscle or gland) and thereby initiates a physiological response (e.g. muscle contraction).

moulting 1. The seasonal loss of hair, fur, or feathers that occurs in mammals and birds. **2.** The periodic loss of the integument of arthropods and reptiles. *See* ecdysis.

mouthparts Modified paired appendages on the head segments of arthropods, used for feeding. A typical insect has a *labium (lower lip), one pair each of *mandibles and *maxillae, and a *labrum (upper lip), although in many the mouthparts are modified to form piercing stylets or a sucking proboscis. Crustaceans, centipedes, and millipedes have one pair of mandibles and two pairs of maxillae used for cutting and holding the food. Crustaceans also have several pairs of *maxillipedes.*

mucilage A gumlike substance frequently present in the cell walls of aquatic plants and in the seed coats of certain other species. Mucilages are hard when dry and slimy when wet. Like *gums they probably have a general protective function or serve to anchor the plant. Some organisms (e.g. certain bacteria) are completely covered with mucilage and in such cases it probably prevents water loss.

mucous membrane A layer of tissue comprising an epithelium supported on connective tissue. Within the epithelium are special *goblet cells*, which secrete *mucus onto the surface, and the epithelium often bears cilia. Mucous membranes line body cavities communicating with the exterior, including the alimentary and respiratory tracts. *Compare* serous membrane.

mucus The slimy substance secreted by goblet cells onto the surface of a *mucous membrane to protect and lubricate it and to trap bacteria, dust particles, etc. Mucus consists of water, the glycoprotein *mucin*, cells, and salts.

multicellular Describing tissues, organs, or organisms that are composed of a number of cells. *Compare* unicellular.

multifactorial inheritance The determination of a particular characteristic, e.g. height or skin colour, by many genes, each having a small effect individually. Characteristics controlled in this way show *continuous variation.

Musci A class of plants within the *Bryophyta comprising the mosses, which are found in both damp (including freshwater) and drier situations. Mosses possess erect or prostrate leafy stems, which give rise to leafless stalks bearing capsules. Spores formed in the capsules are released and grow to produce new plants. A common moss is *Funaria*.

muscle A tissue consisting of sheets or bundles of cells (*muscle fibres*) that are capable of contracting, so producing movement or tension in the body. There are three types of muscle. *Voluntary muscle produces voluntary movement (e.g. at joints); *involuntary muscle mainly effects the movements of hollow organs (e.g. intestine and bladder); and *cardiac muscle occurs only in the heart.

mutagen An agent that causes an increase in the number of mutants (*see* mutation) in a population. Mutagens operate either by causing changes in the DNA of the *genes, so interfering with the coding system, or by causing chromosome damage. Various chemicals (e.g. *colchicine) and forms of radiation (e.g. X-rays) have been identified as mutagens.

mutation A sudden random change in the genetic material of a cell that may cause it and all cells derived from it to differ in appearance or behaviour from the normal type. An organism affected by a mutation (especially one with visible effects) is described as a *mutant*. *Somatic mutations* affect the nonreproductive cells and are therefore restricted to the tissues of a single organism but *germ-line mutations*, which occur in the reproductive cells or their precursors, may be transmitted to the organism's descendants and cause abnormal development.

Mutations occur naturally at a low rate but this may be increased by radiation and by some chemicals (*see* mutagen). Most (the *gene mutations*) consist of invisible changes in the DNA of the chromosomes, but some (the *chromosome mutations*) affect the appearance or the number of the chromosomes. An example of a chromosome mutation is that giving rise to *Down's syndrome.

The majority of mutations are harmful, but a very small proportion may increase an organism's *fitness; these spread through the population over successive generations by natural selection. Mutation is therefore essential for evolution, being the ultimate source of genetic variation.

mutualism An interaction between two species in which both species benefit. (The term *symbiosis is often used synonymously with mutualism.) A well-known example of mutualism is the association between termites and the specialized protozoans that inhabit their guts. The protozoans, unlike the termites, are able to digest the cellulose of the wood that the termites eat and release sugars that the termites absorb. The termites benefit by being able to use wood as a foodstuff, while the protozoans are supplied with food and a suitable environment.

mycelium A network of *hyphae that forms the body of a fungus. It consists of feeding hyphae together with reproductive hyphae, which produce *sporangia and *gametangia.

mycology The scientific study of *fungi.

Mycota *See* fungi.

myelin sheath (medullary sheath) The layer of fatty material that surrounds and electrically insulates the axons of most vertebrate and some invertebrate neurones. The myelin sheath enables a more rapid transmission of nerve impulses (at speeds up to 120 $m\,s^{-1}$). It consists of layers of membrane derived from *Schwann cells. The sheath is interrupted at intervals along the axon by *nodes of Ranvier*; myelinated sections of axon are called *internodes*.

myeloid tissue Tissue within red *bone marrow that produces the blood cells. It is found around the blood vessels and contains various cells that are precursors of the blood cells.

myoglobin A globular protein occurring widely in muscle tissue as an oxygen carrier. It comprises a single polypeptide chain and a *haem group, which reversibly binds a molecule of oxygen. This is only relinquished at relatively low external oxygen concentrations, e.g. during strenuous exercise when muscle oxygen demand outpaces supply from the blood. Myoglobin thus acts as an emergency oxygen store.

myopia Short-sightedness. It results from the lens of the eye refracting the parallel rays of light entering it to a focus in front of the retina generally because of an abnormally long eyeball. The condition is corrected by using diverging spectacle lenses to move the image back to the retina.

Myriapoda A group of terrestrial arthropods generally found in damp conditions, comprising centipedes (class Chilopoda), millipedes (class Diplopoda), pauropods (class Pauropoda), and symphylids (class Symphyla). Myriapods characteristically possess a distinct head with a single pair of antennae and a wormlike body with numerous similar segments, each bearing one or two pairs of legs. Young myriapods may hatch with fewer segments than the adult, gaining more with each successive moult. Centipedes are fast-moving predators with 15–177 body segments, relatively long antennae, and the first pair of appendages specialized as poison fangs; the remaining segments each bear a single pair of legs. Millipedes are slow moving and feed on decaying leaves. The antennae are short and the body consists of 20 to over 60 segments, each bearing two pairs of legs.

Myxomycota *See* slime fungi.

myxovirus One of a group of RNA-containing viruses associated with various diseases of man and other vertebrates. *Orthomyxoviruses* produce diseases of the respiratory tract, e.g. influenza; *paramyxoviruses* include the causal agents of mumps, measles, and fowl pest.

N

NAD (nicotinamide adenine dinucleotide) A *coenzyme, derived from the B vitamin *nicotinic acid, that participates in many biological dehydrogenation reactions. NAD is characteristically loosely bound to the enzymes concerned. It normally carries a positive charge and can accept one hydrogen atom and two electrons to become the reduced form, *NADH*. NADH is generated during the oxidation of food, especially by the reactions of the *Krebs cycle. It then gives up its two electrons (and single proton) to the *electron transport chain, thereby reverting to NAD^+ and generating three molecules of ATP per molecule of NADH.

NADP (*nicotinamide adenine dinucleotide phosphate*) differs from NAD only in possessing an additional phosphate group. It functions in the same way as NAD although anabolic reactions (*see* anabolism) generally use NADPH (reduced NADP) as a hydrogen donor rather than NADH. Enzymes tend to be specific for either NAD or NADP as coenzyme.

nano- Symbol n. A prefix used in the metric system to denote 10^{-9}. For example, 10^{-9} metre = 1 nanometre (nm).

narcotic Any drug that induces stupor and relieves pain, especially morphine and other *opiates. Such narcotics are addictive and cause dependence, and their medical use is strictly controlled.

nares (nostrils) The paired openings of the *nasal cavity in vertebrates. All vertebrates have *external nares*, which open to the exterior; in some species these are situated on a *nose. *Internal nares* (or *choanae*) are present only in air-breathing vertebrates (including lungfish) and open into the mouth cavity. In mammals they open posteriorly, beyond the secondary *palate.

nasal cavity The cavity in the head of a vertebrate that is lined by a membrane rich in sensitive olfactory receptors (*see* olfaction). It is connected to the exterior by external nostrils and (in air-breathing vertebrates) to the respiratory system by internal *nares.

nastic movements Movements of plant organs in response to external stimuli that are independent of the direction of the stimuli. Examples are the opening of crocus and tulip flowers in response to a rise in temperature (*thermonasty*), the opening of evening primrose flowers at night (*photonasty*), and the folding up and drooping of leaves of the sensitive plant (*Mimosa pudica*) when lightly touched (*seismonasty*). *Compare* tropism. *See also* nyctinasty.

natural group A group of organisms of any taxonomic rank that are believed to be descended from a common ancestor (*see* monophyletic). For example, man and the apes are often regarded as a natural group, descended from fossil ancestors, the dryopithecines (or their near relatives). In an ideal natural classification all taxa should be natural groups. *See also* cladistics.

natural order In the classification of plants, the former name for a *family.

natural selection The process that, according to *Darwinism, brings about the evolution of new species of animals and plants. Darwin noted that the size of any population tends to remain constant despite the fact that more offspring are produced than are needed to maintain it. He also saw that variations existed between individuals of the population and concluded that disease, competition, and other forces acting on the population eliminated those individuals less well adapted to their environment. The survivors would pass on any inheritable advantageous characteristics (i.e. characteristics with survival value) to their offspring and in time the composition of the population would change in adaptation to a changing environment. Over a long period of time this process could give rise to organisms so different from the original population that new species are formed. *See also* adaptive radiation. *Compare* punctuated equilibrium.

nature and nurture The combined effects of inherited factors (nature) and environmental factors (nurture) on the development of an organism. The genetic potential of an organism will only be realized under appropriate environmental conditions. *See also* phenotype.

Neanderthal man A form of fossil man that appeared in Europe and Western Asia about 100 000 years ago. He was thought to be a different species (*Homo neanderthalensis*) from modern man but is now generally regarded as a subspecies of *H. sapiens*. The fossil remains indicate that Neanderthals were fairly short and had low brows but that the brain size was the same as modern man's. They were nomadic hunters and are thought to have had a language. *Cromagnon began to replace Neanderthal man about 35 000 years ago. The name is derived from the site in the Neander valley, West Germany, where fossils were found in 1856.

near point The nearest point at which the human eye can focus an object. As the lens becomes harder with age, the extent to which accommodation can bring a near object into focus decreases. Therefore with advancing age the near point recedes – a condition known as *presbyopia.

nectar A sugary liquid produced in plants by *nectaries*, regions of secretory cells on the receptacle or other parts of a flower. It attracts pollinating insects or other animals.

negative feedback *See* feedback.

nekton *Pelagic organisms that actively swim through the water. Examples are fish, jellyfish, turtles, and whales. *Compare* plankton.

nematoblast *See* thread cell.

Nematoda A phylum of invertebrates comprising the roundworms. They are characterized by a smooth narrow cylindrical unsegmented body tapered at both ends. They shed their tough outer cuticle four times during life to allow growth. The microscopic free-living forms are found in all parts of the world, where they play an important role in the destruction and recycling of organic matter. The many parasitic nematodes are much larger; they include the filaria (*Wuchereria*) and Guinea worm (*Dracunculus*), which cause serious diseases in man.

neo-Darwinism (modern synthesis) The current theory of the process of *evolution, formulated between about 1920 and 1950, that combines evidence from classical genetics with the Darwinian theory of evolution by *natural selection (*see* Darwinism). It makes use of modern knowledge of genes and chromosomes to explain the source of the genetic variation upon which selection works. This aspect was unexplained by traditional Darwinism.

neo-Lamarckism Any of the comparatively modern theories of evolution based on Lamarck's theory of the inheritance of acquired characteristics (*see* Lamarckism). These include the unfounded dogma of *Lysenkoism and the recent controversial experiments on the inheritance of acquired immunological tolerance in mice.

Neolithic The New Stone Age, beginning in the Middle East approximately 9000 BC and lasting until 6000 BC, during which man first developed agriculture. Grinding and polishing of stone tools was also practised.

neoplasm Any new abnormal growth of cells, forming either a harmless (benign) tumour or a malignant one (*see* cancer).

neoteny The retention of the juvenile body form, or particular features of it, in a mature animal. For example, the axolotl, a salamander, retains the gills of the larva in the adult. Neoteny is thought to have been an important mechanism in the evolution of certain groups, such as man, who is believed to have developed from the juvenile form of apes.

nephron The excretory unit of the vertebrate *kidney (see illustration). Many constituents of the blood are filtered from the *glomerulus into the Bowman's capsule at one end of the nephron. The glomerular filtrate passes along the length of the nephron and most of its water, plus some salts, glucose, and amino acids, are reabsorbed into the surrounding blood capillaries. The re-

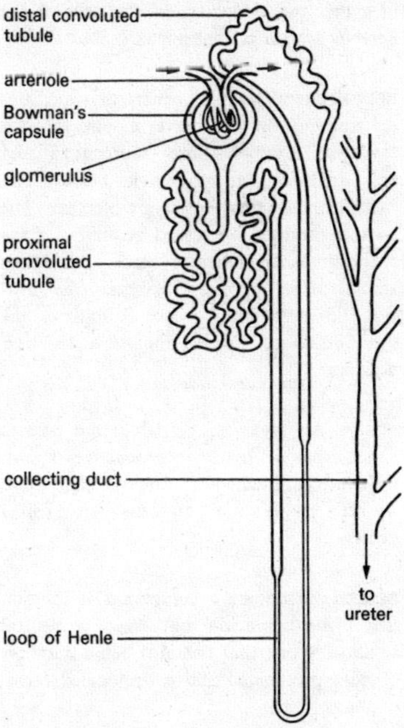

Structure of a single nephron

sulting concentrated solution of nitrogen-containing wastes (*urea) plus inorganic salts passes into the collecting ducts of the nephrons and is discharged as *urine into the *ureter.

neritic zone The region of the sea over the continental shelf, which is less than 200 metres deep (approximately the maximum depth for organisms carrying out photosynthesis). *Compare* oceanic zone.

nerve A strand of tissue comprising many *nerve fibres plus supporting tissues, enclosed in a connective-tissue sheath. Nerves connect the central nervous system with the organs and tissues of the body. A nerve may carry only motor nerve fibres (*motor nerve*) or only sensory fibres (*sensory nerve*) or it may be mixed and carry both types (*mixed nerve*). Although the nerve fibres are in close proximity within the nerve, their physiological responses are independent of each other.

nerve cell *See* neurone.

nerve cord A large bundle of nerve fibres, running down the longitudinal axis of the body, that forms an important part of the *central nervous system. Most invertebrates have a pair of solid nerve cords, situated ventrally and bearing segmentally arranged *ganglia. All animals of the phylum *Chordata have a dorsal hollow nerve cord; in vertebrates this is the *spinal cord.

nerve fibre The *axon of a *neurone together with the tissues associated with it (such as a *myelin sheath). The length and diameter of nerve fibres are very variable, even within the same organism. *See also* giant fibre.

nerve impulse *See* impulse.

nerve net A network of nerve cells connected with each other by synapses or fusion. The nervous system of certain invertebrates (e.g. coelenterates and echinoderms) consists exclusively of a nerve net in the body wall.

nervous system The system of cells and tissues in multicellular animals by which information is conveyed between sensory cells and organs and effectors (such as muscles and glands). It consists of the *central nervous system (in vertebrates the *brain and *spinal cord; in invertebrates the *nerve cord and *ganglia) and the *peripheral nervous system. Its function is to receive, transmit, and interpret information and then to formulate appropriate responses for the effector organs. It also serves to coordinate responses that require more than one physiological process. Nervous tissue consists of *neurones, which convey the information in the form of *impulses, and supporting tissue.

neuroendocrine system Any of the systems of dual control of certain activities in the body of some higher animals by nervous and hormonal stimulation. For example, the posterior *pituitary gland and the medulla of the *adrenal gland receive direct nervous stimulation to secrete their hormones, whereas the anterior pituitary gland is stimulated by factors (*neurosecretions*) released from the hypothalamus.

neurohormone Any hormone that is produced not by an endocrine gland but by a specialized nerve cell and is secreted from nerve endings into the bloodstream or directly to the tissue or organ whose growth or function it controls. Examples of neurohormones are *noradrenaline, *vasopressin, and hormones associated with metamorphosis and moulting in insects (*see* ecdysone, juvenile hormone).

neurone (nerve cell) An elongated branched cell that is the fundamental unit of the *nervous system, being specialized for the conduction of *impulses. A neurone consists of a *cell body, containing the nucleus and *Nissl granules; *dendrites, which receive incoming impulses and pass them towards the cell body; and an *axon, which conducts impulses away from the cell body, sometimes over long distances. Impulses are passed from one neurone to the next via *synapses. *Sensory neurones* transmit information from *receptors to the central nervous system. *Motor neurones* conduct information from the central nervous system to *effectors (e.g. muscles).

neurone theory The hypothesis, now accepted, that the nervous system consists of nerve cells (*neurones), which are functionally linked at *synapses but are physically separate. This theory has superseded the idea that the cytoplasm of the cells of the nervous system is continuous.

neurotransmitter A chemical that mediates the transmission of a nerve impulse across a *synapse. Examples are *adrenaline and *noradrenaline (in adrenergic nerves) and *acetylcholine (in cholinergic nerves). The neurotransmitter is released at the tip of the *axon into the synaptic space. It diffuses across to the opposite membrane (the postsynaptic membrane), where it initiates the transmission of a nerve impulse in the next neurone.

neuter An organism that does not possess either male or female reproductive organs. Cultivated ornamental flowers that have neither pistils nor stamens are called neuters.

neutral Describing a compound or solution that is neither acidic nor basic. A neutral solution is one that contains equal numbers of both protonated and deprotonated forms of the solvent.

niacin *See* nicotinic acid.

niche *See* ecological niche.

nicotinamide adenine dinucleotide *See* NAD.

nicotine A colourless poisonous *alkaloid present in tobacco. It is used as an insecticide.

nicotinic acid (niacin) A vitamin of the *vitamin B complex. It can be manufactured by plants and animals from the amino acid tryptophan. The amide derivative, nicotinamide, is a component of the coenzymes *NAD and NADP. These take part in many metabolic reactions as hydrogen acceptors. Deficiency of nicotinic acid causes the disease pellagra in humans. Apart from

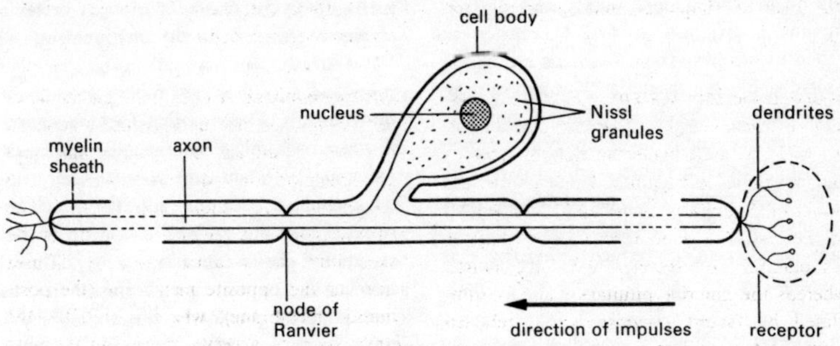

Sensory neurone

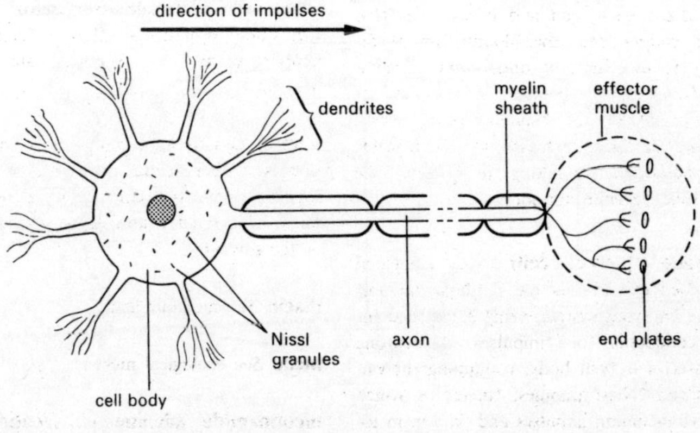

Motor neurone

tryptophan-rich protein, good sources are liver and groundnut and sunflower meals.

nictitating membrane A clear membrane forming a third eyelid in amphibians, reptiles, birds, and some mammals (but not man). It can be drawn across the cornea independently of the other eyelids, thus clearing the eye surface and giving added protection without interrupting the continuity of vision.

nidation *See* implantation.

Nissl granules (Nissl bodies) Particles seen within the cell bodies of *neurones. They are rich in RNA and stain strongly with basic dyes. They are named after F. Nissl (1860–1919), the German neurologist who discovered them.

nitrification A chemical process in which nitrogen (mostly in the form of ammonia) in plant and animal wastes and dead remains is oxidized at first to nitrites and then to nitrates. These reactions are effected mainly by the bacteria *Nitrosomonas* and *Nitrobacter* respectively. Unlike ammonia, nitrates are readily taken up by plant roots; nitrification is therefore a crucial part of the *nitrogen cycle. Nitrogen-containing compounds are often applied to soils deficient in this element, as fertilizer. *Compare* denitrification.

nitrogen Symbol N. A colourless gaseous element that occurs in air (about 78% by volume) and is an essential constituent of proteins and nucleic acids in living organisms (*see* nitrogen cycle).

nitrogen cycle One of the major cycles of chemical elements in the environment. Nitrates in the soil are taken up by plant roots and may then pass along *food chains into animals. Decomposing bacteria convert nitrogen-containing compounds (especially ammonia) in plant and animal wastes and dead remains back into nitrates, which are released into the soil and can again be taken up by plants (*see* nitrification). Though nitrogen is essential to all forms of life, the huge amount present in the atmosphere is not directly available to most organisms (*compare* carbon cycle). It can, however, be assimilated by some specialized bacteria and algae (*see* nitrogen fixation) and is thus made available to other organisms indirect-

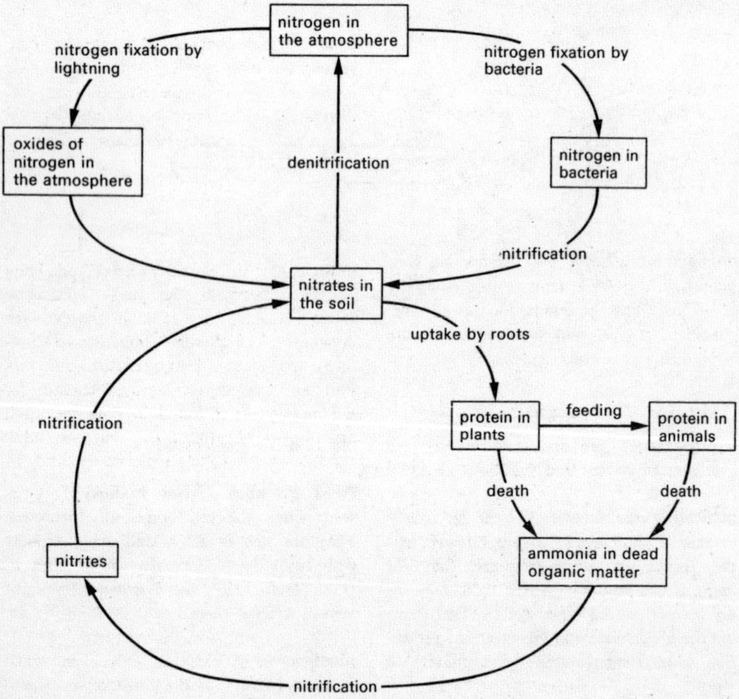

The nitrogen cycle

ly. Lightning flashes also make some nitrogen available to plants by causing the combination of atmospheric nitrogen and oxygen to form oxides of nitrogen, which enter the soil and form nitrates. Some nitrogen is returned from the soil to the atmosphere by denitrifying bacteria (*see* denitrification).

nitrogen fixation A chemical process in which atmospheric nitrogen is assimilated into organic compounds in living organisms and hence into the *nitrogen cycle. The ability to fix nitrogen is limited to certain bacteria (e.g. *Azotobacter*) and blue-green algae (e.g. *Anabaena*). *Rhizobium* bacteria are able to fix nitrogen in association with cells in the roots of leguminous plants such as peas and beans, in which they form characteristic root nodules; cultivation of legumes is therefore one way of increasing soil nitrogen. Various chemical processes are used to fix atmospheric nitrogen in the manufacture of fertilizers. These include the *Birkeland–Eyde process, the cyanamide process (*see* calcium dicarbide), and the *Haber process.

nitrogenous base A basic compound containing nitrogen. The term is used especially of organic ring compounds, such as adenine, guanine, cytosine, and thymine, which are constituents of nucleic acids.

NMR *See* nuclear magnetic resonance.

node 1. (in botany) The part of a plant stem from which one or more leaves arise. The nodes at the stem apex are very close together and remain so in species of monocotyledons that form bulbs. In older regions of the stem they are separated by areas of stem called *internodes*. **2.** (in anatomy) A natural thickening or bulge in an organ or part of the body. Examples are the *sinoatrial node* that controls the heartbeat (*see* pacemaker) and the *lymph nodes.

node of Ranvier *See* myelin sheath.

nomad (in cytology) A cell that migrates or wanders from its site of formation. Certain types of *phagocytes are nomads.

noradrenaline (norepinephrine) A hormone produced by the *adrenal glands and also secreted from nerve endings in the *sympathetic nervous system as a chemical transmitter of nerve impulses (*see* neurotransmitter). Many of its general actions are similar to those of *adrenaline, but it is more concerned with maintaining normal body activity than with preparing the body for emergencies.

norepinephrine *See* noradrenaline.

nose The protuberance on the face of some vertebrates that contains the nostrils (*see* nares) and part of the *nasal cavity. It therefore forms part of the olfactory system (*see* olfaction) and the external opening of the respiratory system.

nostrils *See* nares.

notochord An elastic skeletal rod lying lengthwise beneath the nerve cord and above the alimentary canal in the embryos or adults of all chordate animals (*see* Chordata). Its function is to strengthen and support the body and act as a protagonist for the muscles. It is found in both the adult and larva of *Amphioxus* but in adult vertebrates it is largely replaced by the *vertebral column.

nucellus The tissue that makes up the greater part of the ovule of seed plants. It contains the *embryo sac and nutritive tissue. It is enclosed by the integuments except for a small gap, the *micropyle. In certain flowering plants it may persist after fertilization and provide nutrients for the embryo.

nuclear magnetic resonance (NMR) The absorption of electromagnetic radiation (radio waves) by certain atomic nuclei placed in a strong magnetic field. The main application is in a form of spectroscopy (*NMR spectroscopy*) used for chemical and biochemical analysis and structure determination. The usual technique is to irradiate a sample with radio waves at a fixed frequency and subject it to a strong magnetic field, which can be varied in a controlled way. As the field changes absorption of radiation occurs at certain points, and this produces oscillations in the field, which can be detected. The ^{1}H nucleus is the one commonly studied; other suitable nuclei are ^{13}C, ^{14}N, and ^{19}F, although these have lower natural abundance than hydrogen and produce weaker signals. The spectrum produced is characteristic of the molecule absorbing the radiation. In medicine, NMR *tomography has been developed, in which images of tissue are produced by magnetic-resonance techniques.

nucleic acid A complex organic compound in living cells that consists of a chain of *nucleotides. There are two types: *DNA (deoxyribonucleic acid) and *RNA (ribonucleic acid).

nucleoid (nuclear region) The part of a cell of a bacterium or a blue-green alga (i.e. a prokaryotic *cell) that contains the genetic material *DNA and therefore controls the activity of the cell. It corresponds to the nucleus of the more advanced eukaryotic cells but is not bounded by a membrane.

nucleolus A small dense round body within the nondividing *nucleus of plant and animal cells that consists of protein, DNA, and ribosomal *RNA. It plays an important role in *ribosome manufacture (and therefore protein synthesis).

nucleoprotein Any compound present in cells of living organisms that consists of a nucleic acid (DNA or RNA) combined with a protein. Chromosomes consist of nucleoprotein (DNA and proteins, mostly histones), as do ribosomes (RNA and protein).

nucleoside An organic compound consisting of a nitrogen-containing *purine or *pyrimidine base linked to a sugar (ribose or deoxyribose). An example is *adenosine. *Compare* nucleotide.

nucleotide An organic compound consisting of a nitrogen-containing *purine or *pyrimidine base linked to a sugar (ribose or deoxyribose) and a phosphate group. *DNA and *RNA are made up of long chains of nucleotides (i.e. *polynucleotides*). *Compare* nucleoside.

nucleus The large body embedded in the cytoplasm of all plant and animal *cells (but not the cells of bacteria or blue-green algae) that contains the genetic material *DNA. The nucleus functions as the control centre of the cell. It is bounded by a double membrane (the *nuclear membrane* or *envelope*), which contains many pores for the exchange of material between the nucleus and cytoplasm, and contains a viscous sap (*nucleoplasm*). When the cell is not dividing a *nucleolus is present in the nucleus and the genetic material is dispersed in the sap as *chromatin; in dividing cells the chromatin is organized into *chromosomes and the nucleolus disappears. In certain protozoans there are two nuclei per cell, one concerned with vegetative functions and the other with sexual reproduction.

numerical taxonomy *See* taxonomy.

nut A dry single-seeded fruit that develops from more than one carpel and does not shed its seed when ripe. The fruit wall is woody or leathery. Many nuts are enclosed in a hard or membranous cup-shaped structure, the *cupule. The term nut is often loosely used of any hard fruit. For example, the walnut and coconut are in fact *drupes and the Brazil nut is a seed.

nutation The spiral movement of a plant organ during growth, also known as *circumnutation.* It is seen in climbing plants and helps the plant find a suitable support to twine around. Examples are the coiling movements of the shoot tips of runner beans and of the tendrils of sweet peas.

nyctinasty (sleep movements) *Nastic movements of plant organs in response to the changes in light and temperature that occur between day and night (and vice versa). Examples are the opening and closing of many flowers and the folding together of the leaflets of clover and other plants at night.

nymph The juvenile stage of certain insects, such as dragonflies, grasshoppers, and earwigs, which resembles the adult except that the wings and reproductive organs are undeveloped. There is no pupal stage, and the nymph develops directly into the adult. *Compare* larva.

O

objective The lens or system of lenses nearest to the object being examined through an optical instrument.

occipital condyle A single or paired bony knob that protrudes from the occipital bone of the skull and articulates with the first cervical vertebra (the *atlas). In humans there is a pair of occipital condyles, one on each side of the *foramen magnum. Occipital condyles are absent in most fish, which cannot move their heads.

oceanic zone The region of the open sea beyond the edge of the continental shelf, where the depth is greater than 200 metres. *Compare* neritic zone.

ocellus A simple eye occurring in insects and other invertebrates. It typically consists of light-sensitive cells and a single cuticular lens.

ocular *See* eyepiece.

Odonata An order of insects containing the dragonflies and damselflies, most of which occur in tropical regions. Adult dragonflies have a pair of prominent *compound eyes, a compact thorax bearing two pairs of delicate membranous wings, and a long slender abdomen. They are strong fliers and prey on other insects, either in flight or at rest. The eggs are laid near or in water, and the newly hatched nymphs are aquatic and resemble the adults, with rudimentary wings. They breathe through gills and feed on small aquatic animals. The nymph leaves the water for its final moult into the terrestrial adult.

odontoblast A cell that is responsible for producing the *dentine of vertebrate teeth. Odontoblasts are found around the lining of the *pulp cavity and have processes that extend into the dentine.

oesophagus The section of the *alimentary canal that lies between the *pharynx and the stomach. It is a muscular tube whose function is to transfer food to the stomach by means of wavelike contractions (*peristalsis) along its length.

oestrogen One of a group of female sex hormones, produced principally by the ovaries, that promote the onset of *secondary sexual characteristics (such as breast enlargement and development in women) and control the *oestrous cycle. *Oestradiol* is the most important. Oestrogens are secreted at particularly high levels during ovulation, stimulating the uterus to prepare for pregnancy. They are used in oral contraceptives (with *progestogens) and as treatment for various disorders of the female reproductive organs. Small amounts of oestrogens are produced by the adrenal glands and testes.

oestrous cycle The cycle of reproductive activity shown by most sexually mature nonpregnant female mammals except most primates (*compare* menstrual cycle). There are three phases:
(1) *anoestrus* – no sexual activity, the female is neither receptive nor attractive to the male, reproductive organs inactive;
(2) *pro-oestrus* – reproductive organs become active;
(3) *oestrus* (*heat*) – ovulation normally occurs, the female is ready to mate and becomes sexually attractive to the male.

The length of the cycle depends on the species: larger mammals typically have a single annual cycle with a well-defined breeding season (*monoestrus*). The males have a similar cycle of sexual activity. Other species may have many cycles per year (*polyoestrus*) and the male may be sexually active all the time.

oestrus (heat) *See* oestrous cycle.

offset *See* runner.

oil Any of various viscous liquids that are generally immiscible with water. Natural plant and animal oils are either volatile mixtures of terpenes and simple esters (e.g. *essential oils) or are *glycerides of fatty acids.

oil-immersion lens *See* immersion objective.

oleic acid An unsaturated *fatty acid with one double bond, $CH_3(CH_2)_7CH{:}CH(CH_2)_7$COOH. Oleic acid is one of the most abundant constituent fatty acids of animal and plant fats, occurring in butterfat, lard, tallow, groundnut oil, soya-bean oil, etc. Its systematic chemical name is *cis-octadec-9-enoic acid.*

olfaction The sense of smell or the process of detecting smells. This is achieved by receptors in *olfactory organs* (such as the *nose) that are sensitive to air- or water-borne chemicals. Stimulation of these receptors results in the transmission of information to the brain via the *olfactory nerve.*

Oligocene The third geological epoch of the *Tertiary period. It began about 38 million years ago, following the Eocene epoch, and extended for about 12 million years to the beginning of the Miocene epoch. The epoch was characterized by the continued rise of mammals; the first pigs, rhinoceroses, and tapirs made their appearance.

Oligochaeta A class of hermaphrodite annelid worms that bear only a few bristles (*chaetae). Oligochaetes are very abundant in freshwater and terrestrial habitats. The most familiar members of the class are the earthworm (*Lumbricus*) and the freshwater bloodworm (*Tubifex*).

oligotrophic Describing a body of water (e.g. a lake) with a poor supply of nutrients and a low rate of formation of organic matter by photosynthesis. *Compare* eutrophic.

omnivore An animal that eats both animal and vegetable matter. Pigs, for example, are omnivorous. *Compare* carnivore; herbivore.

oncogenic Describing a chemical, organism, or environmental factor that causes the development of cancer. Some viruses are oncogenic to vertebrates, including the Rous sarcoma virus of chickens, and some are suspected of being oncogenic (e.g. some of the *adenoviruses and *papovaviruses). Many of these viruses contain genes (known as *oncogenes*) that are responsible for the transformation of a normal host cell into a cancerous cell. *See also* retrovirus.

ontogeny The developmental course of an organism from the fertilized egg through to maturity. It has been suggested that "ontogeny recapitulates *phylogeny", i.e. the stages of development, especially of the embryo, reflect the evolutionary history of the organism. This idea is now discredited. *See* recapitulation.

oocyte *See* oogenesis.

oogamy Sexual reproduction involving the formation and subsequent fusion of a large, usually stationary, female gamete and a small motile male gamete. The female gamete may contain nourishment for the development of the embryo, which is often retained and protected by the parent organism.

oogenesis The production and growth of the ova (egg cells) in the animal ovary. Special cells (*oogonia*) within the ovary divide repeatedly by mitosis to produce large numbers of prospective egg cells (*oocytes*). When mature, these undergo meiosis, which halves the number of chromosomes. During the first meiotic division a *polar body* and a secondary oocyte are produced. At the second meiotic division the secondary oocyte produces an ovum and a second polar body. Oocytes may be present in the ovaries at birth and may represent the total number of eggs to be produced.

oosphere (ovum; egg cell) The nonmotile female gamete in plants. In angiosperms it is a cell in the *embryo sac of the ovule. In gymnosperms, pteridophytes, and bryophytes it is situated in an *archegonium. In some lower plants, such as *Fucus*, the oosphere is protected by an *oogonium* until it is shed into the water prior to fertilization. Many oospheres store food in the form of starch or oil droplets.

oospore A zygote that is produced as a result of *oogamy in certain algae and fungi. It contains food reserves, develops a protective outer covering, and enters a resting phase before germination. *Compare* zygospore.

operculum 1. (in zoology) A lid or flap of skin covering an aperture, such as the gill slit cover of fish and larval amphibians and the horny calcareous operculum secreted by many gastropod molluscs, which closes the opening of the shell when the animal is inside. **2.** (in botany) The cone-shaped lid of the *capsule of mosses, which is forcibly detached to release the spores.

operon *See* gene.

opiate One of a group of drugs derived from *opium*, an extract of the poppy plant *Papaver somniferum* that depresses brain function (a *narcotic* action). Opiates include morphine, codeine, and their synthetic derivatives, such as heroin. They are used in medicine chiefly to relieve pain, but the use of morphine and heroin is strictly controlled since they can cause drug dependence and tolerance.

optical activity The ability of certain substances to rotate the plane of plane-polarized light as it passes through a crystal, liquid, or solution. It occurs when the molecules of the substance are asymmetric, so that they can exist in two different structural forms each being a mirror image of the other. The two forms are *optical isomers* or *enantiomers*. The existence of such forms is also known as *enantiomorphism* (the mirror images being *enantiomorphs*). One form will rotate the light in one direction and the other will rotate it by an equal amount in the other. The two possible forms are described as *dextrorotatory or *laevorotatory according to the direction of rotation, and prefixes are used to designate the isomer, as in *d*-tartaric and *l*-tartaric acids. An equimolar mixture of the two forms is not optically active. It is called a *racemic mixture* (or

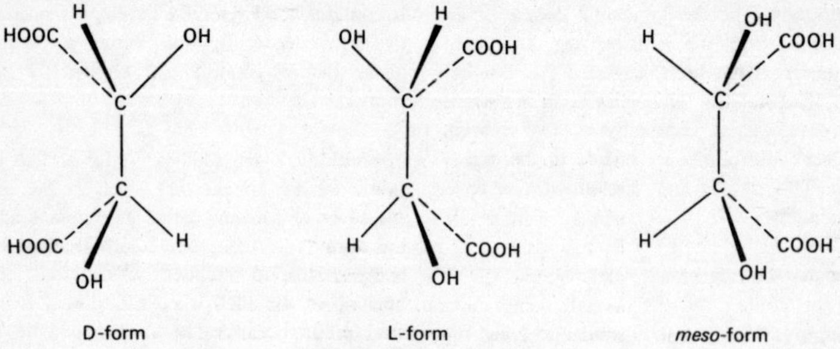

Isomers of tartaric acid

racemate) and designated by *dl*-. In addition, certain molecules can have a *meso form* in which one part of the molecule is a mirror image of the other. Such molecules are not optically active.

Molecules that show optical activity have no plane of symmetry. The commonest case of this is in organic compounds in which a carbon atom is linked to four different groups. An atom of this type is said to be a *chiral centre*. Many naturally occurring compounds show optical isomerism and usually only one isomer occurs naturally. For instance, glucose is found in the dextrorotatory form. The other isomer, *l*-glucose, can be synthesized in the laboratory, but cannot be synthesized by living organisms.

optical fibre A glass fibre through which light can be transmitted with very little leakage through the sidewalls. In the *step-index fibre* a pure glass core, with a diameter between 6 and 250 micrometres, is surrounded by a glass or plastic cladding of lower refractive index. The cladding is usually between 10 and 150 micrometres thick. The interface between core and cladding acts as a cylindrical mirror at which total internal reflection of the transmitted light takes place. This structure enables a beam of light to travel through many kilometres of fibre. In the *graded-index fibre*, each layer of glass, from the fibre axis to its outer wall, has a slightly lower refractive index than the layer inside it. This arrangement also prevents light from escaping through the fibre walls by a combination of refraction and total internal reflection, and can be made to give the same transit time for rays at different angles.

Fibre-optic systems use optical fibres to transmit information, in the form of coded pulses or fragmented images (using bundles of fibres), from a source to a receiver. They are used, for example, in medical instruments (*fibrescopes*) to examine internal body cavities, such as the stomach and bladder.

optical isomers *See* optical activity.

optical microscope *See* microscope.

optic nerve The second *cranial nerve: a paired sensory nerve that runs from each eye to the brain. It is responsible for conveying visual stimuli received by the rods and cones in the retina to the brain for interpretation.

orbit (in anatomy) Either of the two sockets in the skull of vertebrates that house the eyeballs.

order (in taxonomy) A category used in the *classification of organisms that consists of one or several similar or closely related families. Similar orders form a class. Order names typically end in *-ales* in botany, e.g.

Rosales (roses and orchard fruits), and in *-a* in zoology, e.g. Carnivora (flesh eaters).

Ordovician The second geological period of the Palaeozoic era, following the Cambrian and preceding the Silurian periods. It began about 500 million years ago and lasted for about 60 million years. The period was named by the British geologist Charles Lapworth (1842–1920) in 1879. *Graptolites, in deep-water deposits, are the dominant fossils. Other fossils include *trilobites, brachiopods, bryozoans, gastropods, bivalves, echinoids, crinoids, nautiloid cephalopods, and the first corals.

organ Any distinct part of an organism that is specialized to perform one or a number of functions. Examples are ears, eyes, lungs, and kidneys (in animals) and leaves, roots, and flowers (in plants). A given organ will contain many different *tissues.

organ culture The culture of complete living organs (*explants*) of animals and plants outside the body in a suitable culture medium. Animal organs must be small enough to allow the nutrients in the culture medium to penetrate all the cells. Whole plant roots and even root systems can be kept alive in such conditions for a considerable period of time. *See also* explantation.

organelle A minute structure within a plant or animal *cell that has a particular function. Examples of organelles are the nucleus, mitochondria, and lysosomes.

organizer An area of an animal embryo that causes adjacent areas of the embryo to develop in a certain way. The *primary organizer* (blastopore lip or archenteron roof) causes the *gastrula to develop as a complete organism.

origin of life The process by which living organisms developed from inanimate matter, which is generally thought to have occurred on earth between 3500 and 4000 million years ago. It is supposed that the primordial atmosphere was like a chemical soup containing all the basic constituents of organic matter: ammonia, methane, hydrogen, and water vapour. These underwent a process of chemical evolution using energy from the sun and electric storms to combine into ever more complex molecules, such as amino acids, proteins, and vitamins. Eventually self-replicating nucleic acids, the basis of all life, could have developed. The very first organisms may have consisted of such molecules bounded by a simple membrane.

ornithine (orn) An *amino acid, $H_2N(CH_2)_3CH(NH_2)COOH$, that is not a constituent of proteins but is important in living organisms as an intermediate in the reactions of the *urea cycle and in arginine synthesis.

ornithine cycle *See* urea cycle.

orthogenesis An early theory of the nature of evolutionary change, which proposed that organisms evolve along particular paths predetermined by some factor in their genetic make-up. More recent understanding of selection pressure and other external forces that can be shown experimentally to affect the survival of organisms has proved the improbability of the theory.

orthotropism The tendency for a *tropism (growth response of a plant) to be orientated directly in line with the stimulus concerned. An example is the vertical growth of main stems and roots in response to gravity (*orthogeotropism*). *Compare* plagiotropism.

osmium tetroxide (osmium(IV) oxide) A yellow solid, OsO_4, made by heating osmium in air. It is used as a fixative in electron microscopy.

osmometer *See* osmosis.

osmoregulation The control of the water content and the concentration of salts in the body of an animal. In freshwater animals osmoregulation must counteract the tendency for water to pass into the animal by *osmosis. Various methods have been developed to eliminate the excess, such as *contractile vacuoles in protozoans and *kidneys with well-developed glomeruli in freshwater fish. Marine vertebrates have the opposite problem: they prevent excessive water loss and enhance the excretion of salts by having kidneys with few glomeruli and short tubules. In terrestrial vertebrates the dangers of desiccation are reduced by the presence of long convoluted kidney tubules, which increase the reabsorption of water and salts.

osmosis The passage of a solvent through a *semipermeable membrane* separating two solutions of different concentrations. A semipermeable membrane is one through which the molecules of a solvent can pass but the molecules of most solutes cannot. There is a thermodynamic tendency for solutions separated by such a membrane to become equal in concentration, the water (or other solvent) flowing from the weaker to the stronger solution. Osmosis will stop when the two solutions reach equal concentration, and can also be stopped by applying a pressure to the liquid on the stronger-solution side of the membrane. The pressure required to stop the flow from a pure solvent into a solution is a characteristic of the solution, and is called the *osmotic pressure* (symbol Π). Osmotic pressure depends only on the concentration of particles in the solution, not on their nature. For a solution of n moles in volume V at thermodynamic temperature T, the osmotic pressure is given by $\Pi V = RT$, where R is the gas constant. Osmotic-pressure measurements are used in finding the relative molecular masses of compounds, particularly macromolecules. A device used to measure osmotic pressure is called an *osmometer*.

The distribution of water in living organisms is dependent to a large extent on osmosis, water entering the cells through their membranes. A cell membrane is not truly semipermeable as it allows the passage of certain solute molecules; it is described as *differentially permeable*. Animals have evolved various means to counteract the effects of osmosis (*see* osmoregulation); in plant cells, excessive osmosis is prevented by the pressure exerted by the cell wall, which opposes the osmotic pressure.

osmotic pressure *See* osmosis.

ossification The process of *bone formation. It is brought about by the action of special cells called *osteoblasts*, which deposit layers of bone in connective tissue. Some bones are formed directly in connective tissue (*see* membrane bone); others are formed by the replacement of cartilage (*see* cartilage bone).

Osteichthyes The class of vertebrates comprising the bony fishes – marine and freshwater fish with a bony skeleton. All have gills covered with a bony operculum, and a layer of thin overlapping bony *scales covers the entire body surface. Bony fish have a *swim bladder, which acts as a hydrostatic organ enabling the animal to remain suspended in the water at any depth. In some fish this bladder acts as a lung. *See also* Dipnoi; Teleostei. *Compare* Elasmobranchii.

outbreeding Mating between unrelated or distantly related individuals of a species. Outbreeding populations usually show more variation than *inbreeding ones and have a greater potential for adapting to environmental changes. Outbreeding increases the number of *heterozygous individuals, so disadvantageous recessive characteristics tend to be masked by dominant alleles.

outer ear (external ear) The part of the ear external to the *tympanum (eardrum). It is present in mammals, birds, and some reptiles and consists of a tube (the *external auditory meatus*) that directs sound waves onto the tympanum. In mammals it may include an external *pinna, which extends beyond the skull.

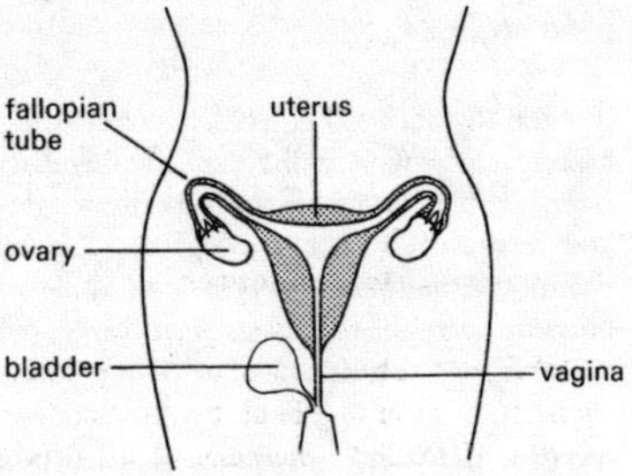

Human female reproductive system

ovary 1. The reproductive organ in female animals in which eggs (ova) are produced. In most vertebrates there are two ovaries (in some fish the ovaries fuse together to form a single structure and in birds the left ovary only is functional). As well as eggs, they produce steroid hormones (*see* oestrogen; progesterone). In mammals each ovary is situated close to the opening of a *fallopian tube; it contains numerous follicles in which the eggs develop and from which they are released in a regular cycle (*see* Graafian follicle; menstrual cycle; oogenesis; ovulation). **2.** The hollow base of the *carpel of a flower, containing one or more *ovules. After fertilization, the ovary wall develops into the fruit enclosing the seeds. In some species, the carpels are fused together to form a complex ovary.

oviduct The tube that conveys an animal egg cell from the ovary to other parts of the reproductive system or to the outside. Eggs are passed along the oviduct by the action of muscles and cilia. *See* fallopian tube.

oviparity Reproduction in which fertilized eggs are laid or spawned by the mother and hatch outside her body. It occurs in most animals except marsupial and placental mammals. *Compare* ovoviviparity; viviparity.

ovipositor An organ at the hind end of the abdomen of female insects through which eggs are laid. It consists of a pair of modified appendages and is often long and piercing, so that eggs can be laid in otherwise inaccessible places. The sting of bees and wasps is a modified ovipositor.

ovoviviparity Reproduction in which fertilized eggs develop and hatch in the oviduct of the mother. It occurs in many invertebrates and in some fish and reptiles (e.g. the viper). *Compare* oviparity; viviparity.

ovulation The release of an egg cell from the ovary. The developing egg cell within its follicle migrates to the ovary surface; when mature, it is released from the follicle (which breaks open) into the body cavity, from where it passes into the oviduct. *See also* menstrual cycle.

ovule The part of the female reproductive organs of seed plants that consists of the *nucellus, *embryo sac, and *integuments. The ovules of gymnosperms are situated on ovuliferous scales of the female cones while those of angiosperms are enclosed in the carpel. After fertilization, the ovule becomes the seed.

ovuliferous scale One of a group of large woody specialized leaves that form the female *cone of conifers and related trees. It bears the ovules, which develop into seeds.

ovum (egg cell) 1. (in zoology) The mature reproductive cell (*see* gamete) of female animals, which is produced by the ovary (*see* oogenesis). It is spherical, has a nucleus, is covered with a vitelline membrane, and is not mobile. **2.** (in botany) The *oosphere of plants.

oxalic acid (ethanedioic acid) A crystalline solid, $(COOH)_2$, that is slightly soluble in water. Oxalic acid is strongly acidic and very poisonous. It occurs in certain plants, e.g. sorrel and the leaf blades of rhubarb.

oxidation *See* oxidation–reduction.

oxidation–reduction (redox) Originally, *oxidation* was simply regarded as a chemical reaction with oxygen. The reverse process – loss of oxygen – was called *reduction*. Reaction with hydrogen also came to be regarded as reduction. Later, a more general idea of oxidation and reduction was developed in which oxidation was loss of electrons and reduction was gain of electrons. This wider definition covered the original one and also applies to reactions that do not involve oxygen. However, it applies only to reactions in which electron transfer occurs – i.e. to reactions involving ions. It can be extended to reactions between covalent compounds by using the concept of *oxidation number* (or *state*). This is a measure of the electron control that an atom has in a compound compared to the atom in the pure element. An oxidation number consists of two parts:

(1) its sign, which indicates whether the control has increased (negative) or decreased (positive);

(2) its value, which gives the number of electrons over which control has changed.

The change of electron control may be complete (in ionic compounds) or partial (in covalent compounds). Oxidation is a reaction involving an increase in oxidation number and reduction involves a decrease. Thus in

$$2H_2 + O_2 \rightarrow 2H_2O$$

the hydrogen in water is +1 and the oxygen −2. The hydrogen is oxidized and the oxygen is reduced. Compounds that tend to undergo reduction readily are *oxidizing agents*; those that undergo oxidation are *reducing agents*.

oxidative phosphorylation A reaction occurring during the final stages of *aerobic respiration, in which ATP is formed from ADP and phosphate coupled to electron transport in the *electron transport chain. The reaction occurs in the mitochondria and takes place at three sites on the electron transport chain.

oxygen Symbol O. A colourless odourless gaseous element. It is the most abundant element in the earth's crust (49.2% by weight) and is present in the atmosphere (28% by volume). Atmospheric oxygen is of vital importance for all organisms that carry out *aerobic respiration.

oxygen debt The physiological state that exists in a normally aerobic animal when insufficient oxygen is available for metabolic requirements (e.g. during a period of strenuous physical activity). To meet the body's increased demand for energy, pyruvate is converted anaerobically (i.e. in the absence of oxygen) to lactic acid, which requires oxygen for its breakdown and accumulates in the tissues. When oxygen is available again lactic acid is oxidized in the liver, thus repaying the debt.

oxyhaemoglobin *See* haemoglobin.

oxytocin A hormone, secreted by the posterior pituitary gland in mammals, that stimulates the ejection of milk in lactating females and the contraction of the uterus during parturition. Like *vasopressin, oxytocin is synthesized in the hypothalamus.

ozone layer (ozonosphere) A layer of the earth's atmosphere in which most of the atmosphere's ozone is concentrated. It occurs 15–50 km above the earth's surface and is virtually synonymous with the stratosphere. In this layer most of the sun's ultraviolet radiation is absorbed by the ozone molecules, causing a rise in the temperature of the stratosphere and preventing vertical mixing so that the stratosphere forms a stable layer. By absorbing most of the solar ultraviolet radiation the ozone layer protects living organisms on earth. The fact that the ozone layer is thinnest at the equator is believed to account for the high equatorial incidence of skin cancer as a result of exposure to unabsorbed solar ultraviolet radiation.

P

P_1 (parental generation) The individuals that are selected to begin a breeding experiment, crosses between which yield the *F_1 generation.

pacemaker 1. (*or* **sinoatrial node**) A small mass of specialized muscle cells in the mammalian heart, found in the wall of the right atrium near the opening for the vena cava. The cells initiate and maintain the heart beat: by their rhythmic and spontaneous contractions they stimulate contractions in the rest of the heart muscle. The cells themselves are controlled by the autonomic nervous system, which determines the heart rate. Similar pacemakers occur in the hearts of other vertebrates. **2.** An electronic or nuclear battery-charged device that can be implanted surgically into the chest to produce and maintain the heart beat. These devices are used when the heart's own pacemaker is defective or diseased.

pachytene The period in the first prophase of *meiosis when paired *homologous chromosomes are fully contracted and twisted around each other.

paedogenesis Reproduction by an animal that is still in the larval or pre-adult form. Paedogenesis is a form of *neoteny and is particularly marked in the axolotl, a larval form of the salamander, which retains its larval features owing to a thyroid deficiency but can breed, producing individuals like itself. If the thyroid hormone thyroxin is given metamorphosis occurs.

pairing (synapsis) The close association between *homologous chromosomes that develops during the first prophase of *meiosis. The two chromosomes move together and an exact pairing of corresponding points along their lengths occurs as they lie side by side. The resulting structure is called a *bivalent*.

palaeobotany The branch of *palaeontology concerned with the study of plants through geological time, as revealed by their *fossil remains (*see also* palynology). It overlaps with other aspects of plant study, including anatomy, ecology, evolution, and taxonomy.

Palaeocene The earliest geological epoch of the *Tertiary period. It began about 65 million years ago, following the Cretaceous period, and extended for about 11.5 million years to the beginning of the *Eocene (the Palaeocene is sometimes included in the Eocene). It was named by the palaeobotanist W. P. Schimper in 1874. A major floral and faunal discontinuity occurred between the end of the Cretaceous and the beginning of the Palaeocene: following the extinction of many reptiles the mammals became abundant on land. By the end of the epoch primates and rodents had evolved.

palaeoclimatology The study of climates of earlier geological periods. This is based largely on the study of sediments that were laid down during these periods and of fossils. The changes in the positions of the continents as a result of *continental drift and *plate tectonics complicate the study.

palaeoecology The study of the relationships of *fossil organisms to each other and to their environments. It involves the study both of the fossils and of the surrounding rock in which they are found. Trace fossils may provide information on the behaviour of the organism.

Palaeolithic The Old Stone Age, lasting in Europe from about 2.5 million to 9000 years ago, during which man used primitive stone tools made by chipping stones and flints.

palaeontology The study of extinct organisms, including their structure, environment, evolution, and distribution, as revealed by their *fossil remains. Palaeontological work also makes important contributions to geology in revealing stratigraphic relationships between rock strata and determining the physical appearance and climate of the earth during past geological ages. *See also* palaeobotany; palaeozoology.

Palaeozoic The first era of *Phanerozoic time. It follows the *Precambrian and is subdivided into the Lower Palaeozoic, comprising the *Cambrian, *Ordovician, and *Silurian periods, and the Upper Palaeozoic, comprising the *Devonian, *Carboniferous, and *Permian periods, It extended from about 570 million years ago to about 225 million years ago, when it was succeeded by the *Mesozoic era.

palaeozoology The branch of *palaeontology concerned with the study of animals throughout geological time, as revealed by their *fossil remains.

palate The roof of the mouth of vertebrates, which separates the *buccal and nasal cavities. In mammals it is divided into two zones, the bony *hard palate* and the *soft palate*, and completely separates the buccal cavity from the air passage to enable simultaneous eating and breathing.

palisade mesophyll *See* mesophyll.

pallium *See* cerebral cortex.

palmitic acid (hexadecanoic acid) A saturated fatty acid, $CH_3(CH_2)_{14}COOH$. Glycerides of palmitic acid occur widely in plant and animal oils and fats.

palp An elongated sensory organ, usually near the mouth, in many invertebrates. Examples are the tactile head appendages of polychaete worms, the ciliated flap of tissue that produces feeding currents in bivalve molluscs, the distal part of the *mandibles of crustaceans, and the olfactory parts of the first and second *maxillae of some insects.

palynology (micropalaeontology) The study of fossil pollen and spores (*pollen analysis*) and various other *microfossils, such as coccoliths and dinoflagellates. Palynology is used in stratigraphy, palaeoclimatology, and archaeology. Pollen and spores are very resistant to decay and therefore their fossils are found in sedimentary rocks. They may be extracted by various methods, including boiling with potassium hydroxide solution, washing with strong oxidizing mixtures, and centrifuging repeatedly. Spores and pollen are classified according to shape, form of aperture, and both internal and external details of the exine (outer coat). They indicate the nature of the dominant flora, and therefore the climate and conditions of the period in which they lived.

pancreas A gland in vertebrates lying between the duodenum and the spleen. Under the influence of the hormone *secretin it secretes digestive enzymes (mainly trypsinogen (*see* trypsin), *amylase, and *lipase) into the duodenum via the pancreatic duct. It also contains groups of cells – the *islets of Langerhans – that function as an *endocrine gland, producing the hormones *insulin and *glucagon, which regulate blood sugar levels.

pancreozymin *See* cholecystokinin.

panicle A type of flowering shoot common in the grass family. The primary axis bears groups of *racemes and is itself racemose, as the youngest groups of flowers are at the top (e.g. oat). The term may be used loosely for any form of branched *racemose inflorescence; for example, the horse chestnut is a raceme of cymes. Both these arrangements are seen in the family Polygonaceae (docks and sorrels).

pantothenic acid A vitamin of the *vitamin B complex. It is a constituent of *coenzyme A, which performs a crucial role in the oxidation of fats, carbohydrates, and certain amino acids. Deficiency rarely occurs because the vitamin occurs in many foods, especially cereal grains, peas, egg yolk, liver, and yeast.

papain A protein-digesting enzyme (*see* protease) occurring in the fruit of the West Indian papaya tree (*Carica papaya*). It is used as a digestant and in the manufacture of meat tenderizers.

paper chromatography A technique for analysing mixtures by *chromatography, in which the stationary phase is absorbent paper. A spot of the mixture to be investigated is placed near one edge of the paper and the sheet is suspended vertically in a solvent, which rises through the paper by capillary action carrying the components with it. The components move at different rates, partly because they absorb to different extents on the cellulose and partly because of partition between the solvent and the moisture in the paper. The paper is removed and dried, and the different components form a line of spots along the paper. Colourless substances are detected by using ultraviolet radiation or by spraying with a substance that reacts to give a coloured spot (e.g. ninhydrin gives a blue coloration with amino acids). The components can be identified by the distance they move in a given time.

papilla Any cone-shaped protuberance projecting from the surface of an organ or organism. Papillae occur, for example, on the tongue and, in plants, on the surface of many petals.

papovavirus One of a group of *DNA-containing viruses that produce tumours in their hosts. *Papilloma* types produce nonmalignant tumours (such as warts) in all vertebrates; *polyoma types* produce malignant tumours in only certain classes of vertebrates (not including man).

pappus A group of modified *sepals, often in the form of a ring of silky hairs. For example, when the fruit of the dandelion matures a pappus of hairs persists at the top of a thin stalk forming a parachute-like structure, which serves to disperse the fruit.

parallel evolution The development of related organisms along similar evolutionary paths due to strong selection pressures acting on all of them in the same way. It is debatable if the phenomenon really exists: many argue that all evolution is ultimately *convergent or divergent (*see* adaptive radiation).

paraphyletic *See* monophyletic.

parasitism An association in which one organism (the *parasite*) lives on (*ectoparasitism*) or in (*endoparasitism*) the body of another (the *host*), from which it obtains its nutrients. Some parasites inflict comparatively little damage on their host, but many cause characteristic diseases (these are, however, never immediately fatal, as killing the host would destroy the parasite's source of food). Parasites are usually highly specialized for their way of life, which may involve one host or several (if the *life cycle requires it). They typically produce vast numbers of eggs, very few of which survive to find their way to another suitable host. *Obligate parasites* can only survive and reproduce as parasites; *facultative parasites* can also live as *saprophytes. The parasites of man include fleas and lice (ectoparasites), various bacteria, protozoans, and fungi (endoparasites causing characteristic diseases), and tapeworms (e.g. *Taenia soleum*, which lives in the gut).

parasympathetic nervous system Part of the *autonomic nervous system. Its nerve endings release acetylcholine as a *neurotransmitter and its actions tend to antagonize those of the *sympathetic nervous system. For example, the parasympathetic nervous system increases salivary gland secretion, decreases heart rate, promotes digestion (by increasing *peristalsis), and dilates blood vessels, while the sympathetic nervous system has opposite effects.

parathyroid glands Two pairs of *endocrine glands situated behind, or embedded within, the thyroid gland in higher vertebrates. They produce *parathyroid hormone* (*parathormone*), which controls the amount of calcium in the blood. A low level of calcium stimulates hormone secretion, which increases transfer of calcium from the bones to the blood (*compare* calcitonin).

parenchyma 1. A plant tissue consisting of roughly spherical relatively undifferentiated cells, frequently with air spaces between them. The cortex and pith are composed of parenchyma cells (*see* ground tissues). **2.** Loose *connective tissue formed of large cells. Its function is to pack the spaces between organs in some simple acoelomate animals, such as flatworms (Platyhelminthes).

parthenocarpy The formation of fruit without prior fertilization of the flower by pollen. The resulting fruits are seedless and therefore do not contribute to the reproduction of the plant; examples are bananas and pineapples. Plant *growth substances may have a role in this phenomenon, which can be induced by auxins in the commercial production of tomatoes and other fruits.

parthenogenesis The development of an organism from an unfertilized egg. This occurs sporadically in many plants (e.g. dandelions and hawkweeds) and in a few animals, but in some species it is the main and sometimes only method of reproduction. For example, in some species of aphid, males are absent or very rare. The eggs formed by the females contain the full (diploid) number of chromosomes and are genetically identical. Variation is consequently very limited in species that reproduce parthenogenetically.

particulate inheritance The transmission from parent to offspring of separate units that determine characteristics. Gregor Mendel observed that *recessive characteristics, absent in the offspring of a cross in which only one parent possessed them, reappeared repeatedly in the progeny of subsequent crosses. This led him to formulate his theory of inherited 'factors' (now called *alleles) that retain their identity through succeeding generations (*see* Mendel's laws). *Compare* blending inheritance.

parturition The act of giving birth to young at the end of the *gestation period. Hormones probably cause the process to start but the mechanism is not fully understood.

pasteurization The treatment of milk to destroy disease-causing bacteria, such as those of tuberculosis, typhoid, and brucellosis. Milk is heated to 65°C for 30 minutes or to 72°C for 15 minutes followed by rapid cooling to below 10°C. The method was devised by the French microbiologist Louis Pasteur (1822–95).

patella (kneecap) A small rounded movable bone that is situated in a tendon in front of the knee joint in most mammals (including humans). The function of the patella is to protect the knee.

pathogen Any disease-causing microorganism. Pathogens include viruses, rickettsiae, and many bacteria, fungi, and protozoans. *See* infection.

patristic Denoting similarity between organisms resulting from common ancestry. *Compare* homoplastic.

peat A mass of dark-brown or black fibrous plant debris produced by the partial disintegration of vegetation in wet places. It may accumulate in depressions. When subjected to burial and hence pressure and heat it may be converted to *coal. Peat is used to improve soil and as a fuel, especially in Ireland and Sweden.

peck order *See* dominant.

pectic substances A group of polysaccharides made up primarily of sugar acids. They are important constituents of plant cell walls and the *middle lamella between adjacent cell walls. Normally they are present in an insoluble form, but in ripening fruits and in tissues affected by certain diseases they change into a soluble form, which is evidenced by softening of the tissues.

pectin A type of *pectic substance. It is used in making jam as it forms a gel with sucrose.

pectoral fins *See* fins.

pectoral girdle (shoulder girdle) The bony or cartilaginous structure in vertebrates to which the anterior limbs (pectoral fins, forelegs, or arms) are attached. In mammals it consists of two dorsal *scapulae (shoulder blades) attached to the backbone and two ventral *clavicles (collar bones) attached to the sternum (breastbone).

pedicel The stalk attaching a flower to the main floral axis (*see* peduncle). Some flowers, described as *sessile*, do not have a pedicel and arise directly from the peduncle.

pedology The science of the study of soils, including their origin and characteristics and their utilization.

peduncle The main stalk of a plant that bears the flowers, which may be solitary or grouped in an *inflorescence. *Compare* pedicel.

pelagic Describing organisms that swim or drift in a sea or a lake, as distinct from those that live on the bottom (*see* benthos). Pelagic organisms are divided into *plankton and *nekton.

pellicle The thin outer covering, composed of protein, that protects and maintains the shape of certain unicellular organisms, e.g. *Euglena*. It is transparent and in ciliated organisms, e.g. *Paramecium*, contains small pores through which the cilia emerge.

pelvic fins *See* fins.

pelvic girdle (pelvis; hip girdle) The bony or cartilaginous structure in vertebrates to which the posterior limbs (pelvic fins or legs) are attached. The pelvic girdle articulates dorsally with the backbone; it is made up of two halves, each produced by the fusion of the *ilium, *ischium, and *pubis.

pelvis **1.** *See* pelvic girdle. **2.** The lower part of the abdomen in the region of the pelvic girdle. **3.** A conical chamber in the *kidney into which urine drains from the kidney tubules before passing to the *ureter.

penicillin An *antibiotic derived from the mould *Penicillium notatum*; specifically it is known as *penicillin G* and belongs to a class of similar substances called penicillins. They are all active against Gram-negative bacteria, producing their effects by disrupting synthesis of the bacterial cell wall, and are used to treat a variety of infections caused by these bacteria.

penis The male reproductive organ of mammals (and also of some birds and reptiles) used to introduce sperm into the female reproductive tract to ensure internal fertilization. It contains a duct (the *urethra) through which the sperms pass. The penis becomes erect during precopulatory activity, either by filling with blood or haemolymph or by the action of muscles, and can be inserted into the vagina (or cloaca). In mammals the urine also leaves the body through the penis.

pentadactyl limb A limb with five digits, characteristic of tetrapod vertebrates (amphibians, reptiles, birds, and mammals). It evolved from the paired fins of primitive fish as an adaptation to locomotion on land and is not found in modern fish. The limb has three parts (see illustration): the upper arm or thigh containing one long bone, the forearm or shank containing two long bones, and the hand or foot, which contains a number of small bones. This basic design is modified in many species, according to the function of the limb, particularly by the loss of fusion of the terminal bones.

pentose A sugar that has five carbon atoms per molecule. *See* monosaccharide.

pepo *See* berry.

pepsin An enzyme, secreted by cells lining the interior of the vertebrate stomach, that catalyses the breakdown of proteins.

peptidase *See* protease.

peptide Any of a group of organic compounds comprising two or more amino acids linked by *peptide bonds*. These bonds are formed by the reaction between adjacent carboxyl ($-COOH$) and amino ($-NH_2$) groups with the elimination of water (see illustration). *Dipeptides* contain two amino acids, *tripeptides* three, and so on. *Polypeptides contain more than ten and usually 100–300. Naturally occurring *oligopeptides* (of less than ten amino acids) include the tripeptide glutathione and the pituitary hormones vasopressin and oxytocin, which are octapeptides. Peptides also result from protein breakdown, e.g. during digestion.

perennation The survival of biennial or perennial plants from one year to the next by vegetative means. In biennials and herbaceous perennials the aerial parts of the plant die down and the plants survive by means of underground storage roots (e.g. carrot), *rhizomes (e.g. couch grass, Solomon's seal), *tubers (e.g. dahlia), *bulbs (e.g. daffodil, snowdrop), or *corms (e.g. crocus, gladiolus). These *perennating organs* are also frequently responsible for *vegetative propagation. Woody perennials survive the winter by reducing their metabolic activity (e.g. by leaf loss in deciduous trees and shrubs).

perennial A plant that lives for a number of years. Woody perennials (trees and shrubs) have a permanent aerial form, which continues to grow year after year. Herbaceous (nonwoody) perennials have aerial shoots that die down each autumn and are replaced in spring by new shoots from an underground structure (*see* perennation). Lupin and rhubarb are examples of herbaceous perennials. *Compare* annual; biennial; ephemeral.

perianth The part of a flower situated outside the stamens and carpels. In dicotyledons it consists of two distinct whorls, the outer of sepals (*see* calyx) and the inner of

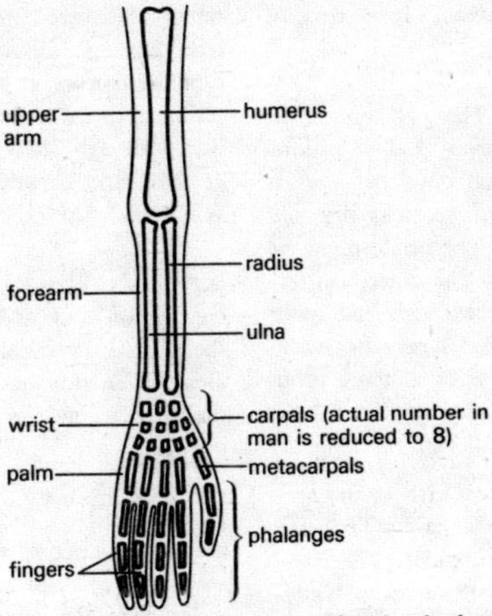

A basic pentadactyl forelimb, as exemplified by the human arm

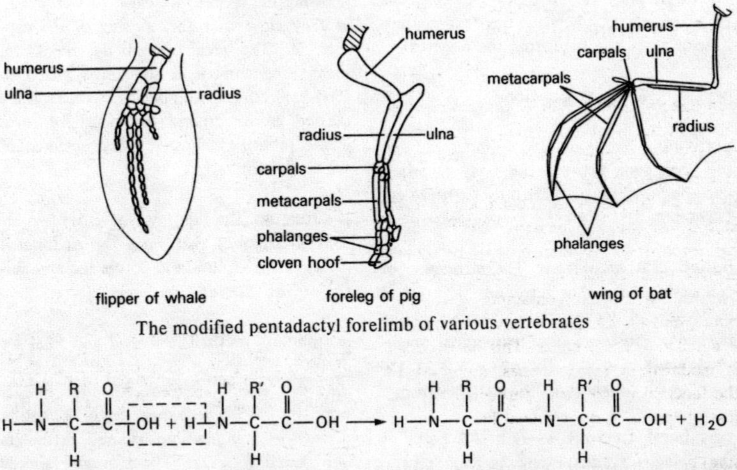

flipper of whale

foreleg of pig

wing of bat

The modified pentadactyl forelimb of various vertebrates

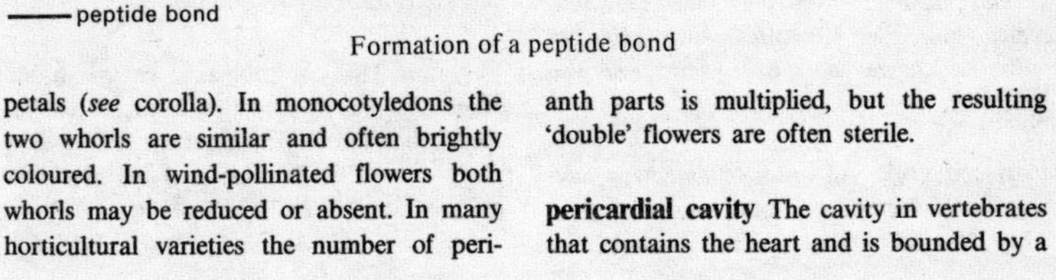

Formation of a peptide bond

petals (*see* corolla). In monocotyledons the two whorls are similar and often brightly coloured. In wind-pollinated flowers both whorls may be reduced or absent. In many horticultural varieties the number of perianth parts is multiplied, but the resulting 'double' flowers are often sterile.

pericardial cavity The cavity in vertebrates that contains the heart and is bounded by a membrane (the *pericardium*). It is part of the *coelom.

pericarp (fruit wall) The part of a fruit that develops from the ovary wall of a flower. The type of fruit that develops depends on whether the pericarp becomes dry and hard or soft and fleshy. The pericarp can be made up of three layers. The outer skin (*epicarp* or *exocarp*) may be tough and hard; the middle layer (*mesocarp*) may be succulent as in peach, hard as in almond, or fibrous as in coconut; and the inner layer (*endocarp*) may be hard and stony as in many *drupes, membranous as in citrus fruits, or indistinguishable from the mesocarp, as in many *berries.

pericycle A plant tissue comprising the outermost layer of the root vascular tissue, lying immediately beneath the *endodermis. Lateral roots originate from the pericycle.

periderm *See* cork cambium.

perilymph The fluid of the *inner ear that fills the space between the bony labyrinth and the membranous labyrinth. *Compare* endolymph.

periodontal membrane The membrane of connective tissue that surrounds the root of a *tooth and anchors it to its socket in the jawbone. Fibres of the periodontal membrane pass into the *cement covering the root, which provides a firm attachment.

peripheral nervous system All parts of the nervous system excluding the *central nervous system. It consists of all the *cranial and *spinal nerves and their branches, which link the *receptors and *effectors with the central nervous system. *See also* autonomic nervous system.

Perissodactyla An order of mammals having hoofed feet with an odd number of toes. They are all herbivores and include the tapirs, rhinoceros, and horse. The teeth are large and specialized for grinding. Cellulose digestion occurs in the caecum and large intestine. Fossils of the Eocene epoch, 60 million years ago, show that these animals were at that time already distinct from the cloven-hoofed *Artiodactyla.

peristalsis Waves of involuntary muscular contraction and relaxation that pass along the alimentary canal, forcing food contents along. It is brought about by contraction of the circular muscles of the gut wall in sequence.

peristome **1.** A ring of toothlike structures around the opening of a moss *capsule. The teeth tend to bend and twist in dry weather, so opening the mouth of the capsule and allowing the spores to escape. In wet weather they close over the opening of the capsule. **2.** The area around the mouth in many invertebrates. It sometimes assists in food collecting. Examples are the spirally ciliated groove around the mouth of some ciliate protozoans and the first segment of the earthworm.

peritoneum The thin layer of tissue (*see* serous membrane) that lines the abdominal cavity of vertebrates and covers the abdominal organs. *See also* mesentery.

permanent teeth The second and final set of teeth that mammals produce after shedding the *deciduous teeth. An adult human normally has 32 permanent teeth, consisting of incisors, canines, molars, and premolars (*see* dental formula). These usually appear between the ages of approximately 6 and 21 years. *See also* diphyodont.

Permian The last geological period in the Palaeozoic era. It extended from the end of the Carboniferous period, about 280 million years ago, to the beginning of the Mesozoic era, about 225 million years ago. It was named by the British geologist Roderick

Murchison (1792–1871) in 1841 after the Perm province in Russia. In some areas continental conditions prevailed, which continued into the following period, the Triassic. These conditions resulted in the deposition of the New Red Sandstone. During the period a number of animal groups became extinct, including the trilobites, tabulate and rugose corals, and blastoids. Amphibians and reptiles continued to be the dominant land animals and gymnosperms replaced pteridophytes as the dominant plants.

peta- Symbol P. A prefix used in the metric system to denote one thousand million million times. For example, 10^{15} metres = 1 petametre (Pm).

petal One of the parts of the flower that make up the *corolla. Petals of insect-pollinated plants are usually brightly coloured and often scented. Those of wind-pollinated plants are usually reduced or absent. Petals are considered to be modified leaves but their structure is simpler. Epidermal hairs may be present and the cuticle is often covered by lines or dots known as *honey guides*, which direct insects to the *nectar.

petiole The stalk that attaches a *leaf blade to the stem. Leaves without petioles are described as *sessile*.

Petri dish A shallow circular flat-bottomed dish made of glass or plastic and having a fitting lid. It is used in laboratories chiefly for culturing bacteria and other microorganisms. It was invented by the German bacteriologist J. R. Petri (1852–1921).

petrification *See* fossil.

pH *See* pH scale.

Phaeophyta (brown algae) A division of *algae in which the green chlorophyll pigments are usually masked by the brown pigment fucoxanthin. Brown algae are usually marine (being abundant in cold water) and many species, such as the wracks (*Fucus*), inhabit intertidal zones. They vary in size from small branched filaments to ribbon-like bodies (known as kelps) many metres long.

phage *See* bacteriophage.

phagocyte A cell that is able to engulf and break down foreign particles, cell debris, and disease-producing microorganisms (*see* phagocytosis). Many protozoans and certain mammalian cells (e.g. *macrophages and *monocytes) are phagocytes. Phagocytes are important elements in the natural defence mechanism of most animals.

phagocytosis The process by which foreign particles invading the body or minute food particles are engulfed and broken down by certain animal cells (known as *phagocytes). The cell membrane of the phagocyte invaginates to capture the particle and then closes around it to form a sac or *vacuole. The vacuole coalesces with a *lysosome, which contains enzymes that break down the particle. *Compare* pinocytosis.

phalanges The bones that make up the *digits of the hand or foot in vertebrates. They articulate with the *metacarpals of the hand or with the *metatarsals of the foot. In the basic *pentadactyl limb there are two phalanges for the first digit (the thumb or big toe in humans) and three for each of the others.

Phanerozoic Geological time since the end of the Precambrian, represented by rock strata containing clearly recognizable fossils. It comprises the *Palaeozoic, *Mesozoic, and *Cenozoic eras and has extended for about 570 million years from the beginning of the Cambrian period. Fossils are extremely rare in Precambrian rocks.

pharmacology The study of the properties of drugs and their effects on living organisms. Clinical pharmacology is concerned with the effects of drugs in treating disease.

pharynx 1. The cavity in vertebrates between the mouth and the *oesophagus and windpipe (*trachea), which serves for the passage of both food and respiratory gases. The presence of food in the pharynx stimulates swallowing (*see* deglutition). In fish and aquatic amphibians the pharynx is perforated by *gill slits. **2.** The corresponding region in invertebrates.

phase-contrast microscope A type of *microscope that is widely used for examining cells and tissues. It makes visible the changes in phase that occur when nonuniformly transparent specimens are illuminated. In passing through an object the light is slowed down and becomes out of phase with the original light. With transparent specimens having some structure diffraction occurs, causing a larger phase change in light outside the central maximum of the pattern. The phase-contrast microscope provides a means of combining this light with that of the central maximum by means of an annular diaphram and a *phase-contrast plate*, which produces a matching phase change in the light of the central maximum only. This gives greater contrast to the final image, due to constructive interference between the two sets of light waves. This is *bright contrast*; in *dark contrast* a different phase-contrast plate is used to make the same structure appear dark, by destructive interference of the same waves.

phellem *See* cork.

phelloderm *See* cork cambium.

phellogen *See* cork cambium.

phenetic Describing a system of *classification of organisms based on similarities and differences in as many observable characteristics as possible. A phenetic system does not aim to reflect evolutionary descent, although it may well do so. *Compare* phylogenetic.

phenolphthalein A dye used as an acid-base indicator. It is colourless below pH 8 and red above pH 9.6. It is used in titrations involving weak acids and strong bases. It is also used as a laxative.

phenotype The observable characteristics of an organism. These are determined by its genes (*genotype*), the dominance relationships between the *alleles, and by the interaction of the genes with the environment.

phenylalanine *See* amino acid.

pheromone (ectohormone) A chemical substance emitted by an organism into the environment as a specific signal to another organism, usually of the same species. Pheromones play an important role in the social behaviour of certain animals, especially insects and mammals. They are used to attract mates, to mark trails, and to promote social cohesion and coordination in colonies. Pheromones are usually highly volatile organic acids or alcohols and can be effective at minute concentrations.

phloem (bast) A tissue that conducts food materials in vascular plants from regions where they are produced (notably the leaves) to regions, such as growing points, where they are needed. It consists of hollow tubes (sieve tubes) that run parallel to the long axis of the plant organ and are formed from elongated cells (sieve elements) joined end to end. The end walls of these cells are broken down to a greater or lesser extent to allow passage of materials. In young plants and in newly formed tissues of mature plants the phloem is formed by the activity of the *apical meristem. In most plants secondary phloem is later differentiated by the vascular *cambium and this replaces the earlier formed phloem in older regions. *See also* companion cell. *Compare* xylem.

phloroglucinol A red dye (usually acidified with hydrochloric acid) that stains lignin in plant cells red.

phosphagen A compound found in animal tissues that provides a reserve of chemical energy in the form of high-energy phosphate bonds. The most common phosphagens are *creatine phosphate, occurring in vertebrate muscle and nerves, and arginine phosphate, found in most invertebrates. During tissue activity (e.g. muscle contraction) phosphagens give up their phosphate groups, thereby generating *ATP from ADP. The phosphagens are then reformed when ATP is available.

phosphatide *See* phospholipid.

phospholipid (phosphatide) One of a group of lipids having both a phosphate group and one or more fatty acids. *Glycerophospholipids* are based on *glycerol; the three hydroxyl groups are esterified with two fatty acids and a phosphate group, which may itself be bound to one of a variety of simple organic groups (e.g. in *lecithin it is choline). *Sphingophospholipids* are based on the alcohol sphingosine and contain only one fatty acid linked to an amino group. With their hydrophilic polar phosphate groups and long hydrophobic hydrocarbon 'tails', phospholipids readily form membrane-like structures in water. They are a major component of cell membranes.

phosphorus Symbol P. A nonmetallic element that is a major *essential element for living organisms. It is an important constituent of tissues (especially bones and teeth) and of cells, being required for the formation of nucleic acids and energy-carrying molecules (e.g. ATP) and also involved in various metabolic reactions.

photic zone The upper layer of a sea or a lake, in which there is sufficient light for photosynthesis. The limit of the photic zone varies from less than a metre to more than 200 metres, depending on the turbidity of the water.

photomicrography The use of photography to obtain a permanent record (a *photomicrograph*) of the image of an object as viewed through a microscope.

photoperiodism The response of an organism to changes in day length (*photoperiod*). Many plant responses are controlled by day length, the most notable being flowering in many species (*see also* day-neutral plant; long-day plant; short-day plant). In plants the pigment *phytochrome is responsible for the regulation of photoperiodic responses. Activities in animals that are determined by photoperiod include breeding, migration, and other seasonal events.

photopic vision The type of vision that occurs when the cones in the eye are the principal receptors, i.e. when the level of illumination is high. Colours can be identified with photopic vision. *Compare* scotopic vision.

photoreceptor A sensory cell or group of cells that reacts to the presence of light. It usually contains a pigment that undergoes a chemical change when light is absorbed, thus stimulating a nerve. *See* eye.

photosynthesis The chemical process by which green plants synthesize organic compounds from carbon dioxide and water in the presence of sunlight. It occurs in the *chloroplasts (most of which are in the leaves) and there are two principal series of reactions. In the *light reactions*, which require the presence of light, energy from sunlight is absorbed by *photosynthetic pigments (chiefly the green pigment *chlorophyll) and converted into chemical energy. In the ensuing *dark reactions*, which can take place either in light or darkness, this chemical energy is used in the production of simple organic compounds from carbon dioxide and water. Further chemical reactions convert these compounds into chemicals useful to the plant. Photosynthesis can be summarized by the equation:

$$CO_2 + 2H_2O \rightarrow [CH_2O] + H_2O + O_2$$

Since virtually all other forms of life are directly or indirectly dependent on plants for food, photosynthesis is the basis for all life on earth. Furthermore virtually all the atmospheric oxygen has originated from oxygen released during photosynthesis.

photosynthetic pigments The plant pigments responsible for the capture of light energy during the light reactions of *photosynthesis. The green pigment *chlorophyll is the principal light receptor, absorbing blue and red light. However the *carotenoids and various other pigments also absorb light energy and pass this on to the chlorophyll molecules.

phototaxis The movement of a cell (e.g. a gamete) or a unicellular organism in response to light. For example, certain algae (e.g. *Chlamydomonas*) can perceive light by means of a sensitive eyespot and move to regions of higher light concentration to enhance photosynthesis. *See* taxis.

phototropism (heliotropism) The growth of plant organs in response to light. Aerial shoots usually grow towards light, while some aerial roots grow away from light. The phototropic response is thought to be controlled by *auxin, a plant growth substance. *See* tropism.

pH scale A logarithmic scale for expressing the acidity or alkalinity of a solution. To a first approximation, the pH of a solution can be defined as $-\log_{10}c$, where c is the concentration of hydrogen ions in moles per cubic decimetre. A neutral solution at 25°C has a hydrogen-ion concentration of 10^{-7} mol dm^{-3}, so the pH is 7. A pH below 7 indicates an acid solution; one above 7 indicates an alkaline solution. pH stands for 'potential of hydrogen'. The scale was introduced by S. P. Sørensen in 1909.

Phycomycetes A class of relatively primitive *fungi, many of which are found in water (e.g. the water moulds, which may be parasitic on fish) or in damp areas. Many are unicellular but those that form mycelia generally have hyphae lacking cross walls, which distinguishes them from the *Ascomycetes and *Basidiomycetes.

phyletic *See* phylogenetic.

phyllotaxis (phyllotaxy) The arrangement of leaves on a plant stem. The leaves may be inserted in whorls or pairs at each node or singly up the stem. When arranged in pairs the two leaves arise on opposite sides of the stem and are usually at right angles to the leaf pairs above and below them. Single leaves may be inserted alternately or in a spiral pattern up the stem. Phyllotaxis generally results in the minimum of shading of leaves by those above them.

phylogenetic (phyletic) Describing a system of *classification of organisms that aims to show their evolutionary history. *Compare* phenetic.

phylogeny The evolutionary history of an organism or group of related organisms. *Compare* ontogeny.

phylum A category used in the *classification of animals that consists of one or several similar or closely related classes. Examples of phyla are the Protozoa, Porifera, Nematoda, Arthropoda, Mollusca, and Chordata. Phyla are grouped into the kingdom Animalia. In plant classification, the *division is usually used instead of the phylum.

physiological saline A liquid medium in which animal tissues may be kept alive for a few hours during experiments without pathological changes or distortion of the cells taking place. Such fluids are salt solutions that are isotonic with and have the same pH as the body fluids of the animal. A well-known example is *Ringer's solution*, formulated by the British physiologist S. Ringer (1835–1910), which is a mixture of sodium chloride, calcium chloride, sodium bicarbonate, and potassium chloride solutions.

physiological specialization The occurrence within a species of several forms that are identical in appearance but differ in physiology: these are termed *physiological races*. For example, many pathogenic fungi develop new physiological races in response to the strong selection pressure exerted when disease-resistant crop varieties are sown over large areas.

physiology The branch of biology concerned with the vital functions of plants and animals, such as nutrition, respiration, reproduction, and excretion.

physisorption *See* adsorption.

phyto- Prefix denoting plants. For example, phytopathology is the study of plant diseases.

phytochrome A protein-based plant pigment present in small quantities in many plant organs. It exists in two interconvertible forms: a physiologically active form, which forms when the plant is illuminated with red light or normal daylight; and an inactive form, formed when the plant is exposed to far-red light or darkness. The active form regulates many plant processes, such as seed germination and the initiation of flowering.

phytogeography *See* plant geography.

phytohormone *See* growth substance.

phytoplankton The plant component of *plankton, consisting chiefly of microscopic algae, such as diatoms and dinoflagellates. Near the surface of the sea there may be many millions of such plants per cubic metre. Members of the phytoplankton are of great importance as they form the basis of food for all other forms of aquatic life, being the primary *producers. *Compare* zooplankton.

pia mater The innermost of the three membranes (*meninges) that surround the brain and spinal cord of vertebrates. The pia mater lies immediately adjacent to the central nervous system, and the *choroid plexus, which secretes cerebrospinal fluid, is an extension of it.

pico- Symbol p. A prefix used in the metric system to denote 10^{-12}. For example, 10^{-12} farad = 1 picofarad (pF).

picornavirus One of a group of small RNA-containing viruses (*pico* = small; hence pico-RNA-virus) commonly present in the alimentary and respiratory tracts of vertebrates. They cause mild infections of these tracts but the group also includes the polioviruses, which attack the central nervous system causing poliomyelitis; and the causal agent of foot and mouth disease in cattle, sheep, and pigs.

pie chart A diagram in which percentages are shown as sectors of a circle. If x percent of the prey of a carnivore comprises species X, y percent species Y, and z percent species Z, a pie chart would show three sectors having central angles $3.6x°$, $3.6y°$, and $3.6z°$.

pileus The umbrella-shaped cap of certain fungi, such as mushrooms. Spores are produced from *gills or pores on the lower surface.

piliferous layer The part of the root epidermis that bears *root hairs. It extends over a region about 4–10 mm behind the root tip. Beyond this the piliferous layer is sloughed off to reveal the hypodermis.

Piltdown man Fossil remains, purported to have been found by Charles Dawson at Piltdown, Sussex, in 1912, that were named *Eoanthropus dawsoni* and described as a representative of the true ancestors of modern man. The skull resembled that of a man but the jaw was apelike. In 1953 dating techniques showed the specimen to be a fraud.

pineal eye *See* median eye.

pineal gland An outgrowth of the *forebrain. In man its functions are obscure, but in other vertebrates it acts as an endocrine gland, secreting hormones that affect reproductive function and behaviour.

pinna (auricle) The visible part of the *outer ear, present in some mammals. It is made of cartilage and its function is to channel sound waves into the external auditory meatus. In some species the pinna is movable and aids in detecting the direction from which a sound originates.

pinocytosis The process by which a living cell engulfs a minute droplet of liquid. It involves a technique similar to *phagocytosis.

pipette A graduated tube used for transferring measured volumes of liquid.

Pisces The fishes: a group of cold-blooded aquatic vertebrates that have *gills and *fins (a few species can move on land and breathe air). The whole body is covered with a tough, usually scaly, skin (*see* scales), which extends over the eye and contains colour pigments and sometimes slime glands. The circulatory system is a single circuit with blood passing through two sets of capillaries, one at the gills and the other in the body tissues. Three *semicircular canals are present in the inner ear. The sense of smell is particularly well developed and pressure waves are detected by the *lateral line system.*

Fossils of fish date back to the Ordovician period, 500–440 million years ago. Modern fish are divided into the *Elasmobranchii (cartilaginous fishes) and *Osteichthyes (bony fishes).

pistil The female part of a flower, consisting either of a single *carpel (*simple pistil*) or a group of carpels (*compound pistil*).

pith 1. (*or* **medulla**) The cylinder of *parenchyma tissue found in the centre of plant stems to the inside of the vascular tissue. It is light in weight and has been put to various commercial uses, notably the manufacture of pith helmets. **2.** (not in scientific usage) The white tissue below the rind of many citrus fruits. **3.** To destroy the central nervous system of an animal, especially a laboratory animal such as a frog, by severing its spinal cord.

Pithecanthropus *See Homo.*

pituitary gland (pituitary body; hypophysis) A pea-sized endocrine gland attached by a thin stalk to the *hypothalamus at the base of the brain. It consists of two lobes: the anterior and the posterior. The *anterior pituitary* secretes such hormones as *growth hormone, the *gonadotrophins, *prolactin, thyroid-stimulating hormone (*see* thyroid gland) and *ACTH. Because these hormones regulate the growth and activity of several other endocrine glands, the anterior pituitary is often referred to as the *master endocrine gland.* Activity of the anterior pituitary itself is regulated by specific hormone-releasing factors (neurosecretions) produced by the hypothalamus (*see also* neuroendocrine system). The *posterior pitui-*

tary secretes the hormones *oxytocin and *vasopressin.

placenta 1. The organ in mammals and other viviparous animals by means of which the embryo is attached to the wall of the uterus. It is composed of embyronic and maternal tissues: extensions of the *chorion and *allantois grow into the uterine wall so that materials (e.g. oxygen, nutrients) can pass between the blood of the embryo and its mother (there is, however, no direct connection between the maternal and embryonic blood). The placenta is eventually expelled as part of the *afterbirth. **2.** A ridge of tissue on the ovary wall of flowering plants to which the ovules are attached. The arrangement of ovules on the placenta (*placentation*) is variable, depending on the number of carpels and whether they are free (*see* apocarpy) or fused (*see* syncarpy).

Placentalia *See* Eutheria.

placoid scale (denticle) *See* scales.

plagiotropism The tendency for a *tropism (growth response of a plant) to be orientated at an angle to the line of action of the stimulus concerned. For example, the growth of lateral branches and lateral roots is at an oblique angle to the stimulus of gravity (*plagiogeotropism*). *Compare* orthotropism.

planarians *See* Turbellaria.

plankton Minute *pelagic organisms that drift or float passively with the current in a sea or lake. Plankton includes many microscopic animals and plants, such as algae, protozoans, various animal larvae, and some worms. It forms an important food source for many other members of the aquatic community and is divided into *zooplankton and *phytoplankton. *Compare* nekton.

plant Any living organism of the kingdom Plantae, characterized by an ability to manufacture carbohydrates by *photosynthesis, in which simple inorganic substances are built up into organic compounds. The radiant energy needed for this process is absorbed by *chlorophyll, a complex pigment not found in animals. Most plants, except for certain microscopic algae, are immobile as there is no necessity to search for food. Other characteristics of plants include the possession of *cell walls (usually composed of *cellulose) and slow response to external stimuli. For a classification of the plant kingdom, see Appendix.

plant geography (phytogeography) The study of the distribution of world vegetation, with particular emphasis on the influence of the environmental factors that determine this distribution.

plantigrade Describing the gait of many mammals, including man, in which the whole lower surface of the foot is on the ground. *Compare* digitigrade; unguligrade.

planula The ciliated free-swimming larva of many coelenterates, consisting of a solid mass of cells. It eventually settles on a suitable surface and develops into a *polyp.

plaque A thin layer of organic material covering all or part of the exposed surface of a tooth. It contains dissolved food (mostly sugar) and bacteria. The bacteria in plaque metabolize the sugar and produce acid, which eats into the surface of the enamel of the tooth and eventually causes tooth decay (caries).

plasma *See* blood plasma.

plasma cells Antibody-producing cells found in the epithelium of the lungs and gut and also in bone-forming tissue. They develop in the lymph nodes, spleen, and bone marrow when antigens stimulate lymphocytes to form the precursor cells that give rise to them.

plasmagel The specialized outer gel-like *cytoplasm of living cells (such as *Amoeba*) that move by extruding part of the cell (known as a *pseudopodium) in the direction of motion. A reversible conversion of plasmagel to the more fluid *plasmasol is believed to be responsible for the continuous flow forward of cytoplasm necessary for forming a pseudopodium.

plasmalemma *See* cell membrane.

plasma membrane *See* cell membrane.

plasmasol The specialized inner sol-like *cytoplasm of living cells that move by producing *pseudopodia. *Compare* plasmagel.

plasmid A structure in cells consisting of DNA that can exist and replicate independently of the chromosomes. In bacterial cells plasmids provide genetic instructions for certain cell activities (e.g. resistance to antibiotic drugs). They can be transferred from cell to cell in a bacterial colony. Bacterial plasmids are used to produce recombinant DNA in techniques of *genetic engineering. Similar blocks of DNA that appear to move from one chromosome to another have been identified in the cells of a number of other species, including man. In organisms (other than man) that have been studied, it appears that they interfere with gene activity. In man this has not been established, although they are almost identical to *retroviruses, which are known to affect gene action.

plasmolysis The loss of water from a plant cell to the extent that the protoplasm shrinks away from the cell wall. Continual plasmolysis results in wilting. *Compare* turgor.

plastid An *organelle within a plant cell, often occurring in large numbers. Apart from the nucleus, plastids are the largest solid inclusions in a plant cell. For convenience they are classified into those containing pigments (*chromoplasts) and those that are colourless (*leucoplasts), although changes from one to the other frequently occur. Plastids develop from *proplastids*, colourless bodies found in meristematic and immature cells; they also arise by division of existing plastids. *See also* chloroplast.

plastron *See* carapace.

platelet (thrombocyte) A small disc of red tissue in the blood. Platelets are formed as fragments of larger cells (*megakaryocytes*) found in red bone marrow; they have no nucleus. They play an important role in *blood clotting (*see also* clotting factors) and release serotonin, which causes blood vessels to constrict, so reducing bleeding. There are about 250 000 per cubic millimetre of blood.

plate tectonics The theory that the surface of the earth is made of lithospheric plates, which have moved throughout geological time resulting in the present-day positions of the continents. The theory explains the locations of mountain building as well as earthquakes and volcanoes. The rigid lithospheric plates consist of continental and oceanic crust together with the upper mantle, which lie above the weaker plastic asthenosphere. These plates move relative to each other across the earth. Six major plates (Eurasian, American, African, Pacific, Indian, and Antarctic) are recognized, together with a number of smaller ones. The plate margins coincide with zones of seismic and volcanic activity.

A *constructive* (or *divergent*) plate margin occurs when two plates move away from each other. It is marked by a mid-oceanic ridge where basaltic material wells up from the mantle to form new oceanic crust, in a process known as *sea-floor spreading*. The production of new crust at constructive plate margins is compensated for by the destruction of material along a *destructive* (or *divergent*) plate margin. Along these margins, which are also known as *subduction zones* and marked by an oceanic trench, one plate (usually oceanic) is forced to plunge down beneath the other (which may be continental or oceanic). The crust becomes partially melted and rises to form a chain of volcanoes in the upper plate parallel to the trench. When two continental plates collide the compression results in the formation of mountain chains. A third type of plate margin – the *transform plate margin* – occurs where two plates are slipping past each other.

Platyhelminthes A phylum of invertebrates comprising the flatworms, characterized by a flattened unsegmented body. The simple nervous system shows some concentration of cells at the head end. The mouth leads to a simple branched gut without an anus. Flatworms are hermaphrodite but self-fertilization is unusual. Many species are parasitic. The phylum contains the classes *Turbellaria (planarians), *Trematoda (flukes), and *Cestoda (tapeworms).

pleiomorphism The existence of distinctly different forms during the life cycle of an individual, e.g. the caterpillar, pupa, and winged adult of a butterfly.

Pleistocene The earlier epoch of the *Quaternary period (*compare* Holocene). It extended from the end of the Pliocene, about 2 million years ago, to the beginning of the Holocene, about 10 000 years ago. The Pleistocene is often known as the *Ice Age* as it was characterized by a series of glacials, in which ice margins advanced towards the equator, separated by interglacials when the ice retreated. *See also* ice age.

pleura The double membrane that lines the thoracic cavity and covers the exterior surface of the lungs. It is a *serous membrane forming a closed sac, with a small space (the *pleural cavity*) between the two layers.

plexus A compact branching network of nerves or blood vessels, such as the *brachial plexus* – a network of spinal nerves that supply branches to the forelimbs in vertebrates. *See also* choroid plexus.

Pliocene The fifth and final epoch of the *Tertiary period. Preceded by the Miocene and followed by the Pleistocene, it began about 7 million years ago and lasted for about 5 million years. Mammals similar to modern forms existed during the epoch and the australopithecines (*see Australopithecus*), early forerunners of man, appeared.

plumule **1.** (in zoology) A *down feather. **2.** (in botany) The part of a plant embryo that develops into the shoot system. It consists of the stem apex and first leaves. In seedlings showing *epigeal germination the plumule grows above the soil surface together with the cotyledons; in seeds showing *hypogeal germination, the plumule alone emerges. *Compare* radicle.

pod *See* legume.

poikilothermy (cold-bloodedness) The passive variation in the internal body temperature of an animal, which depends on the temperature of the environment. All animals except birds and mammals are cold-blooded. Although unable to maintain a constant body temperature, they can respond to compensate for very low or very high temperatures. For example, the tissue composition (especially cell osmotic pressure) can change to regulate the blood flow to peripheral tissues (and thus increase heat loss or heat absorption), and the animals can actively seek sun or shade. Seasonal changes in metabolism are usually under hormonal control. In particularly hot climates, cold-blooded animals may undergo *aestivation to escape the heat. *Compare* homoiothermy.

polar body *See* oogenesis.

pollen The mass of grains containing the male gametes of seed plants, which are produced in large numbers in the *pollen sacs. The pollen grains of insect-pollinated plants may be spiny or pitted and are usually larger than those of wind-pollinated plants, which are usually smooth and light. The pollen grain represents the male *gametophyte generation. The wall of the mature pollen grain consists of the tough outer wall (*exine*) and the more delicate narrower *intine*. The latter gives rise to the *pollen tube. *See also* pollination.

pollen analysis *See* palynology.

pollen sac The structure in seed plants in which pollen is produced. In angiosperms there are usually four pollen sacs in each *anther. In gymnosperms variable numbers of pollen sacs are borne on the microsporophylls that make up the male *cone.

pollen tube An outgrowth of a pollen grain, which transports the male gametes to the ovule. It will only grow if the pollen grain is compatible with the female tissue. In angiosperms, the pollen grain is deposited on the stigma and the pollen tube grows down through the style and into the ovule. In some gymnosperms, e.g. *Pinus* (pines), the pollen tube penetrates the *nucellus but does not develop further until the following year, when the female part of the plant is mature. *See also* embryo sac; fertilization.

pollex The innermost digit on the forelimb of a tetrapod vertebrate. It contains two phalanges (*see* pentadactyl limb) and in man and higher primates it is the thumb, which is opposable (i.e. capable of facing and touching the other digits) and gives the hand greater manipulating ability. In some mammals a pollex is absent. *Compare* hallux.

pollination The transfer of pollen from an anther (the male reproductive organ) to a stigma (the receptive part of the female reproductive organ), either of the same flower (*self-pollination*) or of a different flower of the same species (*cross-pollination*). Cross-pollination involves the action of a pollinating agent to effect transfer of the pollen (*see* anemophily; entomophily; hydrophily). *See also* fertilization; incompatibility.

pollution An undesirable change in the physical, chemical, or biological characteristics of the natural environment, brought about by man's activities. It may be harmful to human or nonhuman life. Pollution may affect the soil, rivers, seas, or the atmosphere. There are two main classes of pollutants: those that are *biodegradable* (e.g. sewage), i.e. can be rendered harmless by natural processes and need therefore cause no permanent harm if adequately dispersed or treated; and those that are *nonbiodegradable* (e.g. heavy metals (such as lead) and *DDT), which eventually accumulate in the environment and may be concentrated in *food chains. Other forms of pollution in the environment include noise (e.g. from jet aircraft, traffic, and industrial processes) and thermal pollution (e.g. the release of excessive waste heat into lakes or rivers causing harm to wildlife). Recent pollution problems include the disposal of *radioactive waste; *acid rain* resulting from industrial emissions of sulphates; increasing levels of human waste; and high levels of carbon dioxide in the atmosphere. Attempts to contain or prevent pollution include strict regulations concerning factory emissions, the use of smokeless fuels, and the banning of certain pesticides.

Polychaeta A class of annelid worms in which each body segment has a pair of flattened fleshy lobes (*parapodia*) bearing numerous bristles (*chaetae). All polychaetes are aquatic and most of them are marine. They include the fanworms (*Sabella*), which construct tubes of sand, etc., in which they live; and the lugworms (*Arenicola*) and ragworms (*Nereis*), which burrow in sand or mud.

polyembryony **1.** The formation of more than one embryo in a plant seed. Often one embryo develops from the fertilized egg cell, while the others have formed asexually from other tissues in the ovule. **2.** The formation of more than one embryo from a single animal zygote. *Identical twins are produced in this way.

polygene Any of a group of genes influencing a quantitative characteristic, e.g. height in man. *See* multifactorial inheritance.

polymer A substance having large molecules consisting of repeated units (the monomers). There are a number of natural polymers, such as polysaccharides.

polymorphism The existence of three or more distinctly different forms within a plant or animal species. An example is the *caste system of social insects, in which there are the workers, drones, and queens. This is an *environmental polymorphism*, i.e. the differences are caused by environmental rather than genetic factors, in this case by the larvae receiving different types of food. There are also *heritable* or *genetic polymorphisms*. An example is the occurrence of sickle-cell anaemia, a genetic disease that principally affects Negro populations of central Africa and is characterized by an abnormal form of the blood pigment haemoglobin and sickle-shaped red blood cells. Three different types of individual occur in such populations: those who have two genes (*AA*) for normal haemoglobin and therefore do not suffer from the disease; those with one normal and one abnormal gene (*AS*), who suffer from a relatively mild form; and those with two abnormal genes (*SS*), who suffer a chronic and eventually fatal form of anaemia. Normally such a harmful gene would have been eliminated from the population by the process of natural selection, but it is maintained in this case because people with the mild form of the disease are resistant to a severe form of malaria endemic in central Africa.

polyp The sedentary stage in the life cycle of the *Coelenterata, consisting of a cylindrical body fixed at one end to a firm base and having a mouth surrounded by a ring of tentacles at the other. Some polyps (e.g. *Hydra*) are single; others (e.g. the corals and *Obelia*) form colonies. Polyps typically reproduce asexually by budding to form either new polyps or *medusae. The latter reproduce sexually giving rise to new polyps. Sea anemones are solitary polyps that reproduce sexually to form new polyps.

polypeptide A *peptide comprising ten or more amino acids. Polypeptides that constitute proteins usually contain 100–300 amino acids. Shorter ones include certain antibiotics, e.g. gramicidin, and some hormones, e.g. *ACTH, which has 39 amino acids. The properties of a polypeptide are determined by the type and sequence of its constituent amino acids.

polyphyletic Denoting any group of organisms the members of which have originated from several different ancestors. An example is the group including all insectivorous animals. Polyphyletic groups are not natural groups and do not have any place in natural classifications. *Compare* monophyletic.

polyphyodont Describing a type of dentition in which the teeth are continuously shed and replaced during the lifetime of the animal. Sharks and frogs have a polyphyodont dentition. *Compare* diphyodont; monophyodont.

polyploid Describing a nucleus that contains more than two sets of chromosomes (*see* diploid) or a cell or organism containing such nuclei. For example, *triploid* plants have three sets of chromosomes and *tetraploid* plants have four. Polyploidy is far more common in plants than in animals; many crops, in particular, are polyploid (bread wheat, for example, is hexaploid, i.e. $6n$).

polysaccharide Any of a group of carbohydrates comprising long chains of monosaccharide (simple sugar) molecules. *Homopolysaccharides* consist of only one type of monosaccharide; *heteropolysaccharides* contain two or more different types. Polysaccharides may have molecular weights of up to several million and are often highly branched. Some important examples are starch, glycogen, and cellulose.

polyspermy The entry of several sperms into the egg during fertilization although only one sperm nucleus actually fuses with the egg nucleus. Polyspermy occurs in animals with yolky eggs (e.g. birds).

Polyzoa (Bryozoa) A phylum of aquatic, mainly marine, invertebrates – the moss animals. They live in colonies, 50 cm or more across, which are attached to rocks, seaweeds, or shells. The individuals making up the colonies are about 1 mm long and superficially resemble coelenterate *polyps, with a mouth surrounded by ciliated tentacles that trap minute particles of organic matter in the water. Some have a horny or calcareous outer skeleton into which the body can be withdrawn.

pome A type of fruit characteristic of apples and pears. The flesh of the fruit develops from the *receptacle of the flower, which completely encloses the fused carpels. After fertilization the carpels form the 'core' of the fruit, which contains the seeds. *See also* pseudocarp.

pons (pons Varolii) A thick tract of nerve fibres in the brain that links the medulla oblongata to the midbrain. Its function is to relay impulses between different parts of the brain. The pons is named after its discoverer, the Italian anatomist C. Varoli (1543–75).

population (in ecology) **1.** A group of individuals of the same species within a *community. The nature of a population is determined by such factors as density, *sex ratio, birth and death rates, emigration, and immigration. **2.** The total number of individuals of a given species or other class of organisms in a defined area, e.g. the population of rodents in Britain.

population dynamics The study of the fluctuations that occur in the numbers of individuals in animal and plant populations and the factors controlling these fluctuations. An important distinction is maintained between those factors that are dependent on population density and have a stabilizing effect (e.g. food supply) and those that are independent of population density (e.g. catastrophes, such as flooding).

population genetics The study of the distribution of inherited *variation among a group of organisms of the same species. The potential for change depends on the sum total of alleles that are available to the organisms (the *gene pool*), and estimates of changes in allele frequency in a population give an indication of its response to a changing condition.

Porifera The phylum of marine and freshwater invertebrates that comprise the sponges, which live permanently attached to rocks or other surfaces. The body of a sponge is hollow and consists basically of an aggregation of cells between which there is little nervous coordination. The body is supported by an internal skeleton of spicules of chalk, silica, or fibrous protein (bath sponges have protein skeletons). Flagellated cells (*choanocytes*) cause water to flow in through openings in the body wall and out through openings at the top; food particles are filtered from the water by the choanocytes.

porphyrin Any of a group of related organic compounds characterized by the possession of a cyclic group of four linked nitrogen-containing rings (a *pyrrole* nucleus). Porphyrins differ in the nature of their side-chain groups. They include the *chlorophylls and the *haem groups of haemoglobin, myoglobin, and the cytochromes.

portal vein (portal circulation; portal system) Any vein that collects blood from one network of capillaries and transports it directly to a second capillary network in another region of the body, without returning to the heart. *See* hepatic portal system.

positive feedback *See* feedback.

postcaval vein *See* vena cava.

posterior 1. Designating the part of an animal that is to the rear, i.e. that follows when the animal is moving. In man and bipedal animals (e.g. kangaroos) the posterior surface is equivalent to the *dorsal surface. **2.** Designating the side of a flower or axillary bud that faces towards the flower stalk or main stem, respectively. *Compare* anterior.

potassium Symbol K. A soft silvery metallic element that is an *essential element for living organisms. The potassium ion, K^+, is the most abundant cation in plant tissues, being absorbed through the roots and being used in such processes as protein synthesis. In animals the passage of potassium and sodium ions across the nerve-cell membrane is responsible for the changes of electrical potential that accompany the transmission of impulses.

potassium–argon dating A *dating technique for certain rocks that depends on the decay of the radioisotope potassium–40 to argon–40, a process with a half-life of about 1.27×10^{10} years. It assumes that all the argon–40 formed in the potassium-bearing mineral accumulates within it and that all the argon present is formed by the decay of potassium–40. The mass of argon–40 and potassium–40 in the sample is estimated and the sample is then dated from the equation:

$$^{40}\mathrm{Ar} = 0.1102\ ^{40}\mathrm{K}(e^{\lambda t} - 1),$$

where λ is the decay constant and t is the time in years since the mineral cooled to about 300°C, when the ^{40}Ar became trapped in the crystal lattice. The method is effective for micas, feldspar, and some other minerals.

potometer An apparatus used to measure the rate of water loss from a shoot (*see* transpiration) under natural or artificial conditions.

poxvirus One of a group of DNA-containing viruses, often enclosed in an outer membrane, that typically produce skin lesions in vertebrates. They include those viruses causing smallpox (variola), cowpox (vaccinia), and myxomatosis. Some poxviruses produce tumours.

Precambrian The division of geological time from the formation of the earth, believed to be about 4600 million years ago, to the beginning of the Cambrian period, some 570 million years ago. The Precambrian thus represents most of geological time. Fossils are rare, although *stromatolites occur, and subsequent metamorphism of Precambrian rocks makes correlation of rocks and events extremely difficult. The Precambrian was formerly divided into the Archaean and the Proterozoic but these divisions are now considered largely irrelevant. The largest areas of exposed Precambrian rocks are the shield areas, such as the Canadian (Laurentian) Shield and the Baltic Shield.

precaval vein *See* vena cava.

precipitin Any *antibody that combines with its specific soluble *antigen to form a precipitate. The term is sometimes applied to the precipitate itself. *See also* agglutination.

premolar A broad ridged tooth in mammals that is situated behind the *canine teeth (when present) and in front of the *molars. Premolars are adapted for grinding and chewing food and are present in both the deciduous and permanent dentitions.

presbyopia A loss of accommodation that normally develops in human eyes over the age of 45–50 years. Vision of distant objects remains unchanged but accommodation of the eye to near objects is reduced as a result of loss of elasticity in the lens of the eye. The defect is corrected by reading glasses using weak converging lenses.

presumptive Describing embryonic tissue that is not yet *determined but which will eventually develop into a certain kind of tissue by virtue of its position in the embryo.

prickle A hard sharp protective outgrowth, many of which may cover the surface of a plant. It contains cortical and vascular tissue and is not regarded as an epidermal outgrowth. *Compare* spine; thorn.

primary growth The increase in size of shoots and roots of plants that results from the activity of the *apical (tip) meristems and subsequent expansion of the cells produced. The tissues thus produced are called *primary tissues* and the resultant plant parts constitute the *primary plant body*. *Compare* secondary growth.

Primates An order of mammals that includes the monkeys, apes, and man. Primates evolved from arboreal insectivores 130 million years ago. They are characterized by thumbs and big toes that are opposable (i.e. capable of facing and touching the other digits), which permits manual dexterity, and forward-facing eyes allowing *binocular vision. The brain, particularly the cerebrum, is relatively large and well-developed, accounting for the intelligence and quick reactions of these mammals. The young are usually produced singly and undergo a long period of growth and development to the adult form.

primordium A group of cells that represents the initial stages in development of a plant organ. Root and shoot primordia are present in a young plant embryo while leaf primordia (or *leaf buttresses*) are seen as small bulges just below the shoot apex.

Proboscidea The order of mammals that comprises the elephants. They are herbivorous, with a muscular trunk (*proboscis) used for drinking, bathing, and collecting food. The tusks are continuously growing upper incisors and the enormous ridged molar teeth are produced in sequence to replace worn teeth throughout life. The order, which evolved in the Eocene epoch, was formerly much larger and more widespread than it is today and included the extinct mammoths. There are ony two species of modern elephants: the African and Indian species.

proboscis **1.** The trunk of an elephant: a muscular and very flexible elongation of the nose, which has a finger-like extremity and is capable of picking up and moving objects, taking in water, collecting food, etc. **2.** The elongated mouthparts of certain invertebrates, such as the two-winged flies (Diptera).

procambium A plant tissue formed by the *apical meristems of shoots and roots. It consists of cells elongated parallel to the long axis of the plant. The procambium subsequently gives rise to the primary *vascular tissue.

procaryote *See* prokaryote.

Proconsul A genus of extinct apes known from fossil remains, 50 million years old, found mostly in Kenya and assigned to the species *P. africanus*. A find made in 1948 was the first discovery of a skull of Miocene ape. It is believed that *Proconsul* was similar and possibly closely related to the genus **Dryopithecus*.

producer An organism considered as a source of energy for those above it in a *food chain (i.e. at the next *trophic level). Green plants, which convert energy from sunlight into chemical energy, are *primary producers*; herbivores are *secondary producers*, as they utilize energy from green plants and supply energy for carnivores. *Compare* consumer.

productivity (in ecology) The rate at which an organism, population, or community assimilates energy (*gross productivity*) or makes energy potentially available (as body tissue) to an animal that feeds on it (*net productivity*). The difference between these two rates is due to the rate at which energy is lost through excretion and respiration. Thus *gross primary productivity* is the rate at which plants assimilate light energy, and *net primary productivity* is the rate at which energy is incorporated as plant tissue. In terrestrial plants, much of the net productivity is not actually available to *consumers, e.g. tree roots are not eaten by herbivores. *See* producer.

profundal Occurring in or designating the deep-water zone of an inland lake. Light intensity, oxygen concentration, and (during summer and autumn) temperature are markedly lower than in the surface layer. *Compare* littoral; sublittoral.

progesterone A hormone, produced primarily by the *corpus luteum of the ovary but also by the placenta, that prepares the inner lining of the uterus for implantation of a fertilized egg cell. If implantation fails, the corpus luteum degenerates and progesterone production ceases accordingly. If implantation occurs, the corpus luteum continues to secrete progesterone, under the influence of *luteinizing hormone and *prolactin, for several months of pregnancy, by which time the placenta has taken over this function. During pregnancy, progesterone maintains the constitution of the uterus and prevents further release of eggs from the ovary. Small amounts of progesterone are produced by the testes. *See also* progestogen.

progestogen One of a group of naturally occurring or synthetic hormones that maintain the normal course of pregnancy. The best known is *progesterone. In high doses progestogens inhibit secretion of *luteinizing hormone, thereby preventing ovulation, and alter the consistency of mucus in the vagina so that conception tends not to occur. They are therefore used as major constituents of oral contraceptives.

prokaryote (procaryote) An organism in which the genetic material is not enclosed in a cell nucleus. Bacteria and blue-green algae are prokaryotes. It is believed that eukaryotic cells (*see* eukaryote) probably evolved as symbiotic associations of prokaryotes. *See* cell.

prolactin (lactogenic hormone; luteotrophic hormone; luteotrophin) A hormone produced by the anterior pituitary gland. In mammals it stimulates the mammary glands to produce milk (*see* lactation) and the corpus luteum of the ovary to secrete the hormone *progesterone. Secretion of prolactin is increased by suckling. In birds prolactin stimulates secretion of crop milk by the crop glands.

proline *See* amino acid.

pronation Rotation of the lower forearm so that the hand faces backwards or downwards with the radius and ulna crossed. *Compare* supination.

prophase The first stage of cell division, during which chromosomes contract and divide along their length (except for the centromeres) into chromatids. In *mitosis, the chromosomes remain separate from each other. In the first division of *meiosis, homologous chromosomes become paired (*see* pairing). By the end of first prophase the two chromosomes begin to move apart.

proplastid *See* plastid.

proprioceptor Any *receptor that is sensitive to movement, pressure, or stretching within the body. Proprioceptors occurring in muscles, tendons, and ligaments are important for the coordination of muscular activity and the maintenance of balance and posture.

prop root Any of the modified roots that arise from the stem of certain plants and provide extra support. Such stems are usually tall and slender and the prop roots develop at successively higher levels as the stem elongates, as in the maize plant. *Buttress roots*, which develop at the base of the trunks of many tropical trees, are similar but tend to have a more flattened appearance. *Stilt roots* are stouter than prop roots. Those formed at the base of the mangrove tree provide firm anchorage in the soft mud of the swamps.

prosencephalon *See* forebrain.

prostaglandin Any of a group of organic compounds derived from *essential fatty acids and causing a range of physiological effects in animals. Prostaglandins have been detected in most body tissues. They act at very low concentrations to cause the contraction of smooth muscle; natural and synthetic prostaglandins are used to induce abortion or labour in humans and domestic animals. Two prostaglandin derivatives have antagonistic effects on blood circulation: *thromboxane* A_2 causes blood clotting while *prostacyclin* causes blood vessels to dilate. Inflammation in allergic reactions and other diseases is also thought to involve prostaglandins.

prostate gland A gland in male mammals that surrounds and opens into the urethra where it leaves the bladder. During ejaculation it secretes a fluid into the semen that activates the sperms and prevents them from sticking together.

prosthetic group A tightly bound nonpeptide inorganic or organic component of a protein. Prosthetic groups may be lipids, carbohydrates, metal ions, phosphate groups, etc. Some *coenzymes are more correctly regarded as prosthetic groups.

protamine Any of a group of proteins of relatively low molecular weight found in association with the chromosomal *DNA of vertebrate male germ cells. They contain a single polypeptide chain comprising about 67% arginine. Protamines are thought to protect and support the chromosomes.

protandry 1. The condition in which the male reproductive organs (stamens) of a flower mature before the female ones (carpels), thereby ensuring that self-fertilization does not occur. Examples of protandrous flowers are ivy and rosebay willowherb. *Compare* protogyny; homogamy. *See also* dichogamy. **2.** The condition in hermaphrodite or colonial invertebrates in which the male gonads or individuals are sexually mature before the female ones. *Compare* protogyny.

protease (peptidase; proteinase; proteolytic enzyme) Any enzyme that catalyses the splitting of proteins into smaller *peptide fractions and amino acids, a process known as *proteolysis*. Examples are *pepsin and *trypsin. Several proteases, acting sequentially, are normally required for the complete digestion of a protein to its constituent amino acids.

protein Any of a large group of organic compounds found in all living organisms. Proteins comprise carbon, hydrogen, oxygen, and nitrogen and most also contain sulphur; molecular weights range from 6000 to several million. Protein molecules consist of one or several long chains (*polypeptides) of *amino acids linked in a characteristic sequence. This sequence is called the *primary structure* of the protein. These polypeptides may undergo coiling or pleating, the nature and extent of which is described as the *secondary structure*. The three-dimensional shape of the coiled or pleated polypeptides is called the *tertiary structure*. *Quaternary structure* specifies the structural relationship of the component polypeptides.

Proteins may be broadly classified into globular proteins and fibrous proteins. Globular proteins have compact rounded molecules and are usually water-soluble. Of prime importance are the *enzymes, proteins that catalyse biochemical reactions. Other globular proteins include the *antibodies, which combine with foreign substances in the body; the carrier proteins, such as *haemoglobin; the storage proteins (e.g. *casein in milk and *albumin in egg white), and certain hormones (e.g. *insulin). Fibrous proteins are generally insoluble in water and consist of long coiled strands or flat sheets, which confer strength and elasticity. In this category are *keratin and *collagen. Actin and myosin are the principal fibrous proteins of muscle, the interaction of which brings about muscle contraction. *Blood clotting involves the fibrous protein called fibrin.

When heated over 50°C or subjected to strong acids or alkalis, proteins lose their specific tertiary structure and may form insoluble coagulates (e.g. egg white). This usually inactivates their biological properties.

protein synthesis The process by which living cells manufacture proteins from their constituent amino acids, in accordance with the genetic information carried in the DNA of the chromosomes. This information is encoded in messenger *RNA, which is transcribed from DNA in the nucleus of the cell (*see* genetic code; transcription): the sequence of amino acids in a particular protein is determined by the sequence of nucleotides in messenger RNA. At the ribosomes the information carried by messenger RNA is translated into the sequence of amino acids of the protein in the process of *translation.

proteolysis The enzymic splitting of proteins. *See* protease.

proteolytic enzyme *See* protease.

prothallus A small flattened multicellular structure that represents the independent *gametophyte generation of pteridophytes. In some pteridophytes a single prothallus bears both male and female sex organs. In others there are separate male and female prothalli.

Protista A kingdom into which all organisms of simple biological organization can be classified. It includes the *algae, *bacteria, *fungi, and *Protozoa. Some authorities restrict membership of the kingdom to unicellular organisms. *Compare* Thallophyta.

protogyny **1.** The condition in which the female reproductive organs (carpels) of a flower mature before the male ones (stamens), thereby ensuring that self-fertilization does not occur. Examples of protogynous flowers are plantain and figwort. *Compare* protandry; homogamy. *See also* dichogamy. **2.** The condition in hermaphrodite or colonial invertebrates in which the female gonads or individuals are sexually mature before the male ones. *Compare* protandry.

protoplasm The granular material comprising the living contents of a *cell, i.e. all the substances in a cell except large vacuoles and material recently ingested or to be excreted. It consists of a *nucleus embedded in a jelly-like *cytoplasm.

protoplast (energid) The living unit (protoplasm) of a cell, consisting of the nucleus and cytoplasm bounded by the cell membrane. Protoplasts of bacterial and plant cells can be prepared by removing the cell wall; they are used to study the processes involved in cell metabolism and reproduction.

Prototheria A subclass of mammals – the monotremes – that lay large yolky eggs. It contains only the duckbilled platypus and the spiny anteater. After hatching, the young feed on milk from simple mammary glands inside a maternal abdominal pouch. In the anteater the eggs are also incubated in this pouch, while the platypus builds an underground nest. Adult monotremes have no true teeth. Their skeleton resembles that of a reptile, and although they are warm-blooded the body temperature is somewhat variable. They are believed to have originated at least 150 million years ago.

Protozoa A phylum comprising unicellular or acellular, usually microscopic, organisms regarded either as simple animals or members of the kingdom *Protista. They are very widely distributed in marine, freshwater, and moist terrestrial habitats; most protozoans are saprophytes, but some are parasites, including the agents causing malaria (*Plasmodium*) and sleeping sickness (*Trypanosoma*), and a few contain chlorophyll and carry out photosynthesis, like plants. Protozoan cells may be flexible or rigid, with an outer *pellicle or protective *test*. In some (such as *Paramecium* and *Trypanosoma*) *cilia or *flagella are present for locomotion; others (such as **Amoeba*) have *pseudopodia for movement and food capture. *Contractile vacuoles occur in freshwater protozoans. Reproduction is usually asexual, by binary *fission, but some protozoans undergo a form of sexual reproduction (*see* conjugation).

proventriculus **1.** The anterior part of the stomach of a bird, where digestive enzymes are secreted. Food passes from the proventriculus to the *gizzard. **2.** *See* gastric mill.

proximal Denoting the part of an organ that is nearest to the organ's point of attachment. For example, the knuckles are at the proximal end of the fingers. *Compare* distal.

pseudocarp (false fruit) A fruit that incorporates, in addition to the ovary wall, other parts of the flower, such as the *receptacle. For example, the fleshy part of the strawberry is formed from the receptacle and the 'pips' on the surface are the true fruits. *See also* composite fruit; pome; sorosis; syconus.

pseudoparenchyma A plant tissue that superficially resembles parenchyma but is made up of an interwoven mass of hyphae (in fungi) or filaments (in algae). Examples of pseudoparenchymatous structures are the fruiting bodies (mushrooms, toadstools, etc.) of certain fungi and the thalli of certain red and brown algae.

pseudopodium A temporary outgrowth of the cell of some protozoans (e.g. *Amoeba*), which serves as a feeding and locomotory organ. Pseudopodia may be blunt or thread-like, form a branching network, or be stiffened with an internal supporting rod. White blood cells also form pseudopodia to engulf invading bacteria.

pseudopregnancy A state resembling pregnancy that may occur in some mammals in which many of the phenomena of pregnancy are present but there is no fetus developing in the uterus.

Pteridophyta A division of the plant kingdom that includes the ferns, horsetails, and clubmosses. Pteridophytes possess true stems, leaves, and roots, and have an organized vascular system. They have a wide (mainly terrestrial) distribution with the maximum development in the tropics. Many genera are now extinct, having formed prominent plant groups in the Carboniferous period. There is a marked alternation of generations between gamete-bearing forms (gametophytes) and spore-bearing forms (sporophytes). Unlike the bryophytes, the generations are often independent and the sporophyte is the dominant generation. Pteridophytes differ from gymnosperms and angiosperms in not producing seeds. The division includes the classes *Lycopsida, *Pteropsida, and *Sphenopsida. In some classification schemes these classes are regarded as subdivisions of the *Tracheophyta.

Pteridospermales (seed ferns) An extinct order of the *Gymnospermae that flourished in the Carboniferous period. They possessed characteristics of both the ferns (*see* Filicinae) and the seed plants (*see* Spermatophyta) in reproducing by means of seeds and yet retaining fernlike leaves. Their internal anatomy combined both fern and seed-plant characteristics.

pterodactyls *See* Pterosauria.

Pteropsida A subdivision of the *Tracheophyta containing the ferns (*see* Filicinae) and seed plants (*see* Gymnospermae; Angiospermae), or a class of the *Pteridophyta containing only the ferns.

Pterosauria An extinct order of flying reptiles – the pterodactyls – that lived in the Jurassic and Cretaceous periods (180–70 million years ago). Pterodactyls had beaked jaws and an elongated fourth finger that supported a membranous wing. They had long jointed tails, no feathers, and could probably only fly by soaring.

ptyalin An enzyme that digests carbohydrates (*see* amylase). It is present in mammalian *saliva and is responsible for the initial stages of starch digestion.

pubis One of the three bones that make up each half of the *pelvic girdle. It is the most anterior of the three pelvic bones. In mammals and many reptiles the pubes are united at a slightly movable joint, the *pubic symphysis*. *See also* ilium; ischium.

pulmonary artery The artery that conveys deoxygenated blood from the right ventricle of the heart to the lungs, where it receives oxygen.

pulmonary vein The vein that conveys oxygenated blood from the lungs to the left atrium of the heart.

pulp cavity The central region of a tooth, which is connected by a narrow channel at the tip of the root with the surrounding tissues. The pulp cavity contains the *pulp* – connective tissue in which blood vessels and nerve fibres are embedded, and it is lined with *odontoblasts, which produce the *dentine.

pulse (in physiology) A series of waves of dilation that pass along the arteries, caused by pressure of blood pumped from the heart through contractions of the left ventricle. In man it can be felt easily where arteries pass close to the skin surface, e.g. at the wrist.

pulvinus A group of cells at the base of a leaf or leaflet in certain plants that, by rapidly losing water, brings about changes in the position of the leaves. In the sensitive plant (*Mimosa pudica*), the pulvinus is responsible for the folding of the leaves that occurs at nightfall or when the plant is touched or injured.

punctuated equilibrium A theory proposing that plant and animal species usually arise very quickly in terms of geological time (in less than 100 000 years) and seldom through a process of gradual change. It thus questions the traditional Darwinian theory of evolution, citing as evidence the discontinuities observed in the fossil records of certain animal groups (e.g. the ammonites).

pupa The third stage of development in the life cycle of some insects. During the pupal stage locomotion and feeding cease and *metamorphosis from the larva to the adult form takes place. There are three types of pupa. The commonest is the *exarate* or free pupa, in which the wings and other appendages are visible and movable. In the *obtect* type the wings are stuck to the body and immovable, as in the *chrysalis* of a butterfly or moth; and in the *co-arctate* type an exarate pupa develops within a hard barrel-shaped *puparium*, as in the housefly and other Diptera.

pupil *See* iris.

pure line A population of plants or animals all having a particular feature that has been retained unchanged through many generations. The organisms are *homozygous and are said to 'breed true' for the feature concerned.

purine An organic nitrogenous base, sparingly soluble in water, that gives rise to a

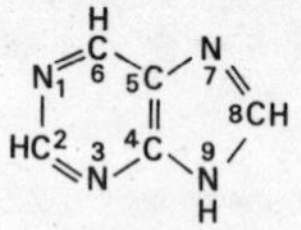

Purine

group of biologically important derivatives, notably *adenine and *guanine, which occur in *nucleotides and nucleic acids (DNA and RNA).

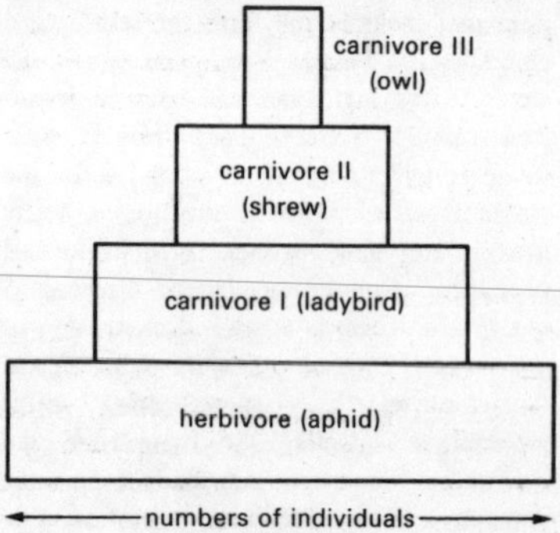

Pyramid of numbers for a woodland food chain

pyramid of numbers The numbers of animals found in an area at ascending *trophic levels of a *food chain. Because only a small proportion of the energy taken in by an organism is converted to tissue and is thus available to consumers at the next trophic level, the number of organisms that can be supported at each level is generally much less than the number at the level that supplies its food (i.e. the level below). A *pyramid of biomass*, representing the *biomass of animals at ascending trophic levels, shows a smaller decrease at each trophic level since the animals at the higher levels are generally larger (although fewer) than those at the lower levels.

pyrenocarp *See* drupe.

pyrenoid A spherical protein body found in the *chloroplasts of many algae. Pyrenoids are associated with the storage of starch: layers of starch are often found around them.

pyridoxine *See* vitamin B complex.

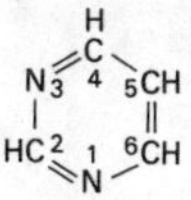

Pyrimidine

pyrimidine An organic nitrogenous base, sparingly soluble in water, that gives rise to a group of biologically important derivatives, notably *uracil, *thymine, and *cytosine, which occur in *nucleotides and nucleic acids (DNA and RNA).

pyruvic acid (2-oxopropanoic acid) A colourless liquid organic acid, $CH_3COCOOH$. It is an important intermediate compound in metabolism, being produced during *glycolysis and converted to acetyl coenzyme A, required for the *Krebs cycle.

Q

quadrat An ecological sampling unit consisting of a small square area of ground within which all species of interest are noted or measurements taken. Quadrats may be spaced over a larger area to form an overall view when a total survey would be impracticable, or they may be used to sample along a *transect.

quadrate A paired bone in the upper jaw of bony fishes, amphibians, reptiles, and birds that articulates with the lower jawbone. It is absent in mammals, being reduced to a small bone (the incus) in the middle ear (*see* ear ossicles).

qualitative variation *See* discontinuous variation.

quantitative variation *See* continuous variation.

Quaternary The second period of the Cenozoic era, which began about 2 million years ago, following the *Tertiary period, and includes the present. It is subdivided into two epochs – the *Pleistocene and *Holocene. The beginning of the Quaternary is usually based on the onset of a worldwide cooling. During the period four principal glacial phases occurred in Europe and North America, in which ice advanced towards the equator, separated by interglacials during which conditions became warmer and the ice sheets and glaciers retreated. The last glacial ended about 10 000 years ago. Man became the dominant terrestrial species during the Quaternary. Among the fauna adapted to the cold conditions were the mammoth and the woolly rhinoceros.

R

race 1. (in biology) A category used in the *classification of organisms that consists of a group of individuals within a species that are geographically, ecologically, physiologically, or chromosomally distinct from other members of the species. The term is frequently used in the same sense as *subspecies. *Physiological races*, for example, are identical in appearance but differ in function. They include strains of fungi adapted to infect different varieties of the same crop species. **2.** (in anthropology) A distinct human type possessing several characteristics that are genetically inherited. The major races of man are Mongoloid, Caucasian, Negroid, and Australoid.

raceme A type of *racemose inflorescence in which the main flower stalk is elongated and bears stalked flowers. An example is the lupin. *See also* panicle.

racemose inflorescence (indefinite inflorescence) A type of flowering shoot (*see* inflorescence) in which the growing region at the tip of the flower stalk continues to produce new flower buds during growth. As a result, the youngest flowers are at the top and the oldest flowers are at the base of the stalk. In a flattened inflorescence, the youngest flowers are in the centre and the oldest flowers are on the outside. Types of racemose inflorescence include the *capitulum, *catkin, *corymb, *raceme, *spadix,

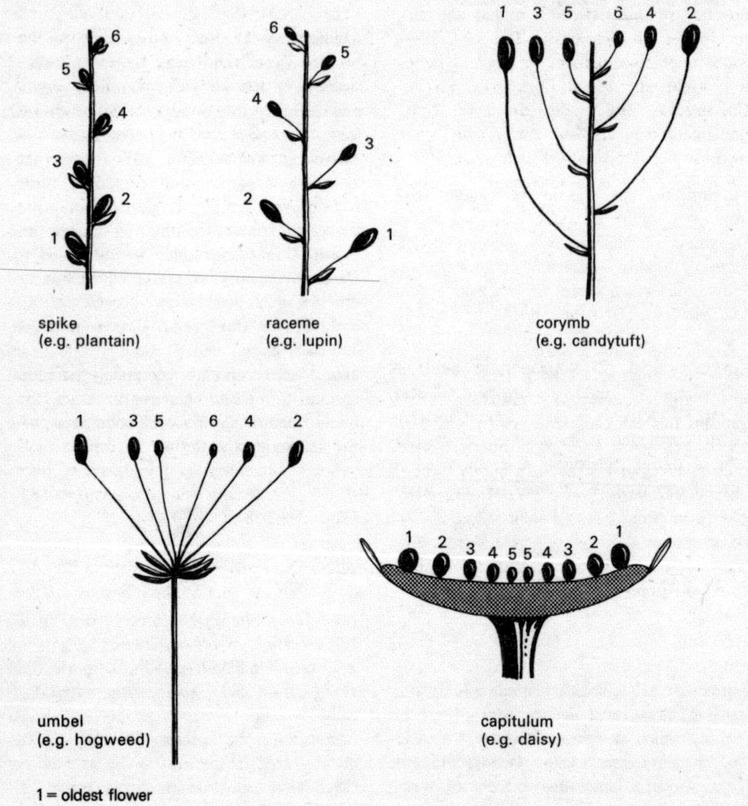

Types of racemose inflorescence

*spike, and *umbel. *Compare* cymose inflorescence.

rachis (rhachis) 1. The main axis of a compound leaf or an inflorescence. **2.** The shaft of a *feather. **3.** The backbone.

rad *See* radiation units.

radial symmetry The arrangement of parts in an organ or organism such that cutting through the centre of the structure in any direction produces two halves that are mirror images of each other. The stems and roots of plants usually show radial symmetry, while all animals belonging to the Coelenterata (e.g. jellyfish) and Echinodermata (e.g. starfish) are radially symmetrical – and typically sessile – in their adult form. The term *actinomorphy* is used to describe radial symmetry in flowers (e.g. a buttercup flower). *Compare* bilateral symmetry.

radiation damage Harmful changes that occur to inanimate materials and living organisms as a result of exposure to energetic electrons, nucleons, fission fragments, or high-energy electromagnetic radiation. In organisms this can cause changes to cells that alter their genetic structure, interfere with their division, or kill them. In humans, these changes can lead to *radiation sickness*, *radiation burns* (from large doses of radiation), or to long-term damage of several kinds, the most serious of which result in various forms of cancer (especially leukaemia).

radiation units Units of measurement used to express the activity of a radionuclide and the *dose of ionizing radiation. The units *curie*, *roentgen*, *rad*, and *rem* are not coherent with SI units but their temporary use with SI units has been approved while the derived SI units *becquerel*, *gray*, and *sievert* become familiar.

The becquerel (Bq), the SI unit of activity, is the activity of a radionuclide decaying at a rate, on average, of one spontaneous nuclear transition per second. Thus 1 Bq = 1 s^{-1}. The former unit, the curie (Ci), is equal to 3.7×10^{10} Bq. The curie was originally chosen to approximate the activity of 1 gram of radium-226.

The gray (Gy), the SI unit of absorbed dose, is the absorbed dose when the energy per unit mass imparted to matter by ionizing radiation is 1 joule per kilogram. The former unit, the rad (rd), is equal to 10^{-2} Gy.

The sievert (Sv), the SI unit of dose equivalent, is the dose equivalent when the absorbed dose of ionizing radiation multiplied by the stipulated dimensionless factors is 1 J kg^{-1}. As different types of radiation cause different effects in biological tissue a weighted absorbed dose, called the *dose equivalent*, is used in which the absorbed dose is modified by multiplying it by dimensionless factors stipulated by the International Commission on Radiological Protection. The former unit of dose equivalent, the rem (originally an acronym for *r*oentgen *e*quivalent *m*an), is equal to 10^{-2} Sv.

In SI units, exposure to ionizing radiation is expressed in coulombs per kilogram, the quantity of X- or gamma-radiation that produces ion pairs carrying 1 coulomb of charge of either sign in 1 kilogram of pure dry air. The former unit, the roentgen (R), is equal to 2.58×10^{-4} C kg^{-1}.

radicle The part of a plant embryo that develops into the root system. The tip of the radicle is protected by a root cap and points towards the micropyle. On germination it breaks through the testa and grows down into the soil. *Compare* plumule.

radioactive age The age of an archaeological or geological specimen as determined by a process that depends on a radioactive decay. *See* carbon dating; fission-track dating; potassium–argon dating; rubidium–strontium dating; uranium–lead dating.

radioactive tracing *See* labelling.

radioactive waste (nuclear waste) Any waste material that contains radionuclides (radioactive atomic nuclei). These wastes are produced in the mining of radioactive ores, the normal running of nuclear power stations and other reactors, the manufacture of nuclear weapons, and in research laboratories. Because high-level radioactive wastes can be extremely dangerous to all living matter and because they may contain radionuclides having half-lives of many thousands of years, their disposal has to be controlled with great stringency.

The first stage in disposal involves extensive reprocessing to retrieve usable material and reduce the activity of the waste. Subsequent final disposal, usually of solid material in strong metal canisters, involves burial of the material. The two currently favoured categories of sites for burial are deep (600 metres) mined cavities in stable geological formations on land and in red clay some 50–100 metres below the seabed away from tectonic plate margins and where the currents at the bottom of the sea are weak.

radiobiology The branch of biology concerned with the effects of radioactive substances on living organisms and the use of radioactive tracers to study metabolic processes (*see* labelling).

radiocarbon dating *See* carbon dating.

radiography The process or technique of producing images of an opaque object on photographic film or on a fluorescent screen by means of radiation (either particles or electromagnetic waves of short wavelength, such as X-rays and gamma-rays). The photograph produced is called a *radiograph*. The process is widely used in diagnostic *radiology, using X-rays. *See also* autoradiography.

radioisotope (radioactive isotope) An isotope of an element that is radioactive. *See* labelling.

radiology The study and use of X-rays, radioactive materials, and other ionizing radiations for medical purposes, especially for diagnosis (*diagnostic radiology*) and the treatment of cancer and allied diseases (*radiotherapy*).

radiometric dating (radioactive dating) *See* dating techniques; radioactive age.

radio-opaque Describing a medium that is opaque to X-rays and gamma rays. Examples are barium salts, used in diagnostic radiology of the digestive tract.

radiotherapy *See* radiology.

radius (in anatomy) The smaller of the two bones in the lower section of the forelimb of a tetrapod vertebrate (*compare* ulna). The radius articulates with some carpal bones and the ulna at the wrist and with the *humerus at the elbow. This sophisticated articulation of the radius enables man (and some other animals) to twist the forearm (*see* pronation; supination).

radula A tonguelike organ of molluscs, consisting of a horny strip whose surface is studded with rows of horny teeth for rasping food. In some species it is modified for scraping or boring.

r.a.m. *See* relative atomic mass.

Ramapithecus A genus of extinct primates that lived between 8 and 14 million years ago. Fossil remains of ramapithecines have been found in India and Pakistan, the Near East, and East Africa. Jaw fragments suggest that they chewed from side to side and had fairly short muzzles, both of which are humanoid features. They were possibly the earliest representatives of the hominid line after this split from the pongids (apes) during the Miocene. *See also Dryopithecus*; *Australopithecus*.

rank (category) The position or status of a *taxon in a *classification hierarchy. Examples of ranks are the family, genus, and species.

Ratitae A group comprising the flightless birds, including the ostrich, kiwi, and emu. They have long legs, heavy bones, small wings, a flat breastbone, and curly feathers. These birds are thought to have descended from a variety of flying birds and are not representatives of a single homologous group.

ray 1. (in optics) A narrow beam of radiation or an idealized representation of such a beam on a *ray diagram*, which can be used to indicate the positions of the object and image in a system of lenses or mirrors. **2.** (in botany) *See* medullary ray.

reaction time (latent period) The period of time between the detection of a stimulus at a sensory receptor and the performance of the appropriate response by the effector organ. This delay is caused by the time taken for the impulse to travel across the *synapses of adjacent neurones. The reaction time for a *reflex response, involving only a single linking synapse, is very short.

Recent *See* Holocene.

receptacle 1. (thalamus *or* **torus)** The tip of a flower stalk, which bears the petals, sepals, stamens, and carpels. The way the receptacle develops determines the position of the flower parts. It can be dilated and dome-shaped, saucer-shaped, or hollow and enclosing the gynaecium. In some plants it may become part of the fruit (*see* pseudocarp). **2.** A swollen part of the thallus of some algae, e.g. *Fucus*, that bears the conceptacles in which the sex organs are situated.

receptor (in physiology) A cell or group of cells specialized to detect a particular stimulus and to initiate the transmission of impulses via the sensory nerves. The eyes, ears, nose, skin, and other sense organs all contain specific receptors responding to external stimuli; other receptors are sensitive to changes within the body. *See also* baroreceptor; chemoreceptor; mechanoreceptor; proprioceptor.

recessive The *allele that does not function when two different alleles are present in the cells of an organism. The aspect of a characteristic controlled by a recessive allele only appears when two such alleles are present, i.e. in the *double recessive* condition. *Compare* dominant.

recipient An individual who receives tissues or organs of the body from another (the *donor).

reciprocal cross A *cross reversing the roles of males and females to confirm the results obtained from an earlier cross. For example, if the pollen (male) from tall plants is transferred to the stigmas (female) of dwarf plants in one cross, the reciprocal cross would use the pollen of dwarf plants to pollinate the stigmas of tall plants.

recombinant DNA DNA that contains genes from different sources that have been combined by the techniques of *genetic engineering rather than by breeding experiments. Genetic engineering is therefore also known as *recombinant DNA technology*.

recombination The rearrangement of genes that occurs when reproductive cells (gametes) are formed (*see* crossing over). Recombination results in offspring that have a combination of characteristics different from that of their parents.

rectum The portion of the *alimentary canal between the *colon and the *anus. Its main function is the storage of *faeces prior to elimination.

red algae *See* Rhodophyta.

red blood cell *See* erythrocyte.

reduction *See* oxidation–reduction.

reduction division *See* meiosis.

reflex An automatic and innate response to a particular stimulus, such as withdrawal of the hand from a painful stimulus (such as fire). A reflex response is extremely rapid. This is because it is mediated by a simple nervous circuit called a *reflex arc*, which at its simplest involves only a receptor linked to a sensory neurone, which synapses with a motor neurone (supplying the effector) in the spinal cord or brain. *See also* conditioned reflex.

regeneration The growth of new tissues or organs to replace those lost or damaged by injury. Many plants can regenerate a complete plant from a shoot segment or a single leaf, this being the basis of many horticultural propagation methods (*see* cutting). The capacity for regeneration in animals is less marked. Some planarians and sponges can regenerate whole organisms from small pieces, and crustaceans (e.g. crabs), echinoderms (e.g. brittlestars), and some reptiles and amphibians can grow new limbs or tails (*see* autotomy), but in mammals regeneration is largely restricted to wound healing.

regma A dry fruit that is characteristic of the geranium family. It is similar to the *carcerulus but breaks up into one-seeded parts, each of which splits open to release a seed.

reinforcement (in animal behaviour) Increasing (or decreasing) the frequency of a particular behaviour through *conditioning, by arranging for some biologically important event (the *reinforcer*) always to follow another event. In instrumental conditioning an *appetitive reinforcer*, or *reward* (e.g. food), given after a response made by the animal, increases that response; an *aversive reinforcer*, or *punishment* (e.g. an electric shock) decreases the response.

relative atomic mass (atomic weight; r.a.m.) Symbol A_r. The ratio of the average mass per atom of the naturally occurring form of an element to 1/12 of the mass of a carbon–12 atom.

relative density (r.d.) The ratio of the density of a substance to the density of some reference substance. For liquids or solids it is the ratio of the density (usually at 20°C) to the density of water (at its maximum density). This quantity was formerly called *specific gravity*.

relative molecular mass (molecular weight) Symbol M_r. The ratio of the average mass per molecule of the naturally occurring form of an element or compound to 1/12 of the mass of a carbon–12 atom. It is equal to the sum of the relative atomic masses of all the atoms that comprise a molecule.

releaser *See* sign stimulus.

rem *See* radiation units.

rennin An enzyme secreted by cells lining the stomach in mammals that is responsible for clotting milk. It acts on a soluble milk protein (*caseinogen*), which it converts to the insoluble form *casein. This ensures that milk remains in the stomach long enough to be acted on by protein-digesting enzymes.

replacing bone *See* cartilage bone.

reproduction The production of new individuals more or less similar in form to the parent organisms. This may be achieved by a number of means (*see* sexual reproduction; asexual reproduction) and serves to perpetuate or increase a species.

Reptilia The class that contains the first entirely terrestrial vertebrates, which can live in dry terrestrial habitats as their skin is covered by a layer of horny scales, preventing water loss. They breathe atmospheric oxygen by means of lungs assisted by respiratory movements principally involving the ribs (there is no diaphragm). Reptiles are cold-blooded (*see* poikilothermy) but behavioural patterns make it possible for them to maintain a fairly even body temperature throughout the day. Fertilization is internal and the majority of reptiles lay eggs on land. These eggs have a porous shell to provide protection from desiccation and allow gas exchange. In some reptiles the eggs are retained within the body of the mother until the young are ready to hatch, thereby greatly reducing juvenile mortality (*see* ovoviviparity).

The class includes the modern crocodiles, lizards and snakes (*see* Squamata), and tortoises and turtles, as well as many extinct forms, such as the *dinosaurs and *Pterosauria.

resin A naturally occurring acidic polymer secreted by many trees (especially conifers) into ducts or canals. Resins are found either as brittle glassy substances or dissolved in essential oils. Their functions are probably similar to those of gums and mucilages, i.e. protective.

resolving power A measure of the ability of an optical instrument to form separable images of close objects or to separate close wavelengths of radiation. The resolving power of a microscope is usually taken as the minimum distance between two points that can be separated; the smaller the resolving power, the better the resolution.

respiration The metabolic process in animals and plants in which organic substances are broken down to simpler products with the release of energy, which is incorporated into special energy-carrying molecules (*see* ATP) and subsequently used for other metabolic processes. In most plants and animals respiration requires oxygen, and carbon dioxide is an end product. The exchange of oxygen and carbon dioxide between the body tissues and the environment is called *external respiration.* In many animals the exchange of gases takes place at *respiratory organs (e.g. *lungs in air-breathing vertebrates) and is assisted by *respiratory movements (e.g. breathing). In plants oxygen enters through pores on the plant surface and diffuses through the tissues via intercellular spaces or dissolved in tissue fluids.

Respiration at the cellular level is known as *internal* (or *tissue*) *respiration* and can be divided into two stages. In the first, *glycolysis, glucose is broken down to pyruvate. This does not require oxygen and is a form of *anaerobic respiration. In the second stage, the *Krebs cycle, pyruvate is broken down by a cyclic series of reactions to carbon dioxide and water. This is the main energy-yielding stage and requires oxygen. The processes of glycolysis and the Krebs cycle are common to all plants and animals that respire aerobically (*see* aerobic respiration).

respiratory chain *See* electron transport chain.

respiratory movement The muscular movement that enables the passage of air to and from the lungs or other *respiratory organs of an animal. The mechanism of the movement varies with the species. In insects abdominal muscles relax and contract rhythmically to encourage the flow of air through the *tracheae. In amphibians air is drawn into the lungs by a pumping action of the muscles in the floor of the mouth. *Breathing* in mammals involves the muscle of the *diaphragm and the *intercostal muscles between the ribs. Contraction of these muscles lowers the diaphragm and raises the ribs, so that the lungs expand and air is drawn in. Relaxation has the opposite effect and forces air out during exhalation.

respiratory organ Any animal organ across which exchange of carbon dioxide and oxygen takes place. The surface membranes of such organs are always moist, thin, and well-supplied with blood. Examples are the *lungs of air-breathing vertebrates, the *gills of fish, and the *tracheae of insects.

respiratory pigment A coloured compound that is capable of reversibly binding with oxygen at high oxygen concentrations and releasing it at low oxygen concentrations. Such pigments are present in the blood, transporting oxygen within the circulatory system from the *respiratory organs to the tissues of the body. In vertebrates the respiratory pigment is *haemoglobin, contained in the erythrocytes (red blood cells).

respiratory quotient (RQ) The ratio of the volume of carbon dioxide produced by an organism during respiration to the volume of oxygen consumed. The RQ is usually about 0.8.

response The physiological, muscular, or behavioural activity that can be elicited by a *stimulus.

resting potential The difference in electrical potential that exists across the membrane of a nerve cell that is not in the process of transmitting a nerve impulse. The resting potential is maintained by means of the *sodium pump. *Compare* action potential.

restriction enzyme (restriction endonuclease) A type of enzyme that can cleave molecules of foreign *DNA. Restriction enzymes are produced by many bacteria and protect the cell by cleaving (and therefore destroying) the DNA of invading viruses. The bacterial cell is protected from attack by its own restriction enzymes by modifying the bases of its DNA during replication. Restriction enzymes are widely used in the techniques of *genetic engineering.

retina The light-sensitive membrane that lines the interior of the eye. The retina consists of two layers. The inner layer contains nerve cells, blood vessels, and two types of light-sensitive cells (*rods and *cones). The outer layer is pigmented, which prevents the back reflection of light and consequent decrease in visual acuity. Light passing through the lens stimulates individual rods and cones, which generates nerve impulses that are transmitted through the optic nerve to the brain, where the visual image is formed.

retinol *See* vitamin A.

retrovirus An RNA-containing virus that causes cancer in animals. Retroviruses contain oncogenes (cancer-causing genes; *see* oncogenic), which are activated when the virus enters its host cell and starts to replicate.

rhachis *See* rachis.

rhesus factor (Rh factor) An *antigen whose presence or absence on the surface of red blood cells forms the basis of the rhesus *blood group system. (The factor was first recognized in rhesus monkeys.) Most people possess the Rh factor, i.e. they are rhesus positive (Rh+). People who lack the factor are Rh−. If Rh+ blood is given to an Rh− patient, the latter develops anti-Rh antibodies. Subsequent transfusion of Rh+

blood results in *agglutination, with serious consequences. Similarly, an Rh− pregnant woman carrying an Rh+ fetus may develop anti-Rh antibodies in her blood; these will react with the blood of a subsequent Rh+ fetus, causing anaemia in the newborn baby.

rhizoid One of a group of delicate and often colourless hairlike outgrowths found in certain algae and the gametophyte generation of bryophytes and some pteridophytes. They anchor the plant to the substrate and absorb water and mineral salts.

rhizome A horizontal underground stem. It enables the plant to survive from one growing season to the next and in some species it also serves to propagate the plant vegetatively. It may be thin and wiry, as in couch grass, or fleshy and swollen, as in *Iris*. Compact upright underground stems, as in rhubarb, strawberry, and primrose, are often called *rootstocks*.

Rhodophyta (red algae) A division of *algae that are often pink or red in colour due to the presence of the pigments phycocyanin and phycoerythrin. Members of the Rhodophyta may be unicellular or multicellular, forming branched flattened thalli or filaments. They are commonly found along the coasts of tropical areas.

rhodopsin (visual purple) The light-sensitive pigment found in the *rods of the vertebrate retina. Light falling on the rod causes a chemical change in rhodopsin, which initiates the transmission of a nerve impulse to the brain. The great sensitivity of rhodopsin allows vision in dim light (night vision).

rhombencephalon *See* hindbrain.

rhytidome *See* bark.

rib One of a series of slender curved bones that form a cage to enclose, support, and protect the heart and lungs (*see* thorax). Ribs occur in pairs, articulating with the *thoracic vertebrae of the spinal column at the back and (in reptiles, birds, and mammals) with the *sternum (breastbone) in front. Movements of the rib cage, controlled by *intercostal muscles* between the ribs, are important in breathing (*see* respiratory movement).

riboflavin *See* vitamin B complex.

ribonucleic acid *See* RNA.

ribose A *monosaccharide, $C_5H_{10}O_5$, rarely occurring free in nature but important as a component of *RNA (ribonucleic acid). Its derivative *deoxyribose*, $C_5H_{10}O_4$, is equally important as a constituent of *DNA (deoxyribonucleic acid), which carries the genetic code in chromosomes.

ribosomal RNA *See* ribosome; RNA.

ribosome A small spherical body within a living cell that is the site of *protein synthesis. Ribosomes consist of a type of RNA (called *ribosomal RNA*) and protein. Usually there are many ribosomes in a cell, either attached to the *endoplasmic reticulum or free in the cytoplasm. During protein synthesis they are associated with messenger RNA in the process of *translation.

rickettsia A very small rod-shaped nonmotile microorganism that resembles a bacterium in its cellular structure and method of asexual reproduction but, like a virus, is totally parasitic, being unable to reproduce outside the cells of its host. Rickettsiae can infect such arthropods as ticks, fleas, lice, and mites, through which they can be transmitted to vertebrates, including man. The group includes the causal agents of trench fever, Rocky Mountain spotted fever, and forms of typhus.

Ringer's solution *See* physiological saline.

ritualization An evolutionary process in which the form or context of an action is altered because it comes to play a role in social communication. For example, many *courtship and greeting ceremonies in animals include ritual food presentation, derived from the action of feeding the young.

RNA (ribonucleic acid) A complex organic compound (a nucleic acid) in living cells that is concerned with *protein synthesis. In some viruses, RNA is also the hereditary material. Most RNA is synthesized in the nucleus and then distributed to various parts of the cytoplasm. An RNA molecule consists of a long chain of *nucleotides in which the sugar is *ribose and the bases are adenine, cytosine, guanine, and uracil (*compare* DNA). *Messenger RNA* (*mRNA*) is re-

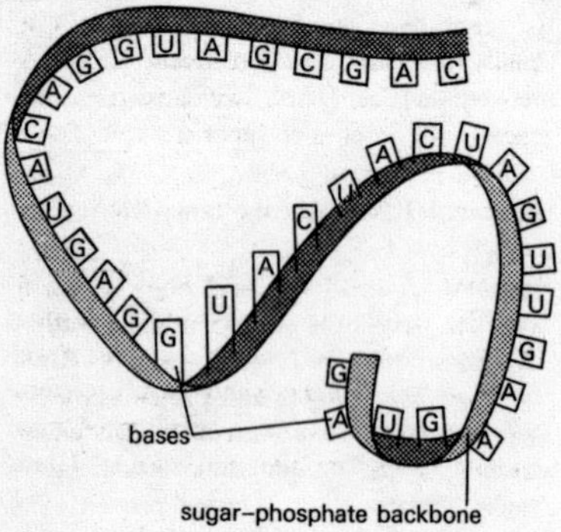

Single-stranded structure of RNA

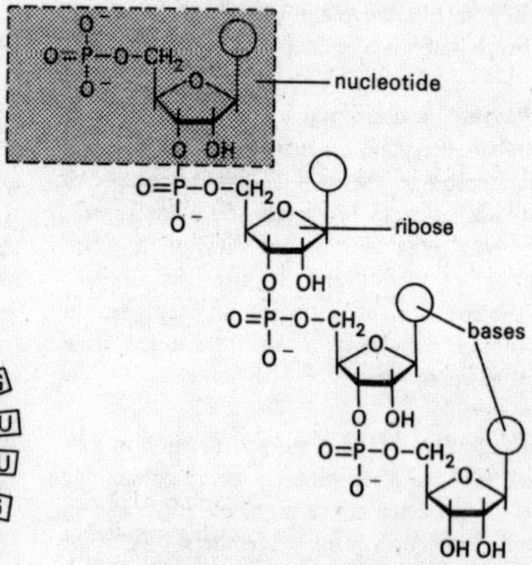

Detail of molecular structure of sugar–phosphate backbone. Each ribose unit is attached to a phosphate group and a base, forming a nucleotide

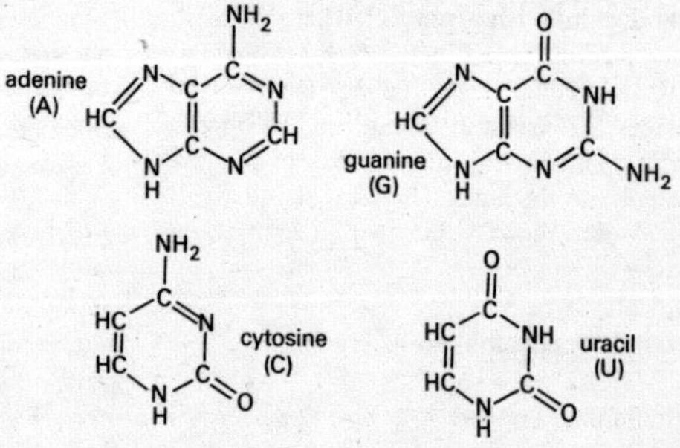

The four bases of RNA

Molecular structure of RNA

sponsible for carrying the *genetic code transcribed from DNA to specialized sites within the cell (known as *ribosomes), where the information is translated into protein composition (*see* transcription; translation). *Ribosomal RNA* (*rRNA*) is present in *ribosomes; it is single-stranded but helical regions are formed by *base pairing within the strand. *Transfer RNA* (*tRNA, soluble RNA, sRNA*) is involved in the assembly of amino acids in a protein chain being synthesized at a ribosome. Each tRNA is specific for an amino acid and bears a triplet of bases complementary with a triplet on mRNA (*see* codon).

rod A type of light-sensitive receptor cell present in the retinas of vertebrates. Rods contain the pigment *rhodopsin and are essential for vision in dim light. They are not evenly distributed on the retina, being absent in the *fovea and occupying all of the retinal margin. *Compare* cone.

Rodentia An order of mammals characterized by a single pair of long curved incisors in each jaw. These teeth are specialized for gnawing: they continue growing throughout life and have enamel only on the front so that they wear to a chisel-shaped cutting edge. Rodents often breed throughout the year and produce large numbers of quickly maturing young. The order includes the squirrels, beavers, rats, mice, and porcupines.

roentgen The former unit of dose equivalent (*see* radiation units). It is named after the discoverer of X-rays, W. K. Roentgen (1845–1923).

root 1. (in botany) The part of a vascular plant that grows beneath the soil surface in response to gravity and water. It anchors the plant in the soil and absorbs water and mineral salts. Unlike the stem, it never produces leaves, buds, or flowers and never contains chlorophyll. The *radicle (embryonic root) may give rise either to a *tap root system* with a single main *tap root* from which lateral roots develop, or a *fibrous root system*, with many roots of equal size. The *apical meristem at the root tip gives rise to a protective sheath, the *root cap, and to the primary tissues of the root. The vascular tissues usually form a central core. This distinguishes roots from stems, in which the vascular tissue often forms a ring. A short distance behind the root tip *root hairs* develop from the epidermis and greatly increase the surface area for the absorption of water and minerals. Beyond this region lateral roots develop.

Roots may be modified in various ways. Some are swollen with food to survive the winter, as in the carrot. Certain plants, such as orchids, have absorptive aerial roots; others, such as ivy, have short clasping roots for climbing. The roots of leguminous plants, such as beans and peas, contain nodules due to the presence of symbiotic bacteria that fix free nitrogen from the soil (*see* nitrogen fixation). Other modifications include *prop roots, stilt roots, and buttress roots, which support the plant.

2. (in dentistry) The portion of a *tooth that is not covered with enamel and is embedded in a socket in the jawbone. Incisors, canines, and premolars have single roots; molars normally have several roots.

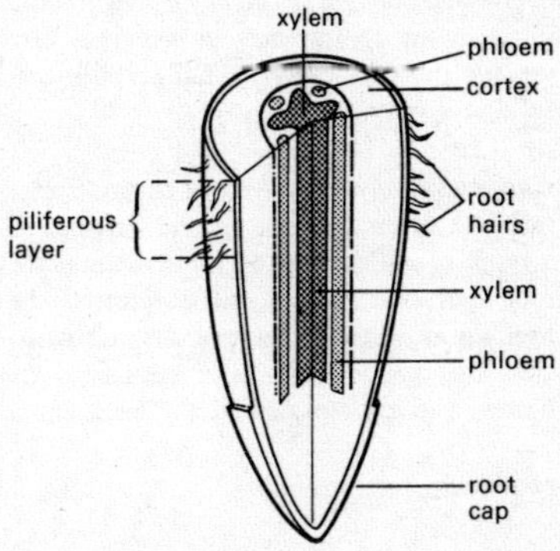

Section through the tip of a plant root

3. (in anatomy) The point of origin of a nerve in the central nervous system. There are two roots for every *spinal nerve, a dorsal and a ventral root.

root cap (calyptra) A cone-shaped structure that covers the root tip and develops as a result of cell division by a meristem at the root apex (*see* calyptrogen). It protects the root tip as it grows between the soil particles. The cells are constantly worn away by friction and are replaced by the meristem.

rootstock *See* rhizome.

roundworms *See* Nematoda.

rubidium–strontium dating A method of dating geological specimens based on the decay of the radioisotope rubidium–87 into the stable isotope strontium–87. Natural rubidium contains 27.85% of rubidium–87, which has a half-life of 4.7×10^{11} years. The ratio $^{87}Rb/^{87}Sr$ in a specimen gives an estimate of its age (up to several thousand million years).

Ruminantia A suborder of hooved mammals (*see* Artiodactyla) comprising the sheep, cattle, goats, deer, and antelopes. They are characterized by a four-chambered stomach, the first part of which is the *rumen*, which contains microorganisms that break down cellulose. Food is temporarily stored in the rumen, then regurgitated and chewed to a pulp before being swallowed and passed to the rest of the stomach where digestion is continued.

runner A stem that grows horizontally along the soil surface and gives rise to new plants either from axillary or terminal buds. Runners are seen in the creeping buttercup and the strawberry. *Offsets*, e.g. those of the houseleek, are short runners.

rusts A group of parasitic fungi belonging to the *Basidiomycetes. Many of these species attack the leaves and stems of cereal crops: characteristic rust-coloured streaks of spores appear on infected plants. The life cycles of some rusts may be complex; many form a number of different types of spore and some require two different host plants. *Compare* smuts.

S

saccharide *See* sugar.

saccharose *See* sucrose.

sacculus (saccule) A chamber of the *inner ear from which the *cochlea arises in reptiles, birds, and mammals. It bears patches of sensory epithelium concerned with balance (*see* macula).

sacral vertebrae The vertebrae that lie between the lumbar and the caudal vertebrae in the *vertebral column. The function of the sacral vertebrae is to articulate securely with the *pelvic girdle, and they are usually fused to form a single bone (the *sacrum*) to provide a firm support. The number of sacral vertebrae varies from animal to animal. Amphibians have a single sacral vertebra, reptiles have two, and mammals have three or more.

safranin A stain used in optical microscopy that colours lignified tissues, cutinized tissues, and nuclei red and chloroplasts pink. It is used mainly for plant tissues, in conjunction with a green or blue counterstain.

salicylic acid (1-hydroxybenzoic acid) A naturally occurring carboxylic acid, HOC_6H_4COOH, found in certain plants. It is used in making aspirin and in the foodstuffs and dyestuffs industries.

saline Describing a chemical compound that is a salt, or a solution containing a salt. *See also* physiological saline.

saliva A watery fluid secreted by the *salivary glands in the mouth. Production of saliva is stimulated by the presence of food in the mouth and also by the smell or thought of food. Saliva contains mucin, which lubricates food and eases its passage into the oesophagus, and in some animals the enzyme *ptyalin, which begins the digestion of starch. The saliva of insects is rich in digestive enzymes, and that of bloodsucking animals contains an anticoagulant.

salivary glands Glands in many terrestrial animals that secrete *saliva into the mouth. In man there are three pairs: the sublingual, submandibular, and the submaxillary glands. The salivary gland cells of some insect larvae produce giant (*polytene*) chromosomes, which are widely used in the study of genetics and protein synthesis.

samara A dry single-seeded indehiscent fruit in which the fruit wall hardens and extends to form a long membranous winglike structure that aids dispersal. Examples are ash and elm fruits. The sycamore fruit is a double samara and technically a *schizocarp. *See also* achene.

sampling The selection of small groups of entities to represent a large number of entities in statistics. In *random sampling* each individual of a population has an equal chance of being selected as part of the sample. In *stratified random sampling*, the population is divided into strata, each of which is randomly sampled and the samples from the different strata are pooled. In *systematic sampling*, individuals are chosen at fixed intervals; for example, every tenth animal in a population. In *sampling with replacement*, each individual chosen is replaced before the next selection is made.

saprophyte Any plant or microorganism that feeds by absorbing dead organic matter. Most saprophytes are bacteria and fungi; the term *saprobe* is frequently used to indicate saprophytic fungi. Saprophytes are important in *food chains as they bring about decay and release nutrients for plant growth. *Compare* parasitism.

sapwood (alburnum) The outer wood of a tree trunk or branch. It consists of living *xylem cells, which both conduct water and provide structural support. *Compare* heartwood.

satellite DNA A portion of the *DNA in plant and animal cells that consists of repeating sequences of *nucleotides. Some 30% of human DNA consists of such satellite regions. They do not appear to form part of any *genes but their function is unclear. The term is sometimes applied to DNA other than that found in the cell nucleus, in particular mitochondrial DNA, which differs in its density.

scales The small bony or horny plates forming the body covering of fish and reptiles. The wings of some insects, e.g. the Lepidoptera (butterflies and moths), are covered with tiny scales that are modified cuticular hairs.

In fish there are three types of scales. *Placoid scales* (*denticles*), characteristic of cartilaginous fish, are small and toothlike, with a projecting spine and a flattened base embedded in the skin. They are made of *dentine, have a pulp cavity, and the spine is covered with a layer of enamel. Teeth are probably modified placoid scales. *Cosmoid scales*, characteristic of lungfish and coelacanths, have an outer layer of hard *cosmin* (similar to dentine) covered by modified enamel (*ganoine*) and inner layers of bone. The scale grows by adding to the inner layer only. In modern lungfish the scales are reduced to large bony plates. *Ganoid scales* are characteristic of primitive ray-finned fishes, such as sturgeons. They are similar to cosmoid scales but have a much thicker layer of ganoine and grow by the addition of material all round. The scales of modern teleost fish are reduced to thin bony plates.

In reptiles there are two types of scales: horny epidermal *corneoscutes* sometimes fused with underlying bony dermal *osteoscutes*.

scanning electron microscope *See* electron microscope.

scapula (shoulder blade) The largest of the bones that make up each half of the *pectoral (shoulder) girdle. It is a flat triangular bone, providing anchorage for the muscles of the forelimb and an articulation for the *humerus at the *glenoid cavity. It is joined to the *clavicle (collar bone) in front.

Schiff's reagent A reagent used for testing for aldehydes and ketones; it consists of a solution of fuchsin dye that has been decolorized by sulphur dioxide. Aliphatic aldehydes restore the pink immediately, whereas aromatic ketones have no effect on the reagent. Aromatic aldehydes and aliphatic ketones restore the colour slowly. It is named after the German chemist Hugo Schiff (1834–1915).

schizocarp A dry indehiscent fruit formed from carpels that develop into separate one-seeded fragments called *mericarps*, which may be dehiscent, as in the *regma, or indehiscent, as in the *cremocarp and *carcerulus.

schizogeny The localized separation of plant cells to form a cavity (surrounded by the intact cells) in which secretions accumulate. Examples are the resin canals of some conifers and the oil ducts of caraway and aniseed fruits. *Compare* lysigeny.

Schwann cell A cell that forms the *myelin sheath of a nerve fibre. Each cell is responsible for a single length of the fibre (called an *internode*); adjacent internodes are separated by small gaps (*nodes of Ranvier*). During its development the cell wraps itself around the fibre, so the sheath consists of concentric layers of Schwann cell membrane. These cells are named after the German anatomist T. Schwann (1810–82).

scion *See* graft.

sclera *See* sclerotic.

scleroprotein Any of a group of proteins found in the exoskeletons of some invertebrates, notably insects. Scleroproteins are formed by conversion of the relatively soft elastic larval protein by a natural tanning process (*sclerotization*) involving orthoquinones. These are secreted and form cross linkages between polypeptides of the proteins, producing a hard rigid covering.

sclerotic (sclera) The tough external layer of the vertebrate eye. At the front of the eye, the sclera is modified to form the *cornea.

scorpions *See* Arachnida.

scotopic vision The type of vision that occurs when the rods in the eye are the principal receptors, i.e. when the level of illumination is low. With scotopic vision colours cannot be identified. *Compare* photopic vision.

scrotum The sac of skin and tissue that contains and supports the *testes in most mammals. It is situated outside the body cavity and allows the testes to develop at the optimum temperature, which is slightly lower than body temperature.

seaweeds Large multicellular *algae living in the sea or in the intertidal zone. They are commonly species of the *Chlorophyta, *Phaeophyta, and *Rhodophyta.

sebaceous gland A small gland occurring in mammalian *skin. Its duct opens into a hair follicle, through which it discharges *sebum onto the skin surface.

sebum The substance secreted by *sebaceous glands onto the surface of the *skin. It is a fatty mildly antiseptic material that protects, lubricates, and waterproofs the skin and hair and helps prevent desiccation.

secondary growth (secondary thickening) The increase in thickness of plant shoots and roots through the activities of the vascular *cambium and *cork cambium. It is seen in most dicotyledons and gymnosperms but not in monocotyledons. The tissues produced by secondary growth are called *secondary tissues* and the resultant plant or plant part is the *secondary plant body*. *Compare* primary growth.

secondary sexual characteristics External features of a sexually mature animal that, although not directly involved in copulation, are significant in reproductive behaviour. The development of such features is controlled by sex hormones (androgens or oestrogens); they may be seasonal (e.g. the antlers of male deer or the body colour of male sticklebacks) or permanent (e.g. breasts in women or facial hair in men).

secondary thickening *See* secondary growth.

secretin A hormone produced by the anterior part of the small intestine (the *duodenum and *jejunum) in response to the presence of hydrochloric acid from the stomach. It causes the pancreas to secrete alkaline pancreatic juice and stimulates bile production in the liver. Secretin, whose function was first demonstrated in 1902, was the first substance to be described as a hormone.

secretion The manufacture and discharge of specific substances into the external medium by cells in living organisms. (The substance secreted is also called the secretion.) Secretory cells are often specialized and organized in groups to form *glands. The substances produced may be released directly into the blood (*endocrine secretion*; *see* endocrine gland) or through a duct (*exocrine secretion*; *see* exocrine gland). Secretions can be classified according to the manner of their discharge. *Merocrine (eccrine) secretion* occurs without the secretory cells sustaining any permanent change; in *apocrine secretion* the cells release a secretory vesicle incorporating part of the secretory cell membrane; and *holocrine secretion* involves the disruption of the entire cell to release its accumulated secretory vesicles.

seed The structure in angiosperms and gymnosperms that develops from the ovule after fertilization. Occasionally seeds may develop without fertilization taking place (*see* apomixis). The seed contains the *embryo and nutritive tissue, either as *endosperm or food stored in the *cotyledons. Angiosperm seeds are contained within a *fruit that develops from the ovary wall. Gymnosperm seeds lack an enclosing fruit and are thus termed *naked*. The seed is covered by a protective layer, the *testa. During development of the testa the seed dries out and enters a resting phase (dormancy) until conditions are suitable for germination.

Annual plants survive the winter or dry season as seeds. The evolution of the seed habit enabled plants to colonize the land, since seed plants do not depend on water for fertilization (unlike the lower plants).

seed coat *See* testa.

seed ferns *See* Pteridospermales.

seed leaf *See* cotyledon.

seed plants *See* Spermatophyta.

segmentation 1. *See* metameric segmentation. **2.** *See* cleavage.

segregation The separation of pairs of *alleles during the formation of reproductive cells so that they contain one allele only of each pair. Segregation is the result of the separation of *homologous chromosomes during *meiosis. *See also* Mendel's laws.

Selachii The major subclass of the Elasmobranchii (cartilaginous fishes), containing the sharks, rays, skates, and similar but extinct forms. Their sharp teeth develop from the toothlike placoid *scales (denticles) and are rapidly replaced as they wear out.

selection pressure The extent to which organisms possessing a particular characteristic are either eliminated or favoured by environmental demands. It indicates the degree of intensity of *natural selection.

self-sterility The condition found in many hermaphrodite organisms in which male and female reproductive cells produced by the same individual will not fuse to form a zygote, or if they do, the zygote is unable to develop into an embryo. In plants this is usually termed *self-incompatibility* (*see* incompatibility).

Seliwanoff's test A biochemical test to identify the presence of ketonic sugars, such as fructose, in solution. It was devised by the Russian chemist F. F. Seliwanoff. A few drops of the reagent, consisting of resorcinol crystals dissolved in equal amounts of water and hydrochloric acid, are heated with the test solution and the formation of a red precipitate indicates a positive result.

semen A fluid containing sperm and nutrients that is produced by a male animal during copulation and is introduced into the body of the female. Spermatozoa are produced by the *testes and the nutrient liquids by the *prostate gland and *seminal vesicles.

semicircular canals The sense organ in vertebrates that is concerned with the maintenance of physical equilibrium (sense of balance). It occurs in the *inner ear and consists of three looped canals set at right angles to each other and attached to the *utriculus. The canals contain a fluid (*endolymph*) that flows in response to movements of the head and body. A swelling (*ampulla*) at one attachment point of each canal contains sensory cells that respond to movement of the endolymph in any of the three planes. These sensory cells initiate nervous impulses to the brain.

seminal receptacle *See* spermatheca.

seminal vesicle 1. A pouch or sac in many male invertebrates and lower vertebrates that is used for storing sperm. **2.** One of a pair of glands in male mammals that secrete seminal fluid into the vas deferens.

seminiferous tubules *See* testis.

semipermeable membrane A membrane that is permeable to molecules of the solvent but not the solute in *osmosis. Semipermeable membranes can be made by supporting a film of material (e.g. cellulose) on a wire gauze or porous pot.

senescence The changes that occur in an organism (or a part of an organism) between maturity and death, i.e. ageing. Characteristically there is a deterioration in functioning as the cells become less efficient in maintaining and replacing vital cell components. In animals this results in a decline in physical ability and, in man, there is also often a reduction in mental ability. Not all the parts of the body necessarily become senescent at the same time or age at the same rate. For example, in deciduous trees the shedding of senescent leaves in the autumn is a normal physiological process.

sense organ A part of the body of an animal that contains or consists of a concentration of *receptors that are sensitive to specific stimuli (e.g. sound, light, pressure, heat). Stimulation of these receptors initiates the transmission of nervous impulses to the brain, where sensory information is analysed and interpreted. Examples of sense organs are the *eye, *ear, *nose, and *taste bud.

sensitization 1. (of a cell) The alteration of the integrity of a cell membrane resulting from the reaction of specific *antibodies with *antigens on the surface of the cell. In the presence of *complement, the cell ruptures. **2.** (of an individual) Initial exposure to a specific antigen such that re-exposure to the same antigen causes a severe immune response (*see* anaphylaxis).

sepal One of the parts of a flower making up the *calyx. Sepals are considered to be modified leaves with a simpler structure. They are usually green and often hairy but in some plants, e.g. monk's hood, they may be brightly coloured.

septum Any dividing wall in a plant or animal. Examples are the septa that separate the chambers of the heart.

serine *See* amino acid.

serology The laboratory study of blood serum and its constituents, particularly *antibodies and *complement, which play a part in the *immune response.

serotonin (5-hydroxytryptamine) A compound, derived from the amino acid tryptophan, that affects the diameter of blood vessels and also functions as a *neurotransmitter. In the brain it is thought to influence mood: drugs, such as LSD, that alter serotonin levels have hallucinogenic effects.

serous membrane A tissue consisting of a layer of *mesothelium attached to a surface by a thin layer of connective tissue. Serous membrane lines body cavities that do not open to the exterior; the *peritoneum and the *pleura are examples.

serum *See* blood serum.

sessile 1. Describing animals that live permanently attached to a surface, i.e. sedentary animals. Many marine animals, e.g. sea anemones and limpets, are sessile. **2.** Describing any organ that does not possess a stalk where one might be expected. For example, the leaves of the oak (*Quercus robur*) are attached directly to the twigs.

seta 1. A bristle or hair in many invertebrates. Setae are produced by the epidermis and consist either of a hollow projection of cuticle containing all or part of an epidermal cell (as in insects) or are composed of chitin (as in the *chaetae of annelid worms). **2.** *See* sporogonium.

sex chromosome A chromosome that operates in the sex-determining mechanism of a species. In most animals there are two kinds of sex chromosomes: the *X chromosome*, similar in size to the other chromosomes, and the smaller *Y chromosome*. In man and many animals a combination of two X chromosomes gives a female while one X and one Y chromosome gives a male. Sex chromosomes carry genes governing the development of sex organs and secondary sexual characteristics. They also carry other genes unrelated to sex (*see* sex linkage). *See also* sex determination.

sex determination The method by which the distinction between males and females is established in a species. It is usually under genetic control. Equal numbers of males and females are produced when sex is determined by *sex chromosomes or by a contrasting pair of alleles. In some species (e.g. bees) females develop from fertilized eggs and males from unfertilized eggs. This does not produce equal numbers.

sex hormones Steroid hormones that control sexual development. The most important are the *androgens and *oestrogens.

sex linkage The tendency for certain inherited characteristics to occur far more frequently in one sex than the other. For example, red–green colour blindness and haemophilia affect men more often than women. This is because the genes governing normal colour vision and blood clotting occur on the X *sex chromosome. Women have two X chromosomes. If one carries an abnormal allele it is likely that its effects will be masked by a normal allele on the other X chromosome. However, men only have one X chromosome and any abnormal alleles therefore will not be masked. *See also* carrier.

sex ratio The ratio of the number of females to the number of males in a *population. Because the mortality rates in the two sexes may be different, the sex ratios in different age classes may differ.

sexual reproduction A form of reproduction that involves the fusion of two reproductive cells (*gametes) in the process of *fertilization. Normally, especially in animals, it requires two parents, one male and the other female. However, most plants bear both male and female reproductive organs and self-fertilization may occur, as it does in hermaphrodite animals. Gametes are formed by *meiosis, a special kind of cell division in the parent reproductive organs that both reassorts the genetic material and halves the chromosome number. Meiosis thus ensures genetic variability in the gametes and therefore in the offspring resulting from their subsequent fusion. Sexual reproduction, unlike *asexual reproduction, therefore generates variability within a species. However, it depends on there being reliable means of bringing together male and female gametes, and many elaborate mechanisms have evolved to ensure this.

sexual selection The means by which it is assumed that certain *secondary sexual characteristics, particularly of male animals, have evolved. Females presumably choose to mate with the male that gives the best courtship display and therefore has the brightest coloration, etc.: these features would be inherited by its male offspring and would thus tend to become exaggerated down the generations.

shoot The aerial part of a vascular plant. It develops from the *plumule and consists of a stem supporting leaves, buds, and flowers.

short-day plant A plant in which flowering can be induced or enhanced by short days, usually of less than 12 hours of daylight. Examples are strawberry and chrysanthemum. *See* photoperiodism. *Compare* day-neutral plant; long-day plant.

short-sightedness *See* myopia.

shoulder girdle *See* pectoral girdle.

siblings Individuals that have both parents in common.

sieve element A type of plant cell occurring within the *phloem. Sieve elements combine to form a series of tubes (*sieve tubes*) connecting the leaves, shoots, and roots in a fine network. Food materials are transported from one element to another via perforations termed *sieve areas* or *sieve plates*. This is an active (energy-requiring) process but the mechanism is uncertain. Sieve elements contain little cytoplasm and no nucleus. It is believed that their metabolic activities may be controlled by *companion cells in angiosperms and by albuminous cells in gymnosperms.

sievert The SI unit of dose equivalent (*see* radiation units).

sieve tube A tube within the *phloem tissue of a plant, composed of joined *sieve elements.

sign stimulus (releaser) The essential feature of a stimulus, which is necessary to elicit a response. For example, a red belly (characteristic of courting male sticklebacks) is the sign stimulus necessary to provoke an attack from a rival male; even a very crude model fish is attacked if it has a red undersurface.

silicula A type of *capsule formed from a bicarpellary ovary. It is longitudinally flattened and divided lengthwise into two cavities (*loculi*). It is broader than a *siliqua. Examples include the fruits of *Alyssum* and candytuft.

siliqua A type of *capsule formed from a bicarpellary ovary. It resembles a *silicula but is longer than it is broad; an example is the fruit of the wallflower. *See also* lomentum.

Silurian A geological period of the Palaeozoic era following the Ordovician period and extending until the beginning of the Devonian period. It began about 440 million years ago and lasted for about 45 million years. The Silurian was named by Roderick Murchison (1792–1871) after an ancient British tribe that inhabited South Wales, where he observed rocks of this period. The majority of Silurian life was marine but during the later part of the period primitive plants began to make their appearance on land. Trilobites and graptolites became less common, brachiopods were numerous and varied, crinoids became common for the first time, and corals also increased. The only known vertebrates during the Silurian were primitive fish; the first jawed fish appeared later in the period. The Caledonian orogeny (mountain-building period) reached its peak towards the end of the Silurian.

sinus A saclike cavity or organ in an animal, e.g. the *sinus venosus* in the heart of lower vertebrates.

sinusoid A tiny blood vessel or blood-filled space in an organ. Sinusoids replace capillaries in certain organs, notably the liver; they allow more direct contact between the blood and the tissue it is supplying.

Siphonaptera An order of wingless insects comprising the fleas. The body of a flea is laterally compressed and bears numerous backward-directed spines. Fleas live as blood-sucking ectoparasites of mammals and birds, having mouthparts adapted to piercing their host, injecting saliva to prevent clotting, and sucking up the blood. The long bristly legs can transmit energy stored in the elastic body wall to leap relatively long distances (over 300 mm horizontally). Apart from causing irritation, fleas can transmit disease organisms, most notably bubonic plague bacteria, which can be carried from rats to humans by the rat flea (*Xenopsylla cheopsis*). The whitish wormlike, legless larvae feed on organic matter. After two moults the larva spins a cocoon and undergoes metamorphosis into the adult.

Siphunculata (Anoplura) An order of wingless insects comprising the sucking lice: blood-sucking ectoparasites of mammals, with piercing and sucking mouthparts forming a snoutlike proboscis. They constitute an irritating pest to humans and domestic animals and can transmit diseases, including typhoid. The human louse (*Pediculus humanus*) exists in two forms: the head louse (*P. humanus capitis*) and the body louse (*P. humanus corporis*).

SI units Système International d'Unités: the international system of units now recommended for all scientific purposes. A coherent and rationalized system of units derived from the *m.k.s. units* (a metric system based on the metre, kilogram, and second), SI units have now replaced *c.g.s. units and *Imperial units. The system has seven *base units* and two *supplementary units* (see Appendix), all other units being derived from these nine units. There are 18 derived units with special names. Each unit has an agreed symbol (a capital letter or an initial capital letter if it is named after a scientist, otherwise the symbol consists of one or two lower-case letters). Decimal multiples of the units are indicated by a set of prefixes; whenever possible a prefix representing 10 raised to a power that is a multiple of three should be used.

skeletal muscle *See* voluntary muscle.

skeleton The structure in an animal that provides mechanical support for the body, protection for internal organs, and a framework for anchoring the muscles. The skeleton may be external (*see* exoskeleton) or internal (*see* endoskeleton). Both types require *joints to allow locomotion. The endoskeleton of higher vertebrates consists of a system of *bones (see illustration).

skin The outer layer of the body of a vertebrate. It is composed of two layers, the *epidermis and *dermis, with a complex nervous and blood supply. The skin may bear a variety of specialized structures, including *hair, *scales, and *feathers. This skin has an important role in protecting the body from mechanical injury, water loss, and the entry of harmful agents (e.g. disease-causing bacteria). It is also a sense organ, containing receptors sensitive to pain, temperature, and pressure. In warm-blooded animals it helps regulate body temperature by means of hair, fur, or feathers and *sweat glands.

skull The skeleton of the head. In mammals it consists of a *cranium enclosing the brain and the bones of the face and jaw. All the joints between the individual bones of the skull are immovable (*see* suture) except for the joint between the mandible (lower jaw) and the rest of the skull. There is a large opening (*foramen magnum*) at the base of the skull through which the spinal cord passes from the brain.

sleep A readily reversible state of reduced awareness and metabolic activity that occurs periodically in many animals. Usually accompanied by physical relaxation, the onset of sleep in humans and other mammals is marked by a change in the electrical activity of the brain, which is recorded by an *electroencephalogram as waves of low frequency and high amplitude (*slow-wave sleep*). This is interspersed by short bouts of high-frequency low-amplitude waves (similar to wave patterns produced when awake) associated with restlessness, dreaming, and rapid eye movement (REM); this is called *REM* (or *paradoxical*) *sleep*. Several regions of the brain are involved in sleep, especially the reticular formation of the *brainstem.

sleep movements *See* nyctinasty.

slime fungi (slime moulds) A group of small simple organisms widely distributed in damp habitats on land. They exist either as free cells or as multinucleate aggregates of cells (plasmodia). Cell walls are generally absent. Slime fungi show amoeboid movement and feed by ingesting small particles of food. They reproduce by means of spores. Since this group of organisms exhibits characteristics of both simple plants and simple animals its taxonomic position is uncertain. They are usually placed in a separate division, the Myxomycota.

small intestine The portion of the *alimentary canal between the stomach and the large intestine. It is subdivided into the *du-

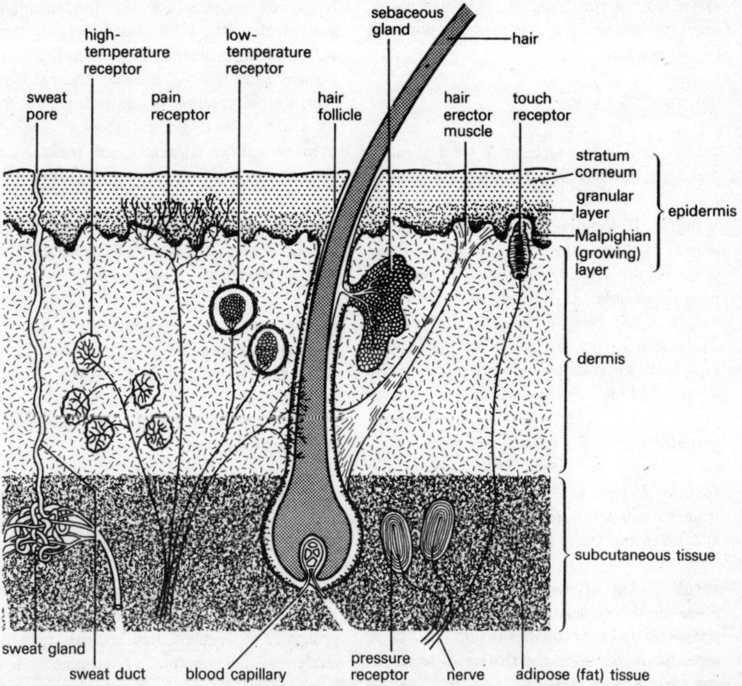

Structure of mammalian skin

odenum, *jejunum, and *ileum. It plays an essential role in the final digestion and absorption of food.

smooth muscle *See* involuntary muscle.

smuts A group of parasitic fungi belonging to the *Basidiomycetes. Many of these species attack the ears of cereal crops, replacing the grain by a mass of dark spores. *Compare* rusts.

snakes *See* Squamata.

sodium Symbol Na. A soft silvery element that is a major *essential element required by animals. It is important in maintaining the *acid–base balance and in controlling the volume of extracellular fluid and functions in the transmission of nerve impulses (*see* sodium pump).

sodium fluoride A crystalline compound, NaF. It is highly toxic but in very dilute solution (less than 1 part per million) it is used in the *fluoridation of water for the prevention of tooth decay on account of its ability to replace OH groups with F groups in the material of dental enamel.

sodium pump A mechanism by which sodium ions are transported out of a neurone across the cell membrane. The process requires energy in the form of ATP, being a form of *active transport. It maintains the different concentration of sodium ions on either side of the neurone membrane necessary for establishing the *resting potential of the neurone.

softwood *See* wood.

sol A *colloid in which small solid particles are dispersed in a liquid continuous phase.

somatic 1. Relating to all the cells of an animal or plant other than the reproductive cells. Thus a somatic *mutation is one that is not heritable. **2.** Relating to organs and tissues of the body other than the gut and its associated structures. The term is applied especially to voluntary muscles, the sense organs, and the nervous system.

somatotrophin *See* growth hormone.

sorosis A type of *composite fruit formed from an entire inflorescence spike. Mulberry and pineapple fruits are examples.

sorus 1. Any of the spore-producing structures on the undersurface of a fern frond, visible as rows of small brown dots. **2.** A reproductive area on the thallus of some algae, e.g. *Laminaria.* **3.** Any of various spore-producing structures in certain fungi.

spadix A flowering shoot (a type of *spike) with a large fleshy floral axis bearing small, usually unisexual, flowers. It is protected by a large petal-like bract, the *spathe*, and is characteristic of plants of the family Araceae (e.g. calla lily).

spathe *See* spadix.

special creation The belief, in accordance with the Book of Genesis, that every species was individually created by God in the form in which it exists today and is not capable of undergoing any change. It was the generally accepted explanation of the origin of life until the advent of *Darwinism. The idea has recently enjoyed a revival, especially among members of the fundamentalist movement in the USA, partly because there still remain problems that cannot be explained entirely by Darwinian theory. However, special creation is contradicted by fossil evidence and genetic studies, and the pseudoscientific arguments of *creation science* cannot stand up to logical examination.

specialization 1. Increasing *adaptation of an organism to a particular environment. **2.** *See* physiological specialization.

speciation The development of one or more species from an existing species. It occurs when *sympatric or *allopatric populations diverge so much from the parent population that interbreeding can no longer occur between them.

species A category used in the *classification of organisms that consists of a group of similar individuals that can usually breed among themselves and produce fertile offspring. Similar or related species are grouped into a genus. Within a species groups of individuals may become reproductively isolated because of geographical or behavioural factors. Such populations may, because of different selection pressures, develop different characteristics from the main population and so form a distinct *subspecies.

spectrum A range of electromagnetic energies arrayed in order of increasing or decreasing wavelength or frequency. The *emission spectrum* of a body or substance is the characteristic range of radiations it emits when it is heated, bombarded by electrons or ions, or absorbs photons. The *absorption spectrum* of a substance is produced by examining, through the substance and through a spectroscope, a continuous spectrum of radiation. The energies removed from the continuous spectrum by the absorbing medium show up as black lines or bands; with a substance capable of emitting a spectrum these are in exactly the same positions in the spectrum as the emission lines and bands would occur in the emission spectrum.

Emission and absorption spectra may show a *continuous spectrum*, a *line spectrum*, or a *band spectrum*. A continuous spectrum contains an unbroken sequence of frequencies over a relatively wide range; it is produced by incandescent solids, liquids, and compressed gases. Line spectra are discontinuous lines produced by excited atoms and ions as they fall back to a lower energy level. Band spectra (closely grouped bands of lines) are characteristic of molecular gases or chemical compounds.

sperm *See* spermatozoon.

spermatheca (seminal receptacle) A sac or receptacle in some female or hermaphrodite animals (e.g. earthworms) in which sperm from the mate is stored until the eggs are ready to be fertilized.

spermatogenesis The series of cell divisions in the testis that results in the production of spermatozoa. Within the seminiferous tubules of the testis germ cells grow and divide by mitosis to produce spermatogonia. These divide by mitosis to produce *spermatocytes*, which divide by meiosis to produce *spermatids*. The spermatids, which thus have half the number of chromosomes of the original germ cells, then develop into spermatozoa.

Spermatophyta A division of the plant kingdom containing plants that reproduce by means of *seeds. These species make up the bulk of the present-day flora – trees, shrubs, grasses, and herbaceous plants. The division is subdivided into the *Angiospermae and the *Gymnospermae. In some classifications seed plants are included within the *Pteropsida.

spermatozoid *See* antherozoid.

spermatozoon (sperm) The mature mobile reproductive cell (*see* gamete) of male animals, which is produced by the testis (*see* spermatogenesis). It consists of a head section containing a *haploid nucleus and an *acrosome*, which allows the sperm to penetrate the egg at fertilization; a middle section containing *mitochondria to provide the energy for movement; and a tail section, which lashes to drive the sperm forward.

Sphenopsida A class of the *Pteridophyta or a subdivision of the *Tracheophyta, the only living members of which are the horsetails (*Equisetum*). Horsetails have a perennial creeping rhizome supporting erect jointed stems bearing whorls of thin leaves. Spores are produced by terminal conelike structures. The group has a fossil record extending back to the Palaeozoic with its greatest development in the Carboniferous period, when giant tree forms were the dominant vegetation with the *Lycopsida.

sphincter A specialized muscle encircling an opening or orifice. Contraction of the sphincter tends to close the orifice. Examples are the anal sphincter (round the opening of the anus) and the pyloric sphincter (at the lower opening of the stomach).

spiders *See* Arachnida.

spike A type of *racemose inflorescence in which stalkless flowers arise from an undivided floral axis, as in plantain and *Orchis*. In the family Gramineae (sedges and grasses) the flowers are grouped in clusters called *spikelets*, which may be arranged to form a compound spike (as in wheat).

spinal column *See* vertebral column.

spinal cord The part of the vertebrate central nervous system that is posterior to the brain and enclosed within the *vertebral column. It consists of a hollow core of *grey matter (H-shaped in cross section) surrounded by an outer layer of white matter; the central cavity contains *cerebrospinal fluid. Paired *spinal nerves arise from the spinal cord.

spinal nerves Pairs of nerves that arise from the *spinal cord (*compare* cranial nerves). In man there are 31 pairs (one from each of the vertebrae). Each nerve contains both motor and sensory fibres (i.e. they are mixed nerves), and the spinal nerves form an important part of the *peripheral nervous system.

spindle A structure formed from fine fibres of protein in the cytoplasm of cells at the beginning of the *metaphase stage of cell division (*see* meiosis; mitosis). Two poles become established at diametrically opposite points close to the nucleus. Fibres radiating from each pole towards the opposite one diverge to their greatest extent midway between. This widest part of the spindle is the *equator*. Chromosomes are attached by their centromeres to the spindle fibres at the equator.

spindle attachment *See* centromere.

spine **1.** *See* vertebral column. **2.** A hard pointed protective structure on a plant that is formed through modification of a leaf, part of a leaf, or a stipule. The edge of the holly leaf is drawn out into spines, but in the gorse the whole leaf is modified. *Compare* prickle; thorn.

spinneret A small tubular appendage from which silk is produced in spiders and some insects. Spiders have four to six spinnerets on the hind part of the abdomen, into which numerous silk glands open. The silk is secreted as a fluid and hardens on contact with the air. Various types of silk are produced depending on its use (e.g. for webs, egg cocoons, etc.). The spinnerets that produce the cocoons of insects are not homologous with those of spiders. For example, the spinneret of the silkworm is in the pharynx and the silk is produced by modified salivary glands.

spiracle **1.** A small paired opening that occurs on each side of the head in cartilaginous fish. It is the reduced first *gill slit, its small size resulting from adaptations of the skeleton for the firm attachment of the jaws. In modern teleosts (bony fish) the spiracle is closed up. In tetrapods the first gill slit develops into the middle ear cavity. **2.** Any of the external openings of the *tracheae along the side of the body of an insect.

spirillum Any rigid spiral-shaped bacterium. Generally, spirilla are Gram-negative (*see* Gram's stain), aerobic, and highly motile, bearing one or more tufts of flagella. They occur in soil and water, feeding on organic matter.

spirochaete Any nonrigid corkscrew-shaped bacterium that moves by means of muscular flexions of the cell. Most spirochaetes are Gram-negative (*see* Gram's stain), anaerobic, and feed on dead organic matter. They are particularly common in sewage-polluted waters. Some, however, cause disease in man; *Treponema*, the agent of syphilis, is an example.

spleen A vertebrate organ, lying behind the stomach, that is basically a collection of *lymphoid tissue. Its functions include producing lymphocytes and destroying foreign particles. It acts as a reservoir for erythrocytes and can regulate the number in circulation. It is also the site for the breakdown of worn-out erythrocytes and it stores the iron they contain.

sponges *See* Porifera.

spongy mesophyll *See* mesophyll.

spontaneous generation The discredited belief that living organisms can somehow be produced by nonliving matter. For example, it was once thought that microorganisms arose by the process of decay and even that vermin spontaneously developed from household rubbish. Controlled experiments using sterilized media by Pasteur and others finally disproved these notions. *Compare* biogenesis. *See also* biopoiesis.

sporangium A reproductive structure in plants that produces asexual spores. *See* sporophyll.

spore A reproductive cell that can develop into an individual without first fusing with another reproductive cell (*compare* gamete). Spores are produced by plants, fungi, bacteria, and some Protozoa. A spore may develop into an organism resembling the parent or into another stage in the life cycle, either immediately or after a period of dormancy. In plants showing *alternation of generations, spores are formed by the *sporophyte generation and give rise to the *gametophyte generation. In ferns, the rows of brown reproductive structures on the undersurface of the fronds are spore-producing bodies.

spore mother cell (sporocyte) A diploid cell that gives rise to four haploid spores by meiosis.

sporocyte *See* spore mother cell.

sporogonium The *sporophyte generation in bryophytes. It is made up of an absorptive *foot*, a stalk (*seta*), and a spore-producing *capsule*. It may be completely or partially dependent on the *gametophyte.

sporophore (fructification) The aerial spore-producing part of certain fungi; for example, the stalk and cap of a mushroom.

sporophyll A leaf that bears *sporangia* (spore-producing structures). In ferns the sporophylls are the normal foliage leaves, but in higher plants the sporophylls are modified and arise in specialized structures such as the strobilus (cone) of certain pteridophytes and gymnosperms and the flower of angiosperms. Angiosperms, gymnosperms, and many pteridophytes produce spores of two different sizes (small *microspores* and large *megaspores*). The sporophylls bearing these are called *microsporophylls* and *megasporophylls* respectively.

sporophyte The generation in the life cycle of a plant that produces spores. The sporophyte is *diploid but its spores are *haploid. It is either completely or partially dependent on the *gametophyte generation in bryophytes but is the dominant plant in the life cycle of pteridophytes and seed plants. *See also* alternation of generations.

Squamata An order of reptiles comprising the lizards and snakes. They appeared at the end of the Triassic period, about 170 million years ago, and have invaded a wide variety of habitats. Most lizards have four legs and a long tail, eardrums, and movable eyelids. Snakes are limbless reptiles that lack eardrums; the eyes are covered by transparent immovable eyelids and the articulation of the jaws is very loose, enabling a wide gape to facilitate swallowing prey whole.

staining A technique in which cells or thin sections of biological tissue that are normally transparent are immersed in one or more coloured dyes (*stains*) to make them more clearly visible through a microscope. Staining heightens the contrast between the various cell or tissue components. Stains are usually organic salts with a positive and negative ion. If the colour comes from the negative ion (organic anion), the stain is described as *acidic*, e.g. *eosin. If the colour comes from the positive ion (organic cation), the stain is described as *basic*, e.g. *haematoxylin. *Neutral stains* have a coloured cation and a coloured anion; an example is *Leishman's stain. Cell constituents are described as being *acidophilic* if they are stained with acidic dyes, *basophilic* if receptive to basic dyes, and *neutrophilic* if receptive to neutral dyes. *Vital stains* are used to colour the constituents of living cells without harming them (*see* vital staining); *nonvital stains* are used for dead tissue.

Counterstaining involves the use of two or more stains in succession, each of which colours different cell or tissue constituents. *Temporary staining* is used for immediate microscopical observation of material, but the colour soon fades and the tissue is subsequently damaged. *Permanent staining* does not distort the cells and is used for tissue that is to be preserved for a considerable period of time.

Electron stains, used in the preparation of material for electron microscopy, are described as *electron-dense* as they interfere with the transmission of electrons. Examples are lead citrate, phosphotungstic acid (PTA), and uranyl acetate (UA).

stamen One of the male reproductive parts of a flower. It consists of an upper fertile part (the *anther) on a thin sterile stalk (the *filament*).

staminode A sterile stamen.

standing crop The total amount of living material in a specified population, expressed as *biomass or its equivalent in terms of energy. The standing crop may vary at different times of the year; for example, in a population of deciduous trees between summer and winter.

stapes (stirrup) The third of the three *ear ossicles of the mammalian middle ear.

starch A *polysaccharide consisting of various proportions of two glucose polymers, *amylose and *amylopectin. It occurs widely in plants, especially in roots, tubers, seeds, and fruits, as a carbohydrate energy store. Starch is therefore a major energy source for animals. When digested it ultimately yields glucose. Starch granules are insoluble in cold water but disrupt if heated to form a gelatinous solution.

statocyst (otocyst) A balancing organ found in many invertebrates. It consists of a fluid-filled sac lined with sensory hairs and contains granules of calcium carbonate, sand, etc. (*statoliths* or *otoliths*). As the animal moves the statoliths stimulate different hairs, giving a sense of the position of the body or part of it. The *semicircular canals in the ears of vertebrates act on the same principle and have a similar function.

stearic acid (octadecanoic acid) A solid saturated *fatty acid, $CH_3(CH_2)_{16}COOH$, that occurs widely (as *glycerides) in animal and vegetable fats.

stele The vascular tissue (i.e. *xylem and *phloem) of seed plants and pteridophytes, together with the endodermis and pericycle (when present). The arrangement of stelar tissues is very variable. In roots the stele often forms a solid core, which better enables the root to withstand tension and compression. In stems it is often a hollow cylinder separating the cortex and pith. This arrangement makes the stem more resistant to bending stresses. Monocotyledons and dicotyledons can usually be distinguished by the pattern of their stelar tissue. In monocotyledons the vascular bundles are scattered throughout the stem whereas in dicotyledons (and gymnosperms) they are arranged in a circle around the pith.

stem The part of a plant that usually grows vertically upwards towards the light and supports the leaves, buds, and reproductive structures. The leaves develop at the *nodes and side or branch stems develop from buds at the nodes. The stems of certain species are modified as bulbs, corms, rhizomes, and tubers. Some species have twining stems;

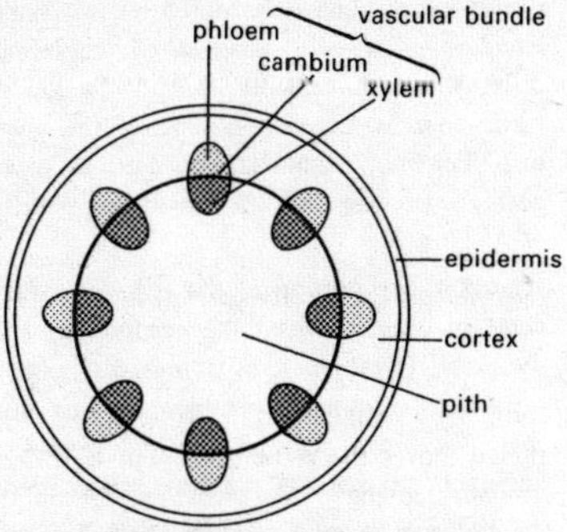

Transverse section through a herbaceous stem

others have horizontal stems, such as *runners. Another modification is the *cladode. Erect stems may be cylindrical or angular; they may be covered with hairs, prickles, or spines and many exhibit secondary growth and become woody (*see* growth ring). In addition to its supportive function, the stem contains *vascular tissue that conducts food, water, and mineral salts between the roots and leaves. It may also contain chloroplasts and carry out photosynthesis.

stereoisomerism The existence of chemical compounds (*stereoisomers*) that have the same molecular formulae and functional groups, but differ in the arrangement of groups in space. Optical isomerism is one form of this (*see* optical activity).

sternum (breastbone) 1. A shield-shaped or rod-shaped bone in terrestrial vertebrates, on the ventral side of the thorax, that articulates with the *clavicle (collar bone) of the pectoral girdle and with most of the ribs. It is absent in fish, and in birds it bears a *keel. **2.** The ventral portion of each segment of the exoskeleton of arthropods.

steroid Any of a group of lipids derived from a saturated compound called cyclopentanoperhydrophenanthrene, which has a nucleus of four rings. Some of the most important steroid derivatives are the steroid alcohols, or *sterols. Other steroids include the *bile acids, which aid digestion of fats in the intestine; the sex hormones (*androgens and *oestrogens); and the *corticosteroid hormones, produced by the adrenal cortex. *Vitamin D is also based on the steroid structure.

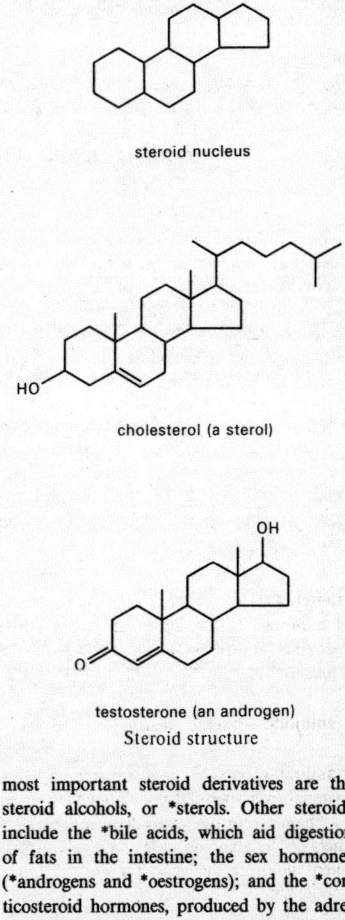

Steroid structure

sterol Any of a group of *steroid-based alcohols having a hydrocarbon side-chain of 8–10 carbon atoms. Sterols exist either as free sterols or as esters of fatty acids. Animal sterols (*zoosterols*) include *cholesterol and lanosterol. The major plant sterol (*phytosterol*) is beta-sitosterol, while fungal sterols (*mycosterols*) include *ergosterol.

stigma 1. The glandular sticky surface at the tip of a carpel of a flower, which receives the pollen. In insect-pollinated plants the stigmas are held within the flower, whereas in wind-pollinated species they hang outside it. **2.** *See* eyespot.

stilt root *See* prop root.

stimulus Any change in the external or internal environment of an organism that provokes a physiological or behavioural response in the organism. In an animal specific *receptors are sensitive to stimuli.

stipe 1. The stalk that forms the lower portion of the fruiting body of certain fungi, such as mushrooms, and supports the umbrella-shaped cap. **2.** The stalk between the holdfast and blade (*lamina*) of certain brown algae, notably kelps.

stipule An outgrowth from the petiole or leaf base of certain plants. Those of the garden pea are leaflike photosynthetic organs. The stipules of the lime tree are scalelike and protect the winter buds, whereas those of the false acacia (*Robinia*) are modified as spines.

stock *See* graft.

stolon A long aerial side stem that gives rise to a new daughter plant when the bud at its apex touches the soil. Plants that multiply in this way include blackberry and currant bushes. Gardeners often pin down stolons to the soil to aid the propagation of such plants. This process is termed *layering.*

stoma (*pl.* **stomata**) A pore large numbers of which are present in the epidermis of leaves (especially on the undersurface) and young shoots. Stomata function in gas exchange between the plant and the atmosphere. Each stoma is bordered by two semicircular *guard cells* (specialized epidermal cells), whose movements (due to changes in water content) control the size of the aperture. The term stoma is also used to mean both the pore and its associated guard cells.

stomach The portion of the vertebrate *alimentary canal between the oesophagus and the small intestine. It is a muscular organ, capable of dramatic changes in size and shape, in which ingested food is stored and undergoes preliminary digestion. Cells lining the stomach produce *gastric juice, which is thoroughly mixed with the food by muscular contractions of the stomach. The resultant acidic partly digested food mass (*chyme) is discharged into the *duodenum through the pyloric *sphincter for final digestion and absorption. Some herbivorous animals (ruminants) have multichambered stomachs from which food is regurgitated, rechewed, and swallowed again.

stomium A region of thin-walled cells in certain spore-producing structures that ruptures to release the spores. For example, in the sporangium of the fern *Dryopteris* the stomium ruptures when the annulus dries out.

stratum corneum The layers of dead keratinized cells that form the outermost layers of mammalian *epidermis. The stratum corneum provides a water-resistant barrier between the external environment and the living cells of the *skin.

striated muscle *See* voluntary muscle.

stridulation The production of sounds by insects rubbing one part of the body against another. The parts of the body involved vary from species to species. Stridulation is typical of the Orthoptera (grasshoppers, crickets, cicadas), in which the purpose of the sounds is usually to bring the sexes together, although they are also used in territorial behaviour, warning, etc.

strobilus **1.** A type of *composite fruit that is formed from a complete inflorescence. It produces *achenes enclosed in bracts and when mature becomes cone-shaped. The hop fruit is an example. **2.** *See* cone.

stroma Tissue that forms the framework of an organ; for example, the tissue of the ovary that surrounds the reproductive cells.

stromatolite A rocky cushion-like mass formed by the unchecked growth of millions of lime-secreting blue-green algae. Stromatolites are found only in areas where other organisms that would normally keep down the algal numbers cannot survive, such as extremely salty bays. It is thought that such formations were abundant 2000 million years ago, when blue-green algae were the most advanced form of life on earth, and that the white rings of fossilized microorganisms found in rocks of this age are the remains of stromatolites.

strontium Symbol Sr. A soft yellowish metallic element. The isotope strontium–90 is present in radioactive fallout (half-life 28 years), and can be metabolized with calcium so that it collects in bone.

strychnine A colourless poisonous crystalline alkaloid found in certain plants.

style The stalk of a carpel, between the stigma and the ovary. In many plants it is elongated to aid pollination.

subarachnoid space The space between the *arachnoid membrane and the *pia mater, two of the membranes (*meninges) that surround the brain and spinal cord. It is filled with *cerebrospinal fluid.

subclavian artery A paired artery that passes beneath the collar bone (clavicle) and branches to supply blood to the arm. The left subclavian artery arises from the aorta; the right from the innominate artery.

subcutaneous tissue The tissue that lies immediately beneath the *dermis (*see* skin). It is made up of loose fibrous *connective tissue, muscle, and fat (*see* adipose tissue), which in some animals (e.g. whales and hibernating mammals) forms an insulating layer or an important food store.

suberin A mixture of waxy substances, similar to *cutin, present in the thickened cell walls of many trees and shrubs, particularly in corky tissues. The deposition of suberin (suberization) provides a protective water-impermeable layer.

sublittoral **1.** Designating or occurring in the shallow-water zone of a sea, over the continental shelf and below the low tide mark. **2.** Designating or occurring in the zone of a lake below the littoral zone, to a depth of 6–10 metres.

subspecies A group of individuals within a *species that breed more freely among themselves than with other members of the species and resemble each other in more characteristics. Reproductive isolation of a subspecies may become so extreme that a new species is formed (*see* speciation). Subspecies are sometimes given a third Latin name, e.g. the mountain gorilla, *Gorilla gorilla beringei* (*see also* binomial nomenclature).

substrate **1.** (in biochemistry) The substance upon which an *enzyme acts in biochemical reactions. **2.** (in biology) The material on which a sedentary organism (such as a barnacle or a plant) lives or grows. The substrate may provide nutrients for the organism or it may simply act as a support.

succession (in ecology) The sequence of communities that develops in an area from the initial stages of colonization until a stable mature *climax community* is achieved. Many factors, including climate and changes brought about by the colonizing organisms, influence the nature of a succession; for example, after many years shrubs produce soil deep enough to support trees, which then shade out the shrubs.

succinic acid (butanedioic acid) A crystalline solid, $HOOC(CH_2)_2COOH$, that is soluble in water. It occurs in living organisms as an intermediate in metabolism, especially in the *Krebs cycle.

succulent A plant that conserves water by storing it in fleshy leaves or stems. Succulents are found either in dry regions or in areas where there is sufficient water but it is not easily obtained, as in salt marshes. Such plants are often modified to reduce water loss by transpiration. For example, the leaves of cacti are reduced to spines.

succus entericus A mixture of enzymes produced by glands within the small intestine. These enzymes are responsible for most of the digestive processes that occur in the alimentary canal and include *proteases, *lipases, and *amylases.

sucker (turion) A shoot that arises from an underground root or stem and grows at the expense of the parent plant. Suckers can be dug up with a portion of root attached and used to propagate a plant. If, however, a plant is grafted onto a different rootstock, as many roses are, any suckers will be of the wild rootstock, rather than the ornamental scion, and must be removed.

sucrose (cane sugar; beet sugar; saccharose) A sugar comprising one molecule of glucose linked to a fructose molecule. It occurs widely in plants and is particularly abundant in sugar cane and sugar beet (15–20%), from which it is extracted and refined for table sugar. If heated to 200°C, sucrose becomes caramel.

sugar (saccharide) Any of a group of water-soluble *carbohydrates of relatively low molecular weight and typically having a sweet taste. The simple sugars are called *monosaccharides. More complex sugars comprise between two and ten monosaccharides linked together: *disaccharides contain two, trisaccharides three, and so on. The name is often used to refer specifically to *sucrose (table sugar).

sulpha drugs *See* sulphonamides.

sulphonamides Organic compounds containing the group $-SO_2.NH_2$. The sulphonamides are amides of sulphonic acids. Many have antibacterial action and are also known as *sulpha drugs*, including:
sulphadiazine, $NH_2C_6H_4SO_2NHC_4H_3N_2$,
sulphathiazole, $NH_2C_6H_4SO_2NHC_5H_2NS$,
and several others. They act by preventing bacteria from reproducing and are used to treat a variety of bacterial infections, especially of the gut and urinary system.

sulphur Symbol S. A yellow nonmetallic element that is an *essential element in living organisms, occurring in the amino acids cysteine and methionine and therefore in many proteins. It is also a constituent of various cell metabolites, e.g. coenzyme A. Sulphur is absorbed by plants from the soil as the sulphate ion ($SO_4{}^{2-}$).

summation *See* synergism.

superior Describing a structure that is positioned above or higher than another structure in the body. For example, in flowering plants the ovary is described as superior when located above the other organs of the flower (*see* hypogyny). *Compare* inferior.

supination Rotation of the lower forearm so that the hand faces forwards or upwards with the radius and ulna parallel. *Compare* pronation.

supplementary units *See* SI units.

suprarenal glands *See* adrenal glands.

surface tension Symbol γ. The property of a liquid that makes it behave as if its surface is enclosed in an elastic skin. The property results from intermolecular forces: a molecule in the interior of a liquid experiences a force of attraction from other molecules equally from all sides, whereas a molecule at the surface is only attracted by molecules below it in the liquid. The surface tension is defined as the force acting over the surface per unit length of surface perpendicular to the force. It is measured in newtons per metre. It can equally be defined as the energy required to increase the surface area by one square metre, i.e. it can be measured in joules per metre squared (which is equivalent to $N\,m^{-1}$).

The surface tension of water is very strong, due to the intermolecular hydrogen bonding, and is responsible for the formation of drops, bubbles, and meniscuses, as well as the rise of water in a capillary tube (*capillarity*), the absorption of liquids by porous substances, and the ability of liquids to wet a surface. Capillarity is very important in plants as it is largely responsible for the transport of water, against gravity, within the plant.

surfactant (surface active agent) A substance, such as a *detergent, added to a liquid to increase its spreading or wetting properties by reducing its *surface tension.

suture The line marking the junction of two body structures. Examples are the immovable joints between the bones of the skull and, in plants, the seam along the edge of a pea or bean pod.

swallowing *See* deglutition.

sweat The salty fluid secreted by the *sweat glands onto the surface of the skin. Excess body heat is used to evaporate sweat, thereby resulting in cooling of the skin surface. Small amounts of urea are excreted in sweat.

sweat gland A small gland in mammalian skin that secretes *sweat. The distribution of sweat glands on the body surface varies between species: they occur over most of the body surface in man and higher primates but have a more restricted distribution in other mammals.

swim bladder (air bladder) An air-filled sac lying above the alimentary canal in bony fish that regulates the buoyancy of the animal. Air enters or leaves the bladder either via a pneumatic duct opening into the oesophagus or stomach or via capillary blood vessels, so that the specific gravity of the fish always matches the depth at which it is swimming. This makes the fish weightless, so less energy is required for locomotion. In lungfish it also has a respiratory function. The lungs of tetrapods are homologous with the swim bladder, which has developed its hydrostatic function by specialization.

syconus A type of *composite fruit formed from a hollow fleshy inflorescence stalk inside which tiny flowers develop. Small *drupes, the 'pips', are produced by the female flowers. An example is the fig.

symbiosis An interaction between individuals of different species. The term symbiosis is usually restricted to interactions in which both species benefit (*see* mutualism), but it may be used for other close associations, such as *commensalism. Many symbioses are obligatory (i.e. the participants cannot survive without the interaction); for example, a lichen is an obligatory symbiotic relationship between an alga and a fungus.

sympathetic nervous system Part of the *autononomic nervous system. Its nerve endings release noradrenaline or adrenaline as a neurotransmitter and its actions tend to antagonize those of the *parasympathetic nervous system, thus achieving a balance in the organs they serve. For example, the sympathetic nervous system decreases salivary gland secretion, increases heart rate, and constricts blood vessels, while the parasympathetic nervous system has opposite effects.

sympatric Describing groups of similar organisms that, although in close proximity and theoretically capable of interbreeding, do not interbreed because of differences in behaviour, flowering time, etc. *Compare* allopatric.

symphysis A *joint that is only slightly movable; examples are the joints between the vertebrae of the vertebral column and that between the two pubic bones in the pelvic girdle. The bones at a symphysis articulate by means of smooth layers of cartilage and strong fibres.

sympodium The composite primary axis of growth in such plants as lime and horse chestnut. After each season's growth the shoot tip of the main stem stops growing (sometimes terminating in a flower spike); growth is continued by the tip of one or more of the lateral buds. *Compare* monopodium.

synapse The junction between two adjacent neurones (nerve cells), i.e. between the axon ending of one and the dendrites of the next. At a synapse, the membranes of the two cells are in close contact, with only a minute gap between them. A nerve *impulse is transmitted across the synapse by the release from the tip of the *axon of a *neurotransmitter substance, which diffuses across the synaptic space to the *dendrites of the second neurone. This triggers the propagation of the impulse from the dendrite along the length of the next neurone. Most neurones have more than one synapse.

synapsis *See* pairing.

syncarpy The condition in which the female reproductive organs (*carpels) of a flower are joined to each other. It occurs, for example, in the primrose. *Compare* apocarpy.

synergism (summation) 1. The phenomenon in which the combined action of two substances (e.g. drugs or hormones) produces a greater effect than would be expected from adding the individual effects of each substance. **2.** The combined action of one muscle (the *synergist*) with another (the *agonist*) in producing movement. *Compare* antagonism.

syngamy *See* fertilization.

synovial membrane The membrane that lines the ligament surrounding a freely movable joint (such as that at the hip or elbow). It secretes a fluid (*synovial fluid*) that lubricates the layers of cartilage forming the articulating surfaces of the joint.

syrinx The sound-producing organ of a bird, situated at the lower end of the trachea where it splits into the bronchi. It has a complex structure with a number of vibrating membranes.

systematics The study of the diversity of organisms and their natural relationships. It is sometimes used as a synonym for *taxonomy. The term *biosystematics* describes the experimental study of diversity, especially at the species level. Biosystematic methods include breeding experiments, biochemical work (*chemosystematics*), and cytotaxonomy.

Système International d'Unités *See* SI units.

systole The phase of the heart beat during which the ventricles of the heart contract to force blood into the arteries. *Compare* diastole. *See* blood pressure.

T

2,4,5-T 2,4,5-trichlorophenoxyacetic acid: a synthetic *auxin frequently used as a herbicide and defoliant. There are now moves to restrict its use, as it tends to become contaminated with the toxic chemical dioxin.

tactic movement *See* taxis.

tannin One of a group of complex organic chemicals commonly found in leaves, unripe fruits, and the bark of trees. Their function is uncertain though the unpleasant taste may discourage grazing animals. Some tannins have commercial uses, notably in the production of leather and ink.

tapetum A reflecting layer, containing crystals of guanine, in the *choroid of the eye of many nocturnal vertebrates. It reflects light back onto the retina, thus improving vision and causing the eyes to shine in the dark.

tapeworms *See* Cestoda.

tap root *See* root.

tarsal (tarsal bone) One of the bones that form the ankle (*see* tarsus) in terrestrial vertebrates.

tarsus The ankle (or corresponding part of the hindlimb) in terrestrial vertebrates, consisting of a number of small bones (*tarsals*). The number of tarsal bones varies with the species: humans, for example, have seven.

taste bud A small sense organ in most vertebrates, specialized for the detection of taste. In terrestrial animals taste buds are concentrated on the upper surface of the *tongue. They are sensitive to four types of taste: sweet, salt, bitter, or sour. The taste bud transmits information about a particular type of taste to the brain via nerve fibres. The four types of taste bud show distinct distribution patterns on the surface of the human tongue.

In fishes, taste buds are distributed over the entire surface of the body and provide information about the surrounding water.

taxis (tactic movement) The movement of a cell (e.g. a gamete) or a microorganism in response to an external stimulus. Certain microorganisms have a light-sensitive region that enables them to move towards or away from high light intensities (positive and negative *phototaxis* respectively). Many bacteria move in response to chemical stimuli (*chemotaxis*); a specific example is *aerotaxis*, in which atmospheric oxygen is the stimulus. Tactic movements are restricted to cells that possess cilia, flagella, or some other means of locomotion. The term is usually not applied to the movements of higher animals. *Compare* kinesis; tropism.

taxon Any named taxonomic group of any *rank in the hierarchical *classification of organisms. Thus the taxa Papilionidae, Lepidoptera, Insecta, and Arthropoda are named examples of a family, order, class, and phylum, respectively.

taxonomy The study of the theory, practice, and rules of *classification of living and extinct organisms. The naming, description, and classification of a given organism draws on evidence from a number of fields. *Classical taxonomy* is based on morphology and anatomy. *Biochemical taxonomy* studies similarities in the structure of certain proteins and nucleic acids. *Cytotaxonomy* compares the size, shape, and number of chromosomes of different organisms. *Numerical taxonomy* uses mathematical procedures to assess similarities and differences and establish taxonomic groups. *See also* systematics.

TCA cycle *See* Krebs cycle.

Teleostei The major subclass of the *Osteichthyes (bony fish), containing about 20 000 species. Teleosts have colonized an extensive variety of habitats and show great diversity of form. The group includes the eel, seahorse, plaice, and salmon. They have been the dominant fish since the Cretaceous period (about 70 million years ago).

telophase The fourth stage of cell division. In *mitosis the chromatids that separated from each other at *anaphase collect at the poles of the spindle. A nuclear membrane forms around each group, producing two daughter nuclei with the same number and kind of chromosomes as the original cell nucleus. In the first telophase of *meiosis, complete chromosomes from the pairs that separated at first anaphase form the daughter nuclei. The number of chromosomes in these nuclei is therefore half the number in the original one. In the second telophase, daughter nuclei are formed from chromatids (as in mitosis).

tendon A thick strand or sheet of tissue that attaches a muscle to a bone. Tendons consist of *collagen fibres and are therefore inelastic: they ensure that the force exerted by muscular contraction is transmitted to the relevant part of the body to be moved.

tendril A slender branched or unbranched structure found in many climbing plants. It may be a modified stem, leaf, leaflet, or petiole. Tendrils respond to contact with solid objects by twining around them (*see* haptotropism). The cells that touch the object lose water and decrease in volume in comparison to the outer cells, thus causing the tendril to curve.

tera- Symbol T. A prefix used in the metric system to denote one million million times. For example, 10^{12} volts = 1 teravolt (TV).

terpenes A group of unsaturated hydrocarbons present in plants (*see* essential oil). Terpenes consist of isoprene units, CH_2:C(CH_3)CH:CH_2. Monoterpenes have two units, $C_{10}H_{16}$, sesquiterpenes three units, $C_{15}H_{24}$, diterpenes four units, $C_{20}H_{32}$, etc.

territory A fixed area that an animal or group of animals defends against intrusion from others of its species. Outside the territory (which may contain food sources, hiding places, and nesting sites) others are not threatened. Many mammals indicate their territory boundaries with scent markings, while birds sing territorial songs that repel would-be intruders. Animals in neighbouring territories normally respect each other's boundaries, which reduces overt *aggression. Some animals are territorial only at certain times of the year, usually the breeding season (*see* courtship).

Tertiary The older geological period of the Cenozoic era (*compare* Quaternary). It began about 65 million years ago, following the Cretaceous period, and extended to the beginning of the Quaternary, about 2 million years ago. It is subdivided into the *Palaeocene, *Eocene, *Oligocene, *Miocene, and *Pliocene epochs in ascending order. The Tertiary period was characterized by the rise of the modern mammals and the development of shrubs, grasses, and other flowering plants.

testa (seed coat) The lignified or fibrous protective covering of a seed that develops from the integuments of the ovule after fertilization. *See also* hilum; micropyle.

test cross *See* back cross.

testicle *See* testis.

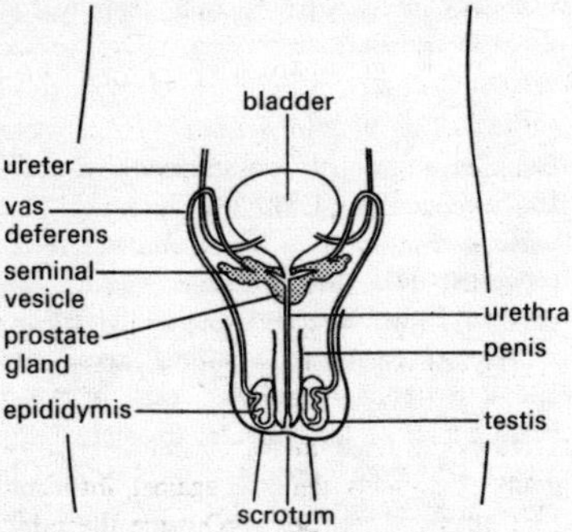

Human male reproductive system

testis (testicle) The reproductive organ in male animals in which spermatozoa are produced. In vertebrates there are two testes; as well as sperm, they produce steroid hormones (*see* androgen). In most animals the testes are within the body cavity but in mammals, although they develop within the body near the kidneys, they come to hang outside the body cavity in a *scrotum. Most of the vertebrate testis is made up of a mass of *seminiferous tubules*, in which the sperms develop. It is connected to the outside by means of the *vas deferens.

testosterone The principal male sex hormone. *See* androgen.

tetrad A group of four *haploid cells formed at the end of the second division of *meiosis.

tetrapod A vertebrate animal with four limbs. Tetrapods include amphibians, reptiles, birds, and mammals. The skeleton of the limbs of all tetrapods is based on the same five-digit pattern (*see* pentadactyl limb).

thalamus 1. (in anatomy) Part of the vertebrate *forebrain that lies above the hypothalamus. It relays sensory information to the cerebral cortex and is also concerned with the translation of impulses into conscious sensations. **2.** (in botany) *See* receptacle.

Thallophyta A former division of the plant kingdom containing relatively simple plants, i.e. those with no leaves, stems, or roots. It included the *algae, *bacteria, *fungi, and lichens (*see* Lichenes). *Compare* Protista.

thallus A relatively undifferentiated plant body with no true roots, stems, leaves, or vascular system. It is found in the algae, lichens, and bryophytes and in the gametophyte generation of pteridophytes.

theca *See* capsule.

thermography A medical technique that makes use of the infrared radiation from the human skin to detect an area of elevated skin temperature that could be associated with an underlying cancer. The heat radiated from the body varies according to the local blood flow, thus an area of poor circulation produces less radiation. A tumour, on the other hand, has an abnormally increased blood supply and is revealed on the *thermogram* (or *thermograph*) as a 'hot spot'. The technique is used particularly in mammography, the examination of the infrared radiation emitted by human breasts in order to detect breast cancer.

thermoluminescence Luminescence produced in a solid when its temperature is raised. It arises when free charge carriers, trapped in a solid as a result of exposure to ionizing radiation, unite and emit photons of light. The process is made use of in *thermoluminescent dating*, which assumes that the number of charge carriers trapped in a sample of pottery is related to the length of time that has elapsed since the pottery was fired. By comparing the luminescence produced by heating a piece of pottery of unknown age with the luminescence produced by heating similar materials of known age, a fairly accurate estimate of the age of an object can be made.

thermoluminescent dating *See* thermoluminescence.

thiamin(e) *See* vitamin B complex.

thigmotropism (haptotropism) The growth of an aerial plant organ in response to localized physical contact. For example, when a tendril of sweet pea touches a supporting structure, it curves in the direction of the support and coils around it. *See* tropism.

thin-layer chromatography A technique for the analysis of liquid mixtures using *chromatography. The stationary phase is a thin layer of an absorbing solid (e.g. alumina) prepared by spreading a slurry of the solid on a plate (usually glass) and drying it in an oven. A spot of the mixture to be analysed is placed near one edge and the plate is stood upright in a solvent. The solvent rises through the layer by capillary action carrying the components up the plate at different rates (depending on the extent to which they are absorbed by the solid). After a given time, the plate is dried and the location of spots noted. It is possible to identify constituents of the mixture by the distance moved in a given time. The technique needs careful control of the thickness of the layer and of the temperature.

thoracic duct The main collecting vessel of the *lymphatic system, running longitudinally in front of the backbone. The thoracic duct drains its lymph into the superior vena cava.

thoracic vertebrae The *vertebrae of the upper back, which articulate with the *ribs. They lie between the *cervical vertebrae and the *lumbar vertebrae and are distinguished by a number of articulating facets for attachment of the ribs. In man there are 12 thoracic vertebrae.

thorax The anterior region of the body trunk of animals. In vertebrates it contains the heart and lungs within the rib cage. It is particularly well-defined in mammals, being separated from the *abdomen by the *diaphragm. In insects the thorax is divided into an anterior *prothorax*, a middle *mesothorax*, and a posterior *metathorax*, each of which bears a pair of legs; the hindmost two segments also both carry a pair of wings. In other arthropods, especially crustaceans and arachnids, the thorax is fused with the head to form a *cephalothorax*.

thorn A hard side stem with a sharp point at the tip, replacing the growing point. In some plants the development of thorns and subsequent suppression of the growing points may be a response to dry conditions. Examples are the thorns of gorse and hawthorn. *Compare* prickle; spine.

thread cell (nematoblast; cnidoblast) A specialized cell found only in the ectoderm of the *Coelenterata. It contains a *nematocyst*, a fluid-filled sac within which lies a long hollow coiled thread. When a small sensory projection (*cnidocil*) on the surface of the thread cell is touched, e.g. by prey, the thread is shot out and adheres to the prey, coils round it, or injects poison into it. Numerous thread cells on the tentacles of jellyfish produce their sting.

threonine *See* amino acid.

threshold (in physiology) The minimum intensity of a stimulus that is necessary to initiate a response.

thrombin An enzyme that catalyses the conversion of fibrinogen to fibrin. *See* clotting factors.

thrombocyte *See* platelet.

thymine A *pyrimidine derivative and one of the major component bases of *nucleotides and the nucleic acid *DNA.

thymus An organ, present only in vertebrates, that is concerned with development of *lymphoid tissue and hence the antibody-producing white blood cells (lymphocytes) and the *immune response. In mammals it is a bilobed organ in the region of the lower neck, above and in front of the heart. It shrinks in size after sexual maturity and is therefore believed to function only during early life.

thyrocalcitonin *See* calcitonin.

thyroid gland A bilobed endocrine gland in vertebrates, situated in the base of the neck. It secretes two iodine-containing hormones, *thyroxine* and *triiodotyrosine*, which control the rate of all metabolic processes in the body and influence physical development. Growth and activity of the thyroid is controlled by a hormone, *thyrotrophin* (or *thyroid-stimulating hormone*), secreted by the anterior *pituitary gland.

thyroxine The principal hormone of the *thyroid gland.

tibia **1.** The larger of the two bones of the lower hindlimb of terrestrial vertebrates (*compare* fibula). It articulates with the *femur at the knee and with the *tarsus at the ankle. The tibia is the major load-bearing bone of the lower leg. **2.** The fourth segment of an insect's leg, which is attached to the femur.

time-lapse photography A form of ciné photography used to record a slow process, such as plant growth. A series of single exposures of the object is made on ciné film at predetermined regular intervals. The film produced is then projected at normal speeds and the process appears to be taking place at a most unusual rate.

tissue A collection of similar cells organized to carry out one or more particular functions. For example, in animals nervous tissue is specialized to perceive and transmit stimuli. An organ, such as a lung or kidney, contains many different types of tissues.

tissue culture The growth of the tissues of living organisms outside the body in a suitable culture medium. Culture (or nutrient) media contain a mixture of nutrients either in solid form (e.g. in *agar) or in liquid form (e.g. in *physiological saline). Tissue culture has proved to be invaluable for gaining information about factors that control the growth and differentiation of cells. *See also* explantation.

titre **1.** The number of infectious virus particles present in a suspension. **2.** A measure of the amount of *antibody present in a sample of serum, given by the highest dilution of the sample that results in the formation of visible clumps with the appropriate antigen (*see* agglutination).

toads *See* Amphibia.

tobacco mosaic virus (TMV) A rigid rod-shaped RNA-containing virus that causes distortion and blistering of leaves in a wide range of plants, especially the tobacco plant. It is transmitted by insects when they feed on plant tissue. TMV was the first virus to be discovered.

tocopherol *See* vitamin E.

Tollen's reagent A reagent used in testing for aldehydes. It is made by adding sodium hydroxide to silver nitrate to give silver(I) oxide, which is dissolved in aqueous ammonia (giving the complex ion $[Ag(NH_3)_2]^+$). The sample is warmed with the reagent in a test tube. Aldehydes reduce the complex Ag^+ ion to metallic silver, forming a bright silver mirror on the inside of the tube (hence the name *silver-mirror test*). Ketones give a negative result.

tomography The use of X-rays to photograph a selected plane of a human body with other planes eliminated. The *CAT* (*computerized axial tomography*) *scanner* is a ring-shaped X-ray machine that rotates through 180° around the horizontal patient, making numerous X-ray measurements every few degrees. The vast amount of information acquired is built into a three-dimensional image of the tissues under examination by the scanner's own computer. The patient is exposed to a dose of X-rays only some 20% of that used in a normal diagnostic X-ray.

tone (tonus) The state of sustained tension in muscles that is necessary for the maintenance of posture. In a tonic muscle contraction, only a certain proportion of the muscle fibres are contracting at any given time; the rest are relaxed and recovering for subsequent contractions. The fibres involved in tone contract more slowly than the fast fibres used for rapid responses by the same muscle. The proportions of slow and fast fibres depends on the function of the muscle.

tongue A muscular organ of vertebrates that in most species is attached to the floor of the mouth. It plays an important role in manipulating food during chewing and swallowing and in terrestrial species it bears numerous *taste buds on its upper surface. In some advanced vertebrates the tongue is used in the articulation of sounds, particularly in human speech.

tonsil A mass of *lymphoid tissue, several of which are situated at the back of the mouth and throat in higher vertebrates. In humans there are the *palatine tonsils* at the back of the mouth, *lingual tonsils* below the tongue, and *pharyngeal tonsils* (or *adenoids*) in the pharynx. They are concerned with the production of *lymphocytes and therefore with defence against infection.

tonus *See* tone.

tooth Any of the hard structures in vertebrates that are used principally for biting and chewing food but also for attack, grooming, and other functions. In fish and amphibians the teeth occur all over the palate, but in higher vertebrates they are concentrated on the jaws. They evolved in cartilaginous fish as modified placoid *scales, and this is reflected in their structure: a body of bony *dentine with a central *pulp cavity and an outer covering of *enamel on the exposed surface (*crown*). The portion of the tooth embedded in the jawbone is the *root* (see illustration).

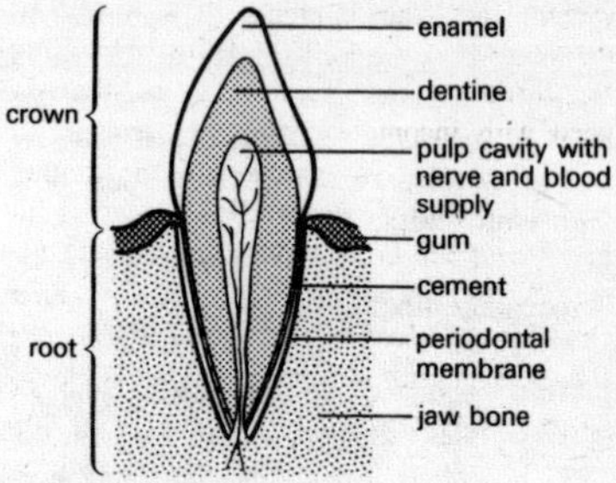

Section through an incisor tooth

In mammals there are four different types of teeth, specialized for different functions (*see* canine; incisor; molar; premolar). Their number varies with the species (*see* dental formula). *See also* deciduous teeth; permanent teeth.

torus *See* receptacle.

toxin A poison produced by a living organism, especially a bacterium. An *endotoxin* is released only when the bacterial cell dies or disintegrates. An *exotoxin* is secreted by a bacterial cell into the surrounding medium.

In the body a toxin acts as an *antigen, producing an *immune response.

trace element *See* essential element.

trace fossil *See* fossil.

trachea **1.** The windpipe in air-breathing vertebrates: a tube that conducts air from the throat to the *bronchi. It is strengthened with incomplete rings of cartilage. **2.** An air channel in insects and most other terrestrial arthropods. Tracheae occur as ingrowths of the body wall. They open to the exterior by *spiracles* and branch into finer channels (*tracheoles*) that terminate in the tissues. Pumping movements of the abdominal muscles cause air to be drawn into and out of the tracheae.

tracheid A type of cell occurring within the *xylem of conifers, ferns, and related plants. Tracheids are elongated and their walls are usually extensively thickened by deposits of lignin. Water flows from one tracheid to another through unthickened regions (pits) in the cell walls. *Compare* vessel element.

Tracheophyta A group (division) of the plant kingdom containing all plants possessing organized *vascular tissue. This classification is an alternative to other classifications, which divide vascular plants into the *Pteridophyta and *Spermatophyta. The Tracheophyta comprises the subdivisions Psilopsida (mostly extinct), *Lycopsida, *Sphenopsida, and *Pteropsida.

tracing (radioactive tracing) *See* labelling.

transamination A biochemical reaction in amino acid metabolism in which an amine group is transferred from an amino acid to a keto acid to form a new amino acid and keto acid. The coenzyme required for this reaction is pyridoxal phosphate.

transcription The process in living cells in which the genetic information of *DNA is transferred to a molecule of messenger *RNA (mRNA) as the first step in *protein synthesis (*see also* genetic code). Transcription takes place in the cell nucleus or nuclear region. An enzyme (transcriptase) proceeds along the DNA strand and assembles the nucleotides necessary to form a complementary strand of mRNA. *Compare* translation.

transect A straight line across an expanse of ground along which ecological measurements are taken, continuously or at regular intervals. Thus an ecologist wishing to study the numbers and types of organisms at different distances above the low-tide line might sample at five-metre intervals along a number of transects perpendicular to the shore.

transfer RNA *See* RNA.

transition zone *See* hypocotyl.

translation The process in living cells in which the genetic information encoded in messenger *RNA (mRNA) in the form of a sequence of base triplets (*codons) is translated into a sequence of amino acids in a polypeptide chain during *protein synthesis. Translation takes place on *ribosomes in the cell cytoplasm. The ribosomes move along the mRNA 'reading' each codon in turn. Molecules of transfer RNA (tRNA), each bearing a particular amino acid, are brought to their correct positions along the mRNA molecule: base pairing occurs between the bases of the codons and the complementary base triplets of tRNA. In this way amino acids are assembled in the correct sequence to form the polypeptide chain.

translocation The movement of minerals and chemical compounds within a plant. There are two main processes. The first is the uptake of soluble minerals from the soil

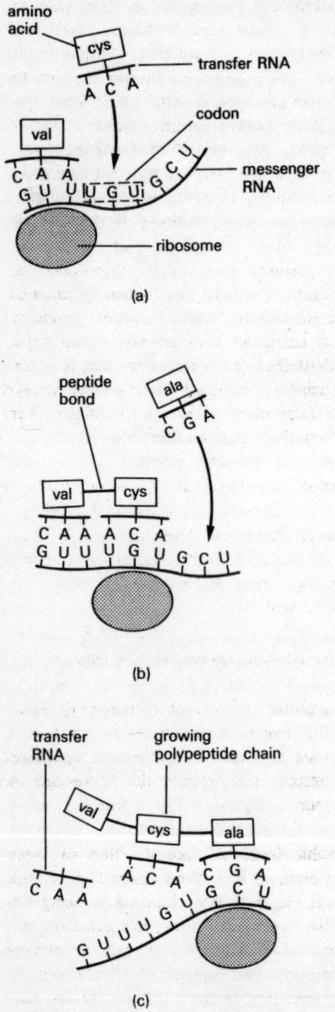

The stages of translation in protein synthesis

and their passage upwards from the roots to various organs by means of the water-conducting vessels (*xylem). The second is the transfer of organic compounds, synthesized by the leaves, both upwards and downwards to various organs, particularly the growing points. This movement occurs within the *phloem tubes.

transmission electron microscope *See* electron microscope.

transpiration The loss of water vapour by plants to the atmosphere. It occurs mainly from the leaves through pores (stomata) whose primary function is gas exchange. The water is replaced by a continuous column of water moving upwards from the roots within the *xylem vessels. *See also* potometer.

transplantation *See* graft.

transplantation antigen *See* antigen; HL-A system.

Trematoda A class of parasitic flatworms (*see* Platyhelminthes) comprising the flukes, such as *Fasciola* (liver fluke). Flukes have suckers and hooks to anchor themselves to the host and their body surface is covered by a protective cuticle. The whole life cycle may either occur within one host or require one or more intermediate hosts to transmit the infective eggs or larvae. *Fasciola hepatica*, for example, undergoes larval development in a land snail (the intermediate host) and infects sheep (the primary host) when contaminated grass containing the larvae is swallowed.

Triassic The earliest period of the Mesozoic era. It began about 225 million years ago, following the Permian, the last period of the Palaeozoic era, and extended until about 190 million years ago when it was succeeded by the Jurassic. It was named, by F. von Alberti in 1834, after the sequence of three divisions of strata that he studied in central Germany – Bunter, Muschelkalk, and Keuper. The Triassic rocks are frequently difficult to distinguish from the underlying

Permian strata and the term *New Red Sandstone* is often applied to rocks of the Permo-Triassic. During the period marine animals diversified: molluscs were the dominant invertebrates – ammonites were abundant and bivalves replaced the declining brachiopods. Reptiles were the dominant vertebrates and included turtles, phytosaurs, dinosaurs, and the marine ichthyosaurs.

tribe A category used in the *classification of plants and animals that consists of several similar or closely related genera within a family. For example the Bambuseae, Oryzeae, Paniceae, and Aveneae are tribes of grasses.

tricarboxylic acid cycle *See* Krebs cycle.

trichome A hairlike projection from a plant epidermal cell. Examples include root hairs and the stinging hairs of nettle leaves.

tricuspid valve A valve, consisting of three flaps, situated between the right atrium and the right ventricle of the mammalian heart. When the right ventricle contracts, forcing blood into the pulmonary artery, the tricuspid valve closes the aperture to the atrium, thereby preventing any backflow of blood. The valve reopens to allow blood to flow from the atrium into the ventricle. *Compare* mitral valve.

triglyceride (triacylglycerol) An ester of glycerol (propane-1,2,3-triol) in which all three hydroxyl groups are esterified with a fatty acid. Triglycerides are the major constituent of fats and oils and provide a concentrated food energy store in living organisms as well as cooking fats and oils, margarines, soaps, etc. Their physical and chemical properties depend on the nature of their constituent fatty acids. In *simple triglycerides* all three fatty acids are identical; in *mixed triglycerides* two or three different fatty acids are present.

trilobite An extinct marine arthropod belonging to the class Trilobita (some 4000 species), fossils of which are found in deposits dating from the Precambrian to the Permian period (590–280 million years ago). Trilobites were typically small (1–7 cm long); the oval flattened body comprised a head (covered by a semicircular dorsal shield) and a thorax and abdomen, which were protected by overlapping dorsal plates with a raised central part and flattened lateral portions, presenting a three-lobed appearance. The head bore a pair of antenna-like appendages and a pair of compound eyes; nearly all body segments bore a pair of Y-shaped (biramous) appendages – one branch for locomotion and the other fringed for respiratory exchange. Trilobites were bottom-dwelling scavengers.

triploblastic Describing an animal having a body composed of three embryonic cell layers: the *ectoderm, *mesoderm, and *endoderm. Most multicellular animals are triploblastic; the coelenterates, which are *diploblastic, are an exception.

tritiated compound *See* labelling.

trochanter 1. Any of several bony knobs on the femur of vertebrates to which muscles are attached. **2.** The second segment of an insect's leg, between the *coxa and the *femur.

trophic level The position that an organism occupies in a *food chain. For example, green plants (which obtain their energy directly from sunlight) are the primary *producers; herbivores are primary *consumers (and secondary producers). A carnivore that eats only herbivores is a secondary consumer and a tertiary producer. Many animals feed at several different trophic levels.

tropism The directional growth of a plant organ in response to an external stimulus, such as light, touch, or gravity. Growth towards the stimulus is a *positive tropism*; growth away from the stimulus is a *negative tropism*. *See also* geotropism; hydrotropism; orthotropism; phototropism; plagiotropism; thigmotropism. *Compare* nastic movements; taxis.

trypsin An enzyme that digests proteins (*see* protease). It is secreted in an inactive form (*trypsinogen*) by the pancreas into the duodenum. There, trypsinogen is acted on by an enzyme (*enterokinase*) produced in the duodenum to yield trypsin. The active enzyme plays an important role in the digestion of proteins in the anterior portion of the small intestine.

trypsinogen *See* trypsin.

tryptophan *See* amino acid.

tuber A swollen underground stem or root in certain plants. It enables the plant to survive the winter or dry season and is also a means of propagation. A *stem tuber*, such as the potato, forms at the end of an underground stem. Each tuber represents several nodes and internodes. The following season several new plants develop from the terminal and axillary buds (eyes). *Root tubers*, such as those of the dahlia, are modified food-storing adventitious roots and may also give rise to new plants.

Turbellaria A class of free-living flatworms (*see* Platyhelminthes) comprising the planarians, which occur in wet soils, fresh water, and marine environments. Their undersurface is covered with cilia, used for gliding over stones and weeds. Planarians can also swim by means of undulations of the body.

turgor The condition in a plant cell when its *vacuole is distended with water, pushing the protoplasm against the cell wall. In this condition the osmotic pressure forcing water in is balanced by pressure exerted by the cell wall. Turgidity assists in maintaining the rigidity of plants; a decrease in turgidity leads to wilting. *Compare* plasmolysis.

turion 1. A winter bud, covered with scale leaves and mucilage, that is produced by certain aquatic plants, such as frogbit. Turions become detached and remain dormant on the pond or lake bottom during the winter before developing into new plants the following season. **2.** *See* sucker.

tympanic cavity *See* middle ear.

tympanum (eardrum) The membrane that separates the *outer ear from the *middle ear. It vibrates in response to sound waves and transmits these vibrations via the *ear ossicles of the middle ear to the site of hearing (the *cochlea of the *inner ear). In amphibians and some reptiles there is no external ear and the tympanum is exposed at the skin surface.

type specimen The specimen used for naming and describing a *species or subspecies. If this is the original specimen collected by the author who named the species it is termed a *holotype*. The type specimen is not necessarily the most characteristic representative of the species. The term *type* is also used of any taxon selected as being representative of the rank to which it belongs. For example, the genus *Solanum* (potato) is said to be the type genus of the family Solanaceae.

tyrosine *See* amino acid.

U

ubiquinone *See* coenzyme Q.

ulna The larger of the two bones in the forearm of vertebrates (*compare* radius). It articulates with the outer carpals at the wrist and with the humerus at the elbow.

ultracentrifuge A high-speed centrifuge used to measure the rate of sedimentation of colloidal particles or to separate macromolecules, such as proteins or nucleic acids, from solutions. Ultracentrifuges are electrically driven and are capable of speeds up to 60 000 rpm.

ultramicroscope A form of microscope that reveals the presence of particles that cannot be seen with a normal optical microscope. Colloidal particles, smoke particles, etc., are suspended in a liquid or gas in a cell with a black background and illuminated by an intense cone of light that enters the cell from the side and has its apex in the field of view. The particles then produce diffraction-ring systems, appearing as bright specks on the dark background.

ultramicrotome *See* microtome.

ultrasonics The study and use of pressure waves that have a frequency in excess of 20 000 Hz and are therefore inaudible to the human ear. Ultrasound is used in medical diagnosis, particularly in conditions such as pregnancy, in which X-rays could have a harmful effect.

ultrastructure The submicroscopic, almost molecular, structure of living cells, which is revealed by the use of an electron microscope.

ultraviolet microscope A *microscope that has quartz lenses and slides and uses *ultraviolet radiation as the illumination. The use of shorter wavelengths than the visible range enables the instrument to resolve smaller objects and to provide greater magnification than the normal optical microscope. The final image is either photographed or made visible on a fluorescent screen by means of an image converter.

ultraviolet radiation (UV) Electromagnetic radiation having wavelengths between that of violet light and long X-rays, i.e. between 400 nanometres and 4 nm. In the range 400–300 nm the radiation is known as the *near ultraviolet*. In the range 300–200 nm it is known as the *far ultraviolet*. Below 200 nm it is known as the *extreme ultraviolet* or the *vacuum ultraviolet*, as absorption by the oxygen in the air makes the use of evacuated apparatus essential. The sun is a strong emitter of UV radiation but only the near UV reaches the surface of the earth as the ozone in the atmosphere absorbs all wavelengths below 290 nm.

Most UV radiation for practical use is produced by various types of mercury-vapour lamps. Ordinary glass absorbs UV radiation and therefore lenses and prisms for use in the UV are made from quartz.

umbel A type of *racemose inflorescence in which stalked flowers arise from the same point on the flower axis, resembling the spokes of an umbrella. An involucre (cluster) of bracts may occur at the point where the stalks emerge. This arrangement is characteristic of the family Umbelliferae (e.g. carrot, hogweed, parsley, parsnip), in which the inflorescence is usually a compound umbel.

umbilical cord The cord that connects the embryo to the *placenta in mammals. It contains a vein and two arteries that carry blood between the embryo and placenta. It is severed after birth to free the newly born animal from the placenta, and shrivels to leave a scar, the navel, on the animal.

ungulate A herbivorous mammal with hoofed feet (*see* unguligrade). Ungulates are grouped into two orders: *Artiodactyla and *Perissodactyla.

unguligrade Describing the gait of ungulates (e.g. horses and cows), in which only the tips of the digits (i.e. the hooves) are on the ground and the rest of the foot is off the ground. *Compare* digitigrade; plantigrade.

unicellular Describing tissues, organs, or organisms consisting of a single cell. For example, the reproductive organs of some algae and fungi are unicellular. Unicellular organisms, e.g. bacteria, protozoans, and certain algae and fungi, are sometimes placed in a separate kingdom, the *Protista. *Compare* acellular; multicellular.

unisexual Describing animals or plants with either male or female reproductive organs but not both. Most of the more advanced animals are unisexual but plants are often *hermaphrodite. Flowers that contain either stamens or carpels but not both are also described as unisexual. *See also* monoecious; dioecious.

unit A specified measure of a physical quantity, such as length, mass, time, etc., specified multiples of which are used to express magnitudes of that physical quantity. For scientific purposes previous systems of units have now been replaced by *SI units.

uracil A *pyrimidine derivative and one of the major component bases of *nucleotides and the nucleic acid *RNA.

uranium–lead dating A group of methods of *dating certain rocks that depends on the decay of the radioisotopes uranium–238 to lead–206 (half-life 4.5×10^9 years) or the decay of uranium–235 to lead–207 (half-life 7.1×10^8 years). One form of uranium–lead dating depends on measuring the ratio of the amount of helium trapped in the rock to the amount of uranium present (since the decay $^{238}U \rightarrow {}^{206}Pb$ releases eight alpha-particles). Another method of calculating the age of the rocks is to measure the ratio of radiogenic lead (^{206}Pb, ^{207}Pb, and ^{208}Pb) present to nonradiogenic lead (^{204}Pb). These methods give reliable results for ages of the order 10^7–10^9 years.

urea (carbamide) A white crystalline water-soluble solid, $CO(NH_2)_2$. Urea is the major end product of nitrogen excretion in mammals, being synthesized by the *urea cycle. Urea is synthesized industrially from ammonia and carbon dioxide for use in urea–formaldehyde resins and pharmaceuticals, as a source of nonprotein nitrogen for ruminant livestock, and as a nitrogen fertilizer.

urea cycle (ornithine cycle) The series of biochemical reactions that converts ammonia to *urea during the excretion of metabolic nitrogen. Urea formation occurs in mammals and, to a lesser extent, in some other animals. The liver converts ammonia to the much less toxic urea, which is excreted in solution in *urine.

ureter The duct in vertebrates that conveys urine from the *kidney to the *bladder.

urethra The duct in mammals that conveys urine from the *bladder to be discharged to the outside of the body. In males the urethra passes through the penis and is joined by the *vas deferens; it therefore also serves as a channel for sperm.

uric acid The end product of purine breakdown in most primates, birds, terrestrial reptiles, and insects and also (except in primates) the major form in which metabolic nitrogen is excreted. Being fairly insoluble, uric acid can be expelled in solid form, which conserves valuable water in arid environments. The accumulation of uric acid in the synovial fluid of joints causes gout.

urine The aqueous fluid formed by the excretory organs of animals for the removal of metabolic waste products. In higher animals, urine is produced by the *kidneys, stored in the *bladder, and excreted through the *urethra or *cloaca. Apart from water, the major constituents of urine are one or more of the end products of nitrogen metabolism – ammonia, urea, uric acid, and creatinine. It may also contain various inorganic ions, the pigments urochrome and urobilin, amino acids, and purines. Precise composition depends on many factors, especially the habitat of a particular species: aquatic animals produce copious volumes; terrestrial animals need to conserve water and produce much less (about 1.0–1.5 litres per day in humans).

uriniferous tubule *See* nephron.

uterus (womb) The organ of female mammals in which the embryo develops. Paired in most mammals but single in humans, it is situated between the bladder and rectum and is connected to the *fallopian tubes and to the *vagina. The lining shows cyclical changes (*see* menstrual cycle; oestrous cycle) associated with egg production and provides a thick spongy layer in which the fertilized egg becomes embedded. The outer wall of the uterus is thick and muscular; by contracting, it forces the fully grown fetus through the vagina to the outside.

utriculus (utricle) A chamber of the *inner ear from which the *semicircular canals arise. It bears patches of sensory epithelium concerned with detecting changes in the direction and speed of movement (*see* macula)

UV *See* ultraviolet radiation.

V

vaccination *See* immunization.

vaccine A liquid preparation of treated disease-producing microorganisms or their products used to stimulate an *immune response in the body and so confer resistance to the disease (*see* immunization). Vaccines are administered orally or by injection (*inoculation*). They take the form of dead viruses or bacteria that can still act as antigens, live but weakened microorganisms (*see* attenuation), specially treated *toxins, or antigenic extracts of the microorganism.

vacuole A space within the cytoplasm of a living *cell that is filled with air, water or other liquid, sap, or food particles. In plant cells there is usually one large vacuole; animal cells usually have several small ones. *See also* contractile vacuole.

vagina The tube leading from the uterus to the outside. Sperm are deposited in the vagina during copulation and the fully developed fetus is born through it. In a number of mammals the vagina may be sealed when the animal is not sexually receptive and only open during oestrus. Its lining produces mucus, which prevents friction and the entry of infective organisms.

vagus nerve The tenth *cranial nerve: a paired nerve that supplies branches to many major internal organs. It carries motor nerve fibres to the heart, lungs, and viscera and sensory fibres from the viscera.

valine *See* amino acid.

valve 1. (in anatomy) Any of various structures for restricting the flow of a fluid through an aperture or along a tube to one direction. Valves in the heart (*see* mitral valve; tricuspid valve), veins, and lymphatic vessels consist of two or three flaps of tissue (*cusps*) fastened to the walls. The cusps are flattened to the walls to allow the normal passage of blood or lymph, but a reverse flow causes them to block the vessel or aperture, so preventing further backflow. **2.** (in botany) **a.** Any of the parts that make up a capsule or other dry fruit that sheds its seeds. **b.** One of the two halves of the cell wall of a diatom.

variation The differences between individuals of a plant or animal species. Variation may be the result of environmental conditions; for example, water supply and light intensity affect the height and leaf size of a plant. Differences of this kind, acquired during the lifetime of an individual, are not transmitted to succeeding generations since the genes are not affected. Variations due to differences in genetic constitution are inherited (*see* continuous variation; discontinuous variation). Wide genetic variation improves the ability of a species to survive in a changing environment, since the chances that some individuals will tolerate a particular change are increased. Such individuals will survive and transmit the advantageous genes to their offspring.

variety A category used in the *classification of plants and animals below the *species level. A variety consists of a group of individuals that differ distinctly from but can interbreed with other varieties of the same species. The characteristics of a variety are genetically inherited. Examples of varieties include breeds of domestic animals and the *races of man. *See also* cultivar. *Compare* subspecies.

varve dating (geochronology) An absolute *dating technique using thin sedimentary layers of clays called *varves*. The varves, which are particularly common in Scandinavia, have alternate light and dark bands corresponding to winter and summer deposition. Most of them are found in the Pleistocene series, where the edges of varve deposits can be correlated with the annual retreat of the ice sheet, although some varve formation is taking place in the present day. By counting varves it is possible to establish an absolute time scale for fossils up to about 20 000 years ago.

vascular bundle (fascicle) A long continuous strand of conducting (vascular) tissue in seed plants and pteridophytes that extends from the roots through the stem and into the leaves. It consists of *xylem and *phloem, which are separated by a *cambium in plants that undergo secondary thickening. *See* vascular system; vascular tissue.

vascular cambium *See* cambium.

vascular plants All plants possessing organized *vascular tissue. They may be grouped in a single division, the *Tracheophyta, or separated into two divisions, the *Pteridophyta and *Spermatophyta.

vascular system 1. A specialized network of vessels for the circulation of fluids throughout the body tissues of an animal. All animals, apart from protozoans and other simple invertebrate groups, possess a *blood vascular system*, which enables the passage of respiratory gases, nutrients, excretory products, and other metabolites into and out of the cells. In vertebrates it consists of a muscular *heart, which pumps blood through major blood vessels (*arteries) into increasingly finer branches until in the *capillaries it is in intimate contact with tissues. It then returns to the heart via another network of vessels (the *veins). This circulation also enables a stable *internal environment for tissue function (*see* homeostasis), the transmission of chemical messengers (*hormones) around the body, and a means of defending the body against pathogens and damage via the immune system (*see* immune response). See illustration.

A *water vascular system* is characteristic of the *Echinodermata.

2. The system of *vascular tissue in plants.

vascular tissue (vascular system) The tissue that conducts water and nutrients through the plant body in higher plants (ferns and seed plants). It consists of *xylem and *phloem. Since the xylem and

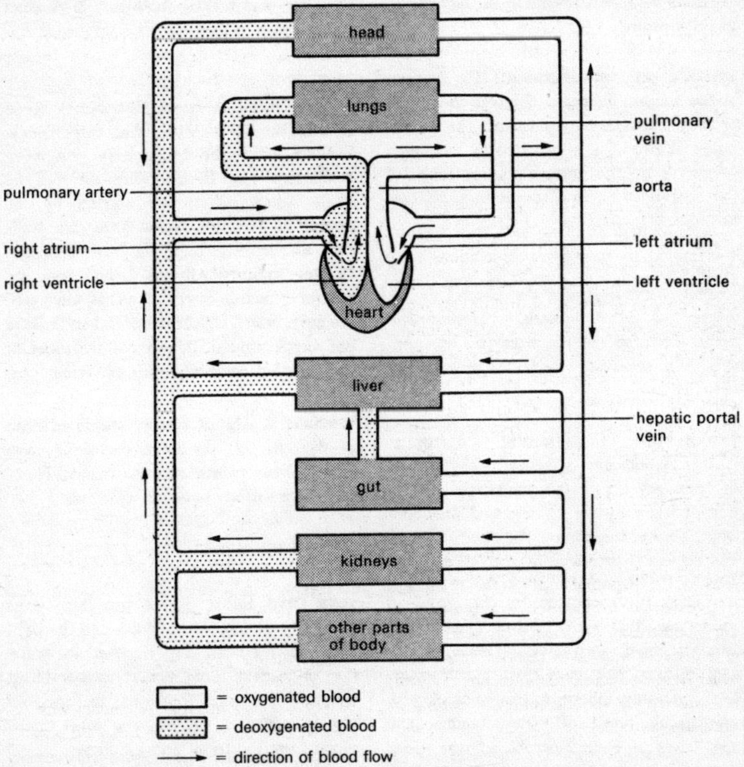

Blood vascular system of a mammal

phloem tissues are always in close proximity to each other, distinct regions of vascular tissue can be identified (*see* vascular bundle). The possession of vascular tissue has enabled the higher plants to attain a considerable size and dominate most terrestrial habitats.

vas deferens One of a pair of ducts carrying sperm from the testis (or *epididymis) to the outside, in mammals through the *urethra.

vas efferens Any of various small ducts carrying sperm. In reptiles, birds, and mammals they convey sperm from the seminiferous tubules of the testis to the *epididymis; in invertebrates they carry sperm from the testis to the vas deferens.

vasoconstriction The reduction in the internal diameter of blood vessels, especially arterioles or capillaries. The constriction of arterioles is mediated by the action of nerves on the smooth muscle fibres of the arteriole walls and results in an increase in blood pressure.

vasodilation (vasodilatation) The increase in the internal diameter of blood vessels, especially arterioles or capillaries. The vasodilation of arterioles is mediated by the action of nerves on the smooth muscle fibres of the arteriole walls and results in a decrease in blood pressure.

vasomotor nerves The nerves of the *autonomic nervous system that control the diameter of blood vessels. *Vasoconstrictor nerves* decrease the diameter (*see* vasoconstriction); *vasodilator nerves* increase it (*see* vasodilation).

vasopressin (antidiuretic hormone; ADH) A hormone, secreted by the posterior *pituitary gland, that stimulates absorption of water by the kidneys and thus controls the concentration of body fluids. Vasopressin is produced by specialized nerve cells in the hypothalamus of the brain and is transported to the posterior pituitary in the bloodstream. Deficiency of ADH results in a disorder known as *diabetes insipidus*, in which large volumes of urine are excreted; it is treated by administration of natural or synthetic hormone.

vector An animal, usually an insect, that passively transmits disease-causing microorganisms from one animal or plant to another or from an animal to man. *Compare* carrier.

vegetative propagation (vegetative reproduction) 1. A form of *asexual reproduction in plants whereby new individuals develop from specialized multicellular structures (e.g. *tubers, *bulbs) that become detached from the parent plant. Examples are the production of strawberry plants from *runners and of gladioli from daughter *corms. Artificial methods of vegetative propagation include grafting (*see* graft), *budding, and making *cuttings. **2.** Asexual reproduction in animals, e.g. budding in *Hydra*.

vein 1. A blood vessel that carries blood towards the heart. Most veins carry deoxygenated blood (the *pulmonary vein is an exception). The largest veins are fed by smaller ones, which are formed by the merger of *venules. Veins have thin walls and a relatively large internal diameter. *Valves within the veins ensure that the flow of blood is always towards the heart. *Compare* artery. **2.** A vascular bundle in a leaf (*see* venation). **3.** Any of the tubes of chitin that strengthen an insect's wing.

velamen A whitish spongy sheath of dead empty cells that surrounds the aerial roots of *epiphytic plants, such as orchids. It absorbs any surface water on the roots.

velum *See* annulus.

vena cava Either of the two large veins that carry deoxygenated blood into the right atrium of the heart. The *precaval vein* (*anterior* or *superior vena cava*) receives blood from the head and forelimbs; the *postcaval vein* (*posterior* or *inferior vena cava*) drains blood from the trunk and hindlimbs.

venation 1. The arrangement of veins (vascular bundles) in a leaf. The leaves of dicotyledons have a central main vein (midrib) with side branches that themselves further subdivide to form a network (*net* or *reticulate venation*). The leaves of monocotyledons have parallel veins (*parallel venation*). **2.** The arrangement of the veins in an insect's wing, which is often important in classification.

ventral Describing the surface of a plant or animal that is nearest or next to the ground or other support, i.e. the lower surface. In bipedal animals, such as man, it is the forward-directed (*anterior) surface. *Compare* dorsal.

ventricle 1. A chamber of the *heart that receives blood from an *atrium and pumps it into the arterial system. Amphibians and fish have a single ventricle, but mammals, birds, and reptiles have two, pumping deoxygenated blood to the lungs and oxygenated blood to the rest of the body, respectively. **2.** Any of the four linked fluid-filled cavities in the brain of vertebrates. One of these cavities is in the *medulla oblongata, two are in the cerebral hemispheres (*see* cerebrum), and the fourth is in the posterior part of the *forebrain. The ventricles contain cerebrospinal fluid filtered from the blood by the *choroid plexus.

venule A small blood vessel that receives blood from the capillaries and transports it to a vein.

vermiform appendix *See* appendix.

vernalization The application of a cold treatment to germinating seeds or seedlings to ensure flowering. Many plants, including winter cereals, will not flower unless subjected to a period of chilling early in their development. Winter cereals are therefore sown in the autumn for flowering the following year. However, if germinating seeds are artificially vernalized they can be sown in the spring for flowering the same year.

vertebra Any of the bones that make up the *vertebral column. In mammals each vertebra typically consists of a main body, or *centrum*, from which arises a *neural arch* through which the spinal cord passes, and *transverse processes* projecting from the side. There are five groups of vertebrae, specialized for various functions and varying in number with the species. In man, for example, there are 7 *cervical vertebrae, 12 *thoracic vertebrae, 5 *lumbar vertebrae, 5 fused *sacral vertebrae, and 5 fused *caudal vertebrae (forming the *coccyx).

vertebral column (backbone; spinal column; spine) A flexible bony column in vertebrates that extends down the long axis of the body and provides the main skeletal support. It also encloses and protects the *spinal cord and provides attachment for the muscles of the back. The vertebral column consists of a series of bones (*see* vertebra) separated by discs of cartilage (*intervertebral discs*). It articulates with the skull by means of the *atlas vertebra, with the ribs at the *thoracic vertebrae, and with the pelvic girdle at the sacrum (*see* sacral vertebrae).

Vertebrata (Craniata) The largest subphylum of the *Chordata, comprising all those animals with backbones (*see* vertebral column). The replacement of the rigid notochord by the more flexible backbone has permitted vertebrates a greater degree of movement and subsequent improvement in the sense organs and enlargement of the brain. The brain is enclosed in a skeletal case, the cranium. There are seven extant classes: *Agnatha (jawless fishes), *Elasmobranchii (cartilaginous fishes), *Osteichthyes (bony fishes), *Amphibia, *Reptilia, *Aves (birds), and *Mammalia.

vesicle A small, usually fluid-filled, membrane-bound sac within the cytoplasm of a living cell. Vesicles form part of the *Golgi apparatus.

vessel 1. (in botany) A tube within the *xylem composed of joined *vessel elements. Vessels facilitate the efficient movement of water from the roots to the shoots and leaves of a plant. **2.** (in zoology) Any of various tubular structures through which substances are transported, especially a blood vessel or a lymphatic vessel.

vessel element A type of cell occurring within the *xylem of flowering plants, many of which, end to end, form water-conducting *vessels*. Vessel elements are frequently very broad and have side walls thickened by deposits of lignin over most of the surface area. However the end walls are broken down to provide connections with the cells both above and below them. *Compare* tracheid.

vestigial organ Any part of an organism that has diminished in size during its evolution because the function it served decreased in importance or became totally unnecessary. Examples are the human appendix and the wings of the ostrich.

vibrio Any comma-shaped bacterium. Generally, vibrios are Gram-negative (*see* Gram's stain), motile, and aerobic. They are widely distributed in soil and water and while most feed on dead organic matter some are parasitic, e.g. *Vibrio cholerae*, the causal agent of cholera.

villus A microscopic outgrowth from the surface of some tissues and organs, which serves to increase the surface area of the organ. Numerous villi line the interior of the small intestine. Their shape may vary from finger-like (in the *duodenum) to spadelike (in the *ileum). Intestinal villi are specialized for the absorption of soluble food material: each contains blood vessels and a lymph vessel (*see* lacteal).

Villi also occur on the chorion of the mammalian placenta, where they increase the surface area for the exchange of materials between the fetal and maternal blood.

virion *See* virus.

virology The scientific study of *viruses. *See* microbiology.

virulence The disease-producing ability of a microorganism. *See also* pathogen.

virus A particle that is too small to be seen with a light microscope or to be trapped by filters but is capable of independent metabolism and reproduction within a living cell. Outside its host cell a virus is completely inert. A mature virus (a *virion*) ranges in size from 20 to 400 nm in diameter. It consists of a core of nucleic acid (DNA or RNA) surrounded by a protein coat (*capsid*). Some bear an outer envelope (*enveloped viruses*). Inside its host cell the virus initiates the synthesis of viral proteins and undergoes replication. The new virions are released when the host cell disintegrates. Viruses are parasites of animals, plants, and some bacteria (*see* bacteriophage). Viral diseases of animals include the common cold, influenza, smallpox, herpes, hepatitis, polio, and rabies (*see* adenovirus; arbovirus; herpesvirus; myxovirus; papovavirus; picornavirus; poxvirus); some viruses are also implicated in the development of cancer (*see* oncogenic). Plant viral diseases include various forms of yellowing and blistering of leaves and stems (*see* tobacco mosaic virus). Antibiotics are ineffective against viral diseases but *vaccines provide good protection.

visual purple *See* rhodopsin.

vital staining A technique in which a harmless dye is used to stain living tissue for microscopical observation. The stain may be injected into a living animal and the stained tissue removed and examined (*intravital staining*) or the living tissue may be removed directly and subsequently stained (*supravital staining*). Microscopic organisms, such as Protozoa, may be completely immersed in the dye solution. Vital stains include trypan blue, vital red, and Janus green, the latter being especially suitable for observing mitochondria.

vitamin One of a number of organic compounds required by living organisms in relatively small amounts to maintain normal health. There are some 14 generally recognized major vitamins: the water-soluble *vitamin B complex (containing 9) and *vitamin C and the fat-soluble *vitamin A, *vitamin D, *vitamin E, and *vitamin K. Most B vitamins and vitamin C occur in plants, animals, and microorganisms; they function typically as *coenzymes. Vitamins A, D, E, and K occur only in animals, especially vertebrates, and perform a variety of metabolic roles. Animals are unable to manufacture many vitamins themselves and must have adequate amounts in the diet. Foods may contain vitamin precursors (called *provitamins*) that are chemically changed to the actual vitamin on entering the body. Many vitamins are destroyed by light and heat, e.g. during cooking.

vitamin A (retinol) A fat-soluble vitamin that cannot be synthesized by mammals and other vertebrates and must be provided in the diet. Green plants contain precursors of the vitamin, notably carotenes, that are converted to vitamin A in the intestinal wall and liver. The aldehyde derivative of vitamin A, *retinal*, is a constituent of the visual pigment *rhodopsin. Deficiency affects the eyes, causing night blindness, xerophthalmia, and eventually total blindness. The role of vitamin A in other aspects of metabolism is less clear; it may be involved in controlling *ATP production and the growth of epithelial cells.

vitamin B complex A group of water-soluble vitamins that characteristically serve as components of *coenzymes. Plants and many microorganisms can manufacture B vitamins but dietary sources are essential for most animals. Heat and light tend to destroy B vitamins.

Vitamin B_1 (thiamin(e)) is a precursor of the coenzyme thiamine pyrophosphate, which functions in carbohydrate metabolism. Deficiency leads to beriberi in humans and to polyneuritis in birds. Good sources include brewer's yeast, wheatgerm, beans, peas, and green vegetables.

Vitamin B_2 (riboflavin) occurs in green vegetables, yeast, liver, and milk. It is a constituent of the coenzymes *FAD and FMN, which have an important role in the metabolism of all major nutrients as well as in the oxidative phosphorylation reactions of the *electron transport chain. Deficiency of B_2 causes inflammation of the tongue and lips and mouth sores.

Vitamin B_6 (pyridoxine) is widely distributed in cereal grains, yeast, liver, milk, etc. It is a constituent of a coenzyme (pyridoxal phosphate) involved in amino acid metabolism. Deficiency causes retarded growth, dermatitis, convulsions, and other symptoms.

Vitamin B_{12} (cyanocobalamin) is manufactured only by microorganisms and natural sources are entirely of animal origin. Liver is especially rich in it. One form of B_{12} functions as a coenzyme in a number of reactions, including the oxidation of fatty acids and the synthesis of DNA. It also works in conjunction with *folic acid (another B vitamin) in the synthesis of the amino acid methionine and it is required for normal production of red blood cells. Vitamin B_{12} can only be absorbed from the gut in the presence of a glycoprotein called intrinsic factor; lack of this factor or deficiency of B_{12} results in pernicious anaemia.

Other vitamins in the B complex include *nicotinic acid, *pantothenic acid, *folic acid, *biotin, and *lipoic acid. *See also* choline.

vitamin C (ascorbic acid) A colourless crystalline water-soluble vitamin found especially in citrus fruits and green vegetables. Most organisms synthesize it from glucose but man and other primates and various other species must obtain it from their diet. It is required for the maintenance of healthy connective tissue; deficiency leads to scurvy. Vitamin C is readily destroyed by heat and light.

vitamin D A fat-soluble vitamin occurring in the form of two steroid derivatives: *vitamin* D_2 (*ergocalciferol*, or *calciferol*), found in yeast; and *vitamin* D_3 (*cholecalciferol*), which occurs in animals. Vitamin D_2 is formed from a steroid by the action of ultraviolet light and D_3 is produced by the action of sunlight on a cholesterol derivative in the skin. Fish-liver oils are the major dietary source. The active form of vitamin D is manufactured in response to the secretion of *parathyroid hormone, which occurs when blood calcium levels are low. It causes increased uptake of calcium from the gut, which increases the supply of calcium for bone synthesis. Vitamin D deficiency causes rickets in growing animals and osteomalacia in mature animals. Both conditions are characterized by weak deformed bones.

vitamin E (tocopherol) A fat-soluble vitamin consisting of several closely related compounds, deficiency of which leads to a range of disorders in different species, including muscular dystrophy, liver damage, and infertility. Good sources are cereal grains and green vegetables. Vitamin E prevents the oxidation of unsaturated fatty acids in cell membranes, so maintaining their structure.

vitamin K A fat-soluble vitamin consisting of several related compounds that act as coenzymes in the synthesis of several proteins necessary for blood clotting. Deficiency of vitamin K, which leads to extensive bleeding, is rare because a form of the vitamin is manufactured by intestinal bacteria. Green vegetables and egg yolk are good sources.

vitelline membrane *See* egg membrane.

vitreous humour The colourless jelly that fills the space between the lens and the retina of the vertebrate eye.

viviparity 1. (in zoology) A form of reproduction in animals in which the developing embryo obtains its nourishment directly from the mother via a *placenta or by other means. Viviparity occurs in some insects and other arthropods, in certain fishes, amphibians, and reptiles, and in the majority of mammals. *Compare* oviparity; ovoviviparity. **2.** (in botany) **a.** A form of *asexual reproduction in certain plants, such as the onion, in which the flower develops into a budlike structure that forms a new plant when detached from the parent. **b.** The development of young plants on the inflorescence of the parent plant, as seen in certain grasses and the spider plant.

vocal cords A pair of elastic membranes that project into the *larynx in air-breathing vertebrates. Vocal sounds are produced when expelled air passing through the larynx vibrates the cords. The pitch of the sound produced depends on the tension of the cords, which is controlled by muscles and cartilages in the larynx.

voluntary muscle (skeletal, striped, or **striated muscle)** Muscle that is under the control of the will and is generally attached to the skeleton. An individual muscle consists of bundles of long *muscle fibres*, each containing many nuclei, the whole muscle being covered with a strong connective tissue sheath (*epimysium*) and attached at each end to a bone by inextensible *tendons. Each fibre contains smaller fibres (*myofibrils*) having alternate light and dark bands (*sarcomeres*), which are responsible for the muscle's contractile ability and its typical striped appearance under the microscope.

The end of the muscle that is attached to a nonmoving bone is called the *origin* of the muscle; the end attached to a moving bone is the *insertion*. As a muscle contracts it becomes shorter and fatter, moving one bone closer to the other. Since a muscle cannot expand, another muscle (the *extensor*, or *antagonist*) is required to move the bone in the opposite direction and stretch the first musscle (known as the *flexor*, or *agonistic muscle*).

vulva The external opening of the *vagina, comprising in women the female external genitalia – two pairs of fleshy folds of tissue (the *labia*).

W

Wallace's line An imaginary line that runs between the Indonesian islands of Bali and Lombok and represents the separation of the Australian and Oriental faunas. It was proposed by the naturalist A. R. Wallace (1823–1913) who had noted that the mammals in Southeast Asia are different from and more advanced than their Australian counterparts. He suggested this was because the Australian continent had split away from Asia before the better adapted placental mammals evolved in Asia. Hence the isolated Australian marsupials and monotremes were able to thrive while those in Asia were driven to extinction by competition from placental mammals. *See also* zoogeography.

warfarin 3-(alpha-acetonylbenzyl)-4-hydroxycoumarin: a synthetic *anticoagulant used both therapeutically in clinical medicine and, in lethal doses, as a rodenticide.

warm-bloodedness *See* homoiothermy.

warning coloration (aposematic coloration) The conspicuous markings of an animal that make it easily recognizable and warn would-be predators that it is a poisonous, foul-tasting, or dangerous species. For example, the yellow-and-black striped abdomen of the wasp warns of its sting. *See also* mimicry.

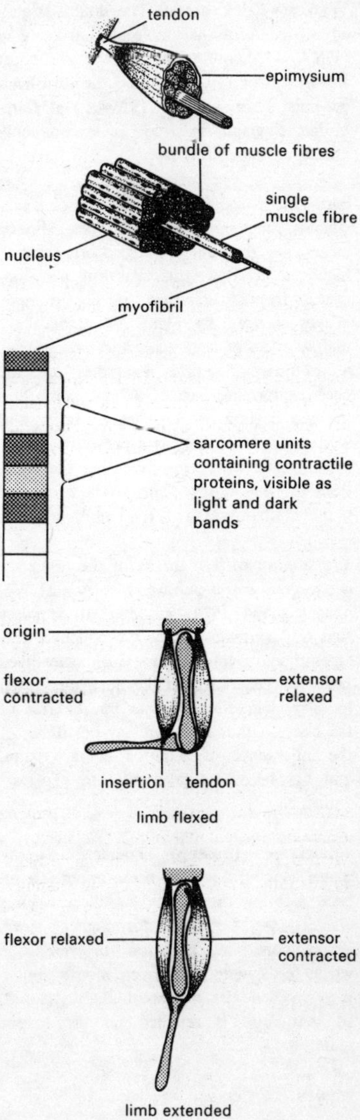

Structure and action of a voluntary muscle

Watson–Crick model The double-stranded twisted ladder-like molecular structure of *DNA as determined by the US biochemist James Watson (1928–) and the British biochemist Francis Crick (1916–) at Cambridge, England, in 1953. It is commonly known as the *double helix*.

wax Any of various solid or semisolid substances. There are two main types. Mineral waxes are mixtures of hydrocarbons with high molecular weights. Paraffin wax, obtained from *petroleum, is an example. Waxes secreted by plants or animals are mainly esters of fatty acids and usually have a protective function. Examples are the beeswax forming part of a honeycomb and the wax coating on some leaves, fruits, and seed coats, which acts as a protective water-impermeable layer supplementing the functions of the cuticle. The seeds of a few plants contain wax as a food reserve.

Weismannism The theory of the *continuity of the germ plasm* published by August Weismann (1834–1914) in 1886. It proposes that the contents of the reproductive cells (sperms and ova) are passed on unchanged from one generation to the next, unaffected by any changes undergone by the rest of the body. It thus rules out any possibility of the inheritance of acquired characteristics, and has become fundamental to neo-Darwinian theory.

whalebone (baleen) Transverse horny plates hanging down from the upper jaw on each side of the mouth of the toothless whales (*see* Cetacea), forming a sieve. Water, containing plankton on which the whale feeds, enters the open mouth and is then expelled with the mouth slightly closed, so that food is retained on the baleen plates.

whales *See* Cetacea.

white blood cell *See* leucocyte.

white matter Part of the tissue that makes up the central nervous system of vertebrates. It consists chiefly of nerve fibres enclosed in whitish *myelin sheaths. *Compare* grey matter.

wild type Describing the form of an *allele possessed by most members of a population in their natural environment. Wild-type alleles are usually *dominant.

womb *See* uterus.

wood The hard structural and water-conducting tissue that is found in many perennial plants and forms the bulk of trees and shrubs. It is composed of secondary *xylem and associated cells, such as fibres. The wood of angiosperms is termed *hardwood*, e.g. oak and mahogany, and that of gymnosperms *softwood*, e.g. pine and fir. New wood is added to the outside of the old wood each growing season by divisions of the vascular cambium (*see* growth ring). Only the outermost new wood (*sapwood) functions in water conduction; the inner wood (*heartwood) provides only structural support.

X

xanthophyll A member of a class of oxygen-containing *carotenoid pigments, which provide the characteristic yellow and brown colours of autumn leaves.

X chromosome *See* sex chromosome.

xeromorphic Describing the structural modifications of certain plants (*xerophytes) that enable them to reduce water loss, particularly from their leaves and stems.

xerophyte A plant that is adapted to live in conditions in which there is either a scarcity of water in the soil, or the atmosphere is dry enough to provoke excessive transpiration, or both. Xerophytes have special structural (*xeromorphic*) and functional modifications, including swollen water-storing stems and specialized leaves that may be hairy, rolled, or reduced to spines or have a thick cuticle to lower the rate of transpiration. Examples of xerophytes are desert cacti and many species growing on sand dunes and exposed moorlands. *Compare* mesophyte; hydrophyte.

X-ray crystallography The use of *X-ray diffraction to determine the structure of crystals or molecules, such as nucleic acids. The technique involves directing a beam of X-rays at a crystalline sample and recording the diffracted X-rays on a photographic plate. The diffraction pattern consists of a pattern of spots on the plate, and the crystal structure can be worked out from the positions and intensities of the diffraction spots. X-rays are diffracted by the electrons in the molecules and if molecular crystals of a compound are used, the electron density distribution in the molecule can be determined.

X-rays Electromagnetic radiation of shorter wavelength than ultraviolet radiation and longer wavelength than gamma radiation. The range of wavelengths is 10^{-11} m to 10^{-9} m. X-rays can pass through many forms of matter and they are therefore used medically and industrially to examine internal structures. X-rays are produced for these purposes by an X-ray tube.

xylem A tissue that transports water and dissolved mineral nutrients in vascular plants. In flowering plants it consists of hollow *vessels* that are formed from cells (*vessel elements) joined end to end. The end walls of the vessel elements are perforated to allow the passage of water. In less advanced vascular plants, such as conifers and ferns, the constituent cells of the xylem are called *tracheids. In young plants and at the shoot and root tips of older plants the xylem is formed by the apical meristems. In plants showing secondary growth this xylem is replaced in most of the plant by secondary xylem, formed by the vascular *cambium. The walls of the xylem cells are thickened with lignin, the extent of this thickening being greatest in secondary xylem. Xylem contributes greatly to the mechanical strength of the plant: *wood is mostly made up of secondary xylem. *See also* fibre. *Compare* phloem.

xylenes *See* dimethylbenzenes.

Y

Y chromosome *See* sex chromosome.

yeasts A group of unicellular fungi many of which belong to the *Ascomycetes. Certain species of the genus *Saccharomyces* are used in the baking and brewing industries.

yolk The food stored in an egg for the use of the embryo. It can consist mainly of protein (*protein yolk*) or of phospholipids and fats (*fatty yolk*). The eggs of oviparous animals (e.g. birds) contain a relatively large yolk.

Z

zone fossil *See* index fossil.

zoogeography The study of the geographical distributions of animals. The earth can be divided into several zoogeographical regions separated by natural barriers, such as oceans, deserts, and mountain ranges. The characteristics of the fauna of each region are believed to depend particularly on the process of *continental drift and the stage of evolution reached when the various land masses became isolated. For example Australia, which has been isolated since Cretaceous times, has the most primitive native mammalian fauna, consisting solely of marsupials and monotremes. *See also* Wallace's line.

zoology The scientific study of animals, including their anatomy, physiology, biochemistry, genetics, ecology, evolution, and behaviour.

zooplankton The animal component of *plankton. All major animal phyla are represented in zooplankton, as adults, larvae, or eggs; some are just visible to the naked eye but most cannot be seen without magnification. Near the surface of the sea there may be many thousands of such animals per cubic metre.

zwitterion (ampholyte ion) An ion that has a positive and negative charge on the same group of atoms. Zwitterions can be formed from compounds that contain both acid groups and basic groups in their molecules. For example, the amino acid glycine has the formula $H_2N.CH_2.COOH$. However, under neutral conditions, it exists in the different form of the zwitterion $^+H_3N.CH_2.COO^-$, which can be regarded as having been produced by an internal neutralization reaction (transfer of a proton from the carboxyl group to the amino group). Glycine therefore has some properties characteristic of ionic compounds, e.g. a high melting point and solubility in water. In acid solutions, the positive ion $^+H_3NCH_2COOH$ is formed. In basic solutions, the negative ion $H_2NCH_2COO^-$ predominates. The name comes from the German *zwei*, two.

zygomorphy *See* bilateral symmetry.

zygospore A zygote with a thick resistant wall, formed by some algae and fungi. It results from the fusion of two gametes, neither of which is retained by the parent in any specialized sex organ (such as an oogonium). It enters a resting phase before germination. *Compare* oospore.

zygote A fertilized female *gamete: the product of the fusion of the nucleus of the ovum or ovule with the nucleus of the sperm or pollen grain. *See* fertilization.

zymogen Any inactive enzyme precursor that, following secretion, is chemically altered to the active form of the enzyme. For example, the protein-digesting enzyme *trypsin is secreted by the pancreas as the zymogen trypsinogen. This is changed in the small intestine by the action of another enzyme, enterokinase, to the active form.

Appendix 1 SI units

Table 1.1 Base and supplementary SI units

Physical quantity	*Name*	*Symbol*
length	metre	m
mass	kilogram	kg
time	second	s
electric current	ampere	A
thermodynamic temperature	kelvin	K
luminous intensity	candela	cd
amount of substance	mole	mol
*plane angle	radian	rad
*solid angle	steradian	sr

*supplementary units

Table 1.2 Derived SI units with special names

Physical quantity	*Name of SI unit*	*Symbol of SI unit*
frequency	hertz	Hz
energy	joule	J
force	newton	N
power	watt	W
pressure	pascal	Pa
electric charge	coulomb	C
electric potential difference	volt	V
electric resistance	ohm	Ω
electric conductance	siemens	S
electric capacitance	farad	F
magnetic flux	weber	Wb
inductance	henry	H
magnetic flux density (magnetic induction)	tesla	T
luminous flux	lumen	lm
illuminance	lux	lx
absorbed dose	gray	Gy
activity	becquerel	Bq
dose equivalent	sievert	Sv

Table 1.3 Decimal multiples and submultiples to be used with SI units

Submultiple	*Prefix*	*Symbol*	*Multiple*	*Prefix*	*Symbol*
10^{-1}	deci	d	10	deca	da
10^{-2}	centi	c	10^2	hecto	h
10^{-3}	milli	m	10^3	kilo	k
10^{-6}	micro	μ	10^6	mega	M
10^{-9}	nano	n	10^9	giga	G
10^{-12}	pico	p	10^{12}	tera	T
10^{-15}	femto	f	10^{15}	peta	P
10^{-18}	atto	a	10^{18}	exa	E

Table 1.4 Conversion of units to SI units

From	*To*	*Multiply by*
in	m	2.54×10^{-2}
ft	m	0.3048
sq. in	m^2	6.4516×10^{-4}
sq. ft	m^2	9.2903×10^{-2}
cu. in	m^3	1.63871×10^{-5}
cu. ft	m^3	2.83168×10^{-2}
l(itre)	m^3	10^{-3}
gal(lon)	m^3	$4.546\ 09 \times 10^{-3}$
gal(lon)	l(itre)	4.546 09
miles/hr	$m\ s^{-1}$	0.477 04
km/hr	$m\ s^{-1}$	0.277 78
lb	kg	0.453 592
$g\ cm^{-3}$	$kg\ m^{-3}$	10^3
lb/in^3	$kg\ m^{-3}$	$2.767\ 99 \times 10^4$
dyne	N	10^{-5}
kgf	N	9.806 65
poundal	N	0.138 255
lbf	N	4.448 22
mmHg	Pa	133.322
atmosphere	Pa	$1.013\ 25 \times 10^5$
hp	W	745.7
erg	J	10^{-7}
eV	J	$1.602\ 10 \times 10^{-19}$
kW h	J	3.6×10^6
cal	J	4.1868

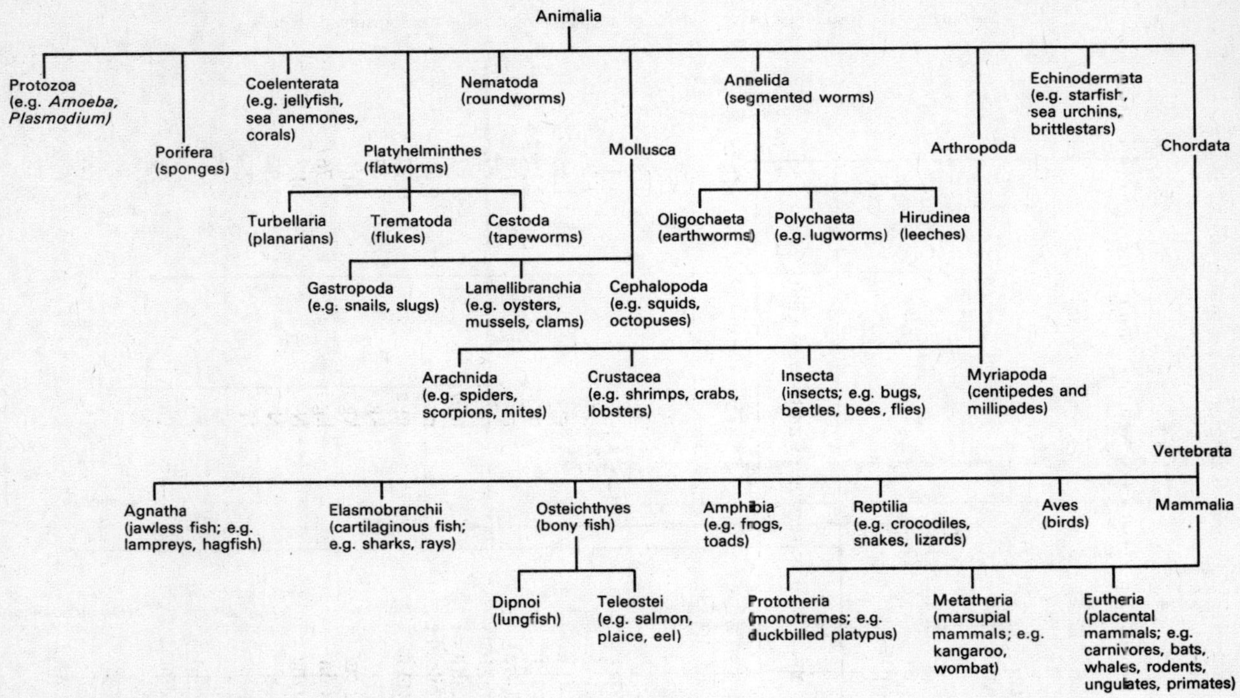

A simplified classification of the animal kingdom (down to class level (subclass for major groups) and including only the major phyla)

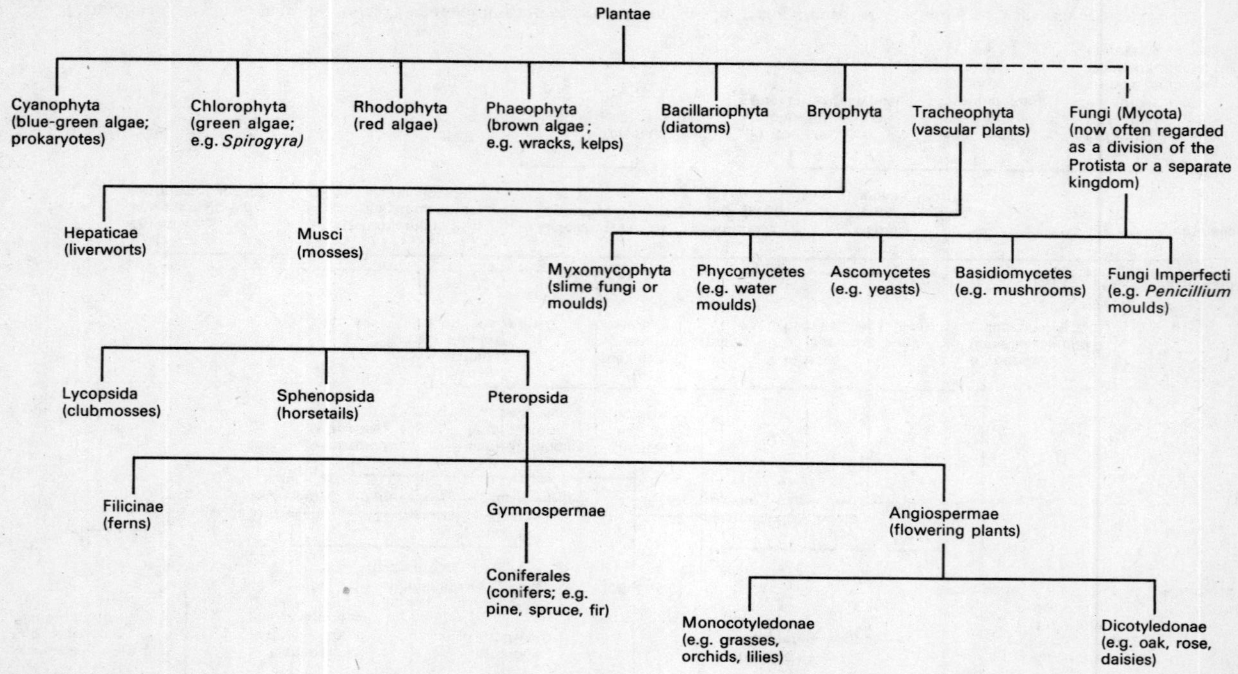

A simplified classification of the plant kingdom (excluding extinct and mostly extinct groups)

Era	Period	Epoch	millions of years ago
Cenozoic	Quaternary	Holocene	
		Pleistocene	1.8
	Tertiary	Pliocene	
		Miocene	
		Oligocene	
		Eocene	
		Palaeocene	65
Mesozoic	Cretaceous		136
	Jurassic		190
	Triassic		225
Palaeozoic	Permian		280
	Carboniferous		345
	Devonian		395
	Silurian		440
	Ordovician		500
	Cambrian		570
Precambrian	Precambrian		4600

Geological time scale